Lecture Notes in Computer Science

Lecture Notes in Computer Science

Edited by G. Goos and J. Hartmanis

425

C.H. Bergman R.D. Maddux
D.L. Pigozzi (Eds.)

Algebraic Logic and Universal Algebra in Computer Science

Conference, Ames, Iowa, USA
June 1–4, 1988
Proceedings

Springer-Verlag
Berlin Heidelberg GmbH

Editors

Clifford H. Bergman
Roger D. Maddux
Don L. Pigozzi
Department of Mathematics, Iowa State University
Ames, Iowa 50011, USA

CR Subject Classification (1987): F.4.1, F.4.3

ISBN 978-0-387-97288-6 ISBN 978-0-387-34804-9 (eBook)
DOI 10.1007/978-0-387-34804-9

Originally published by Springer-Verlag Berlin Heidelberg New York in 1990

2145/3140-543210 – Printed on acid-free paper

Dedicated to the memory of
Evelyn M. Nelson

Preface

Algebraic methods, in particular those of universal algebra and algebraic logic, are playing an increasingly important role in computer science, especially in the areas of algebraic specification of data types, relational database theory, logic of programs, functional and logic programming, and semantics of programming languages. To a large extent this work has been carried forward by computer scientists independently of the very active group of mathematicians who work in universal algebra and algebraic logic. In spite of the fundamental differences between the types of problems the two groups of researchers have been working on, we were convinced that a very substantial area of common interest could be uncovered once the problem of communication was overcome. Thus it was decided to hold a conference at Iowa State University from June 1 to 4, 1988. The main purpose of this conference was to bring together leading researchers from both areas to identify this common ground.

Invited hour addresses were given by Joel Berman (Illinois, Chicago), H. Peter Gumm (SUNY New Paltz), Bjarni Jónsson (Vanderbilt), Dexter Kozen (Cornell), István Németi (Hungarian Academy of Sciences, Budapest), Vaughan Pratt (Stanford), Dana Scott (Carnegie-Mellon), and Eric Wagner (IBM, Yorktown Heights). The program also included the presentation of 23 contributed papers and a round-table discussion on the role of algebra and logic in computer science moderated by George Strawn (Iowa State University). There were 78 participants, about evenly divided between those who characterized themselves as computer scientists and those as mathematicians. Papers submitted to the proceedings of the conference were thoroughly refereed. The 16 papers and extended abstracts included here represent a wide range of topics at the interface of algebra and computer science.

The organizers gratefully acknowledge the support of the Mathematics and Computer Science Departments, the Graduate College and the Computation Center of Iowa State University. The conference was also supported by grants from the National Science Foundation, the Office of Naval Research, and the Institute for Applied Mathematics in Minneapolis. We greatly appreciate the fine job performed by the referees. Finally the organizers want to thank the editorial staff of Springer-Verlag for their help and encouragement.

The editors
Ames, Iowa
November 24, 1989

Table of Contents

Conference Program

Wednesday, June 1

MORNING SESSION — GEORGE MCNULTY, MODERATOR

Invited Address: Eric Wagner, IBM, *All recursive types defined using products and sums can be implemented using pointers*

Ivo Rosenberg, Univ. of Montreal, *Term equations via Mal'cev preiterative algebras*

Ildikó Sain, Hungarian Academy of Sciences, *Comparative study of distinguished program verification methods*

AFTERNOON SESSION — WILLIAM LAMPE, MODERATOR

Invited Address: Joel Berman, Univ. of Illinois, Chicago, *The value of free algebras*

Lawrence Moss, Univ. of Michigan, *Final algebra semantics for insufficiently complete specifications*

Ivan Rival, Univ. of Ottawa, *Graphical data structures for ordered sets*

Hong X. Dang, SUNY Geneseo, *On sublattice lattice varieties*

George McNulty, Univ. South Carolina, *Avoidable words*

Thursday, June 2

MORNING SESSION — DAVID SCHMIDT, MODERATOR

Invited Address: Dana Scott, Carnegie-Mellon University, *Domains and algebras*

Zbigniew Stachniak, York University, *The resolution rule: an algebraic perspective*

Alan Day, Lakehead University, *The interval construction revisited*

Ernie Manes, Univ. of Massachusetts, *Assertions, interleavings and atoms*

AFTERNOON SESSION — STEPHEN COMER, MODERATOR

Invited Address: István Németi, Hungarian Academy of Sciences, *Epimorphisms in algebraic logic with applications to the Beth definability theorem*

Marek Zaionc, Univ. of Alabama, Birmingham, *A characteristic of definable tree operations*

H. Albert Lilly, Univ. of Alabama, Birmingham, *A survey of current research into the use of functional specifications for the compilation of programming languages*

Ivo Düntsch, Univ. of Brunei, Darussalam, *On Galois closed algebras of binary relations*

Richard Thompson, Univ. of California, Berkeley *The manipulatory foundations of non-parallel programming*

EVENING SESSION

Round-table discussion: The role of algebra and logic in computer science. Moderator: George Strawn, Iowa State University

Friday, June 3

MORNING SESSION — DAVID B. BENSON, MODERATOR

Invited Address: Dexter Kozen, Cornell University, *Stone duality in programming language semantics*

H.P. Sankappanavar, SUNY New Paltz, *Linked double weak Stone algebras*

Erzsébet Lukács, Vanderbilt University, *Representability of finite relation algebras with many identity atoms*

Mitsuhiro Okada, Concordia University, *Algebraic proof of normalization theorems for polymorphic lambda calculus and higher order logics*

AFTERNOON SESSION — IVO ROSENBERG, MODERATOR

Invited Address: H.P. Gumm, SUNY New Paltz, *The role of universal algebra in computer science*

Fernando Guzmán, SUNY Binghamton, *Conditional logic*

George Nelson, Univ. of Iowa, *Other logics for equational theories*

Robert W. Quackenbush, Univ. of Manitoba, *The completeness theorem for the universal logic of algebras via congruences*

Irving Anellis, Philosophia Mathematica, *Maslov's inverse method and its application to programming logic*

Saturday, June 4

MORNING SESSION — ERNIE MANES, MODERATOR

Invited Address: Vaughan Pratt, Stanford University, *Dynamic algebras II: the constructive fragment*

Marek Suchenek, Wichita State University, *Incremental models of incomplete information data bases*

Nistala V. Murthy, Univ. of Toledo, *Essentially algebraic categories*

W.D. Maurer, George Washington University, *Three fundamental correctness theorems for the modification index method*

AFTERNOON SESSION — ROBERT QUACKENBUSH, MODERATOR

Invited Address: Bjarni Jónsson, Vanderbilt University, *Relatively free relation algebras*

David Benson, Washington State University, *Interaction automata*

R. Padmanabhan, Univ. of Manitoba, *Equational logic on algebraic curves*

Chihyi Ying, *First-order Boolean-valued semantics*

List of Participants

J.C. Abbott	U.S. Naval Academy
Hajnal Andréka	Hungarian Academy of Sciences
Irving H. Anellis	Philosophia Mathematica
David B. Benson	Washington State University
Clifford Bergman	Iowa State Univeristy
Joel Berman	University of Illinois at Chicago
Jim Bieman	Iowa State University
Willem Blok	University of Illinois at Chicago
Robert Cacioppo	Northeast Missouri State University
William Chao	University of Alabama at Birmingham
Stephen D. Comer	The Citadel
Hong X. Dang	SUNY, Geneseo
Kyung-Goo Doh	Kansas State University
Ivo Duntsch	Universiti Brunei Darussalam
Susan Even	Iowa State University
Isidore Fleischer	University of Iowa
Gebhard Fuhrken	University of Minnesota
Hernando Gaitan	Iowa State University
Abraham Goetz	University of Notre Dame
H.P. Gumm	SUNY New Paltz
Fernando Guzmán	SUNY Binghamton
Lucien Haddad	University of Waterloo
Hartmut Höft	Eastern Michigan University
Margret Höft	University of Michigan
Bjarni Jónsson	Vanderbilt University
David Kelly	University of Manitoba
Dexter Kozen	Cornell University
Peter Ladkin	Kestrel Institute
William Lampe	University of Hawaii
Dean R. Lass	Iowa State University
Albert Lilly	University of Alabama at Birmingham
Erzsébet Lukács	Vanderbilt University
George McNulty	University of South Carolina
Wendy MacCaull	Francis Xavier University
Roger Maddux	Iowa State University
Ernie Manes	University of Massachusetts
W.D. Maurer	George Washington University
J.D. Monk	University of Colorado
David L. Morgan	Kennesaw College

Larry Moss	University of Michigan
Nistala V. Murthy	University of Toledo
David A. Naumann	University of Texas, Austin
George C. Nelson	University of Iowa
István Németi	Hungarian Academy of Sciences
William C. Nemitz	Livonia, MI
Mitsuhiro Okada	Concordia University
R. Padmanabhan	Univ. of Manitoba
John Pedersen	UCLA
J.D. Phillips	Iowa State University
Thomas F. Piatkowski	SUNY Binghamton
Don Pigozzi	Iowa State University
Craig Platt	University of Manitoba
Vaughan Pratt	Stanford University
Robert Quackenbush	University of Manitoba
Ivan Rival	University of Ottawa
Ivo Rosenberg	University of Montreal
Ildikó Sain	Hungarian Academy of Sciences
H. P. Sankappanavar	SUNY New Paltz
David Schmidt	Kansas State University
Dana S. Scott	Carnegie Mellon University
Shigeko Seki	California State University, Fresno
Jacob Shapiro	Baruch College
Jerry Shelton	Computer Sciences Corporation
Giora Slutzki	Iowa State University
Laura Smith	Iowa State University
Zbigniew Stachniak	York University
Michael Stone	University of Calgary
George Strawn	Iowa State University
Marek A. Suchenek	The Wichita State University
Richard J. Thompson	University of California, Berkeley
Steven Tschantz	Vanderbilt University
Eric G. Wagner	IBM
Todd Wilson	Fresno, California
Chihyi Ying	Dubuque, Iowa
Marek Zaionc	University of Alabama at Birmingham

Relatively Free Relation Algebras

H. Andréka[1], B. Jónsson[2], I. Németi[1]

Introduction

Every non-trivial variety of relation algebras contains as a member a non-trivial finite relation algebra, and from this it follows by Jónsson–Tarski [5] that a free algebra on n generators in such a variety is never generated by a set with fewer than n elements. The answers to two related questions, however, are less trivial: (a) Can a free algebra on n generators contain as a subalgebra a free algebra on $n+1$ generators? (b) Can a free algebra on n generators be nonfreely generated by an n element set? For many familiar varieties, such as the variety of all groups and the variety of all lattices, the answer is known to be affirmative for (a) and negative for (b). In the present situation this is reversed. More precisely, the answer to (a) is negative for a large class of varieties of relation algebras, while the answer to (b) is affirmative for every variety of relation algebras that contains an infinite full relation set algebra. We also investigate atoms in free relation algebras.

Our results on (b) make use of properties that are special to relation algebras, and they involve some rather detailed arithmetic arguments. The results on (a), on the other hand, are really special cases of results in universal algebra. Indeed, the only facts that we need are that the variety under consideration has the congruence extension property, and that some non-trivial finite member is an absolute retract. The results about atoms in a free algebra use the universal algebraic condition that the variety under consideration is a discriminator variety. In locally finite discriminator varieties we completely describe the finitely generated free algebras.

1 Embeddability of the $(n+1)$–Generated Free Algebra in the n–Generated One, and Atoms in a Free Algebra

1.1 Preliminaries

We list for easy reference the basic facts about discriminator varieties that

[1]Mathematical Institute of the Hungarian Academy of Sciences, Budapest, PF.127, H-1364, Hungary
[2]Dept. of Mathematics, Vanderbilt University, Nashville, TN 37235, U.S.A.

we will be using. These results are due to many authors, but the primary contributors are A. Pixley and H. Werner. A largely selfcontained exposition can be found in Werner [14], where more detailed credits and references are also provided.

The (ternary) discriminator on a set U is the ternary operation f on U such that, for x, y, z in U,

$$f(x,y,z) = \begin{cases} z & \text{if } x = y \\ x & \text{otherwise.} \end{cases}$$

A variety V of algebras is called a discriminator variety if V is generated by a class K of algebras, with the property that some ternary term t represents the ternary discriminator on each member of K. We call t a discriminator term for K. The theorem below, which is essentially Theorem 2.2 in Werner [14], contains most of the information that we need about discriminator varieties. Given a class K of algebras, we write $S(K)$ for the class of all subalgebras of members of $K, Pu(K)$ for the class of all ultraproducts of members of K, $Si(K)$ for the class of all subdirectly irreducible members of K, and $Si'(K)$ for $Si(K) \cup \{A\}$, where A is the one–element algebra. An algebra is said to be semisimple if it is a subdirect product of simple algebras, and a variety is said to be semisimple if all its members are semisimple. For an algebra A, $\mathrm{Con}(A)$ is the lattice of all congruence relations on A, and if u and v are members of A, then $\mathrm{con}(u, v)$ is the principal congruence generated by (u, v). A variety V of algebras is said to have the congruence extension property if, for every algebra A in V, every congruence relation on a subalgebra B of A is equal to the restriction to B of a congruence relation on A.

Theorem 1.1. *Suppose V is a discriminator variety generated by a class K, with t a discriminator term for K. Then the following statements hold.*

I. V is congruence permutable, congruence distributive and semisimple, and V has the congruence extension property.

II. For every non–trivial algebra A in V, the following conditions are equivalent:

(i) *A is subdirectly irreducible.*
(ii) *A is simple.*
(iii) *A is directly indecomposable.*
(iv) *$A \in SPu(K)$*
(v) *t represents the discriminator on A.*

III. For every algebra A in V:

(i) *For all $u, v \in A$, $\mathrm{con}(u, v) = \{(x, y) : t(u, v, x) = t(u, v, y)\}$.*
(ii) *Every principal congruence on A is a factor congruence.*
(iii) *Every compact congruence on A is principal.*
(iv) *The principal congruences on A form a sublattice of $\mathrm{Con}(A)$.*

IV. $Si'(V)$ is a universal class, and the map $M \longrightarrow \mathrm{Var}(M)$ is an isomorphism from the lattice of all universal subclasses of $Si'(V)$ onto the lattice *of all subvarieties of V.*

V. Let F be the smallest set of formulas in the language of V that contains all equations and is closed under conjunction, disjunction and implication. There is an effective way of associating with each formula α in F an equation α^* in such a way that α and α^* have the same truth set in every member of $Si(V)$.

1.2 Free algebras in discriminator varieties

We denote by $Fr(V, n)$ the V–free algebra on n generators. If V is a locally finite discriminator variety, then the finite algebra $Fr(V, n)$ is a direct product of finite subdirectly irreducible algebras. The factors are the members of $Si(V)$ that are n–generated (are generated by some set of order n or less). In order to describe how often each such algebra S occurs in the direct product, we need the following concept.

Definition 1.2. *Suppose S is an algebra.*

(i) *Two mappings f and g from a set X onto generating sets for S are said to be equivalent if there exists an automorphism h of S such that $g = hf$.*

(ii) *If n is a positive integer, then we denote by $\mu(S, n)$ the number of non–equivalent mappings from an n–element set onto a generating set for S.*

The cardinal $\mu(S, n)$ is of course an invariant of the algebra S, and does not depend on the choice of the set. If S is finite, which is the only case that concerns us, then $\mu(S, n)$ is a natural number; it will be positive iff S is n–generated.

Theorem 1.3. *If V is a locally finite discriminator variety, then $Fr(V, n)$ is isomorphic to the direct product of all the algebras S^m, with S an n–generated member of $Si(V)$ and $m = \mu(S, n)$.*

Corollary 1.4. *Suppose $\mathbf{V}'$ and $\mathbf{V}''$ are subvarieties of a discriminator variety $\mathbf{V}$ of a finite similarity type. If $\mathbf{V}' \cap \mathbf{V}''$ is locally finite, then for every positive integer n,*

$$Fr(\mathbf{V}' + \mathbf{V}'', n) \times Fr(\mathbf{V}' \cap \mathbf{V}'', n) \cong Fr(\mathbf{V}', n) \times Fr(\mathbf{V}'', n).$$

Recall that a subdirectly irreducible member S of a variety V is said to be splitting (in V) if the class of all subvarieties of V that do not contain S as a member has a largest member V^*, and that V^* is then called the conjugate variety of S.

Lemma 1.5. *Every finite, subdirectly irreducible member of a discriminator variety is splitting.*

We will be considering finite, subdirectly irreducible members S of a discriminator variety V, having a very strong additional property: S is maximal in $Si(V)$, in the sense that S is not embeddable in any other member of $Si(V)$. In Jónsson [3] it is shown that in the variety RA of all relation algebras, the full relation algebra on a finite set has this property.

The next two theorems will not be needed here, but they are relevant for the potential applications of our principal results.

We say that the algebra A is an absolute retract in the class K of algebras, if for any embedding $f : A \to B$, $B \in K$ there is a homomorphism $g : B \to A$ such that gf is the identity on A.

Theorem 1.6. *Suppose V is a variety and $S \in Si(V)$. Consider the following conditions.*

(i) *S is maximal in $Si(V)$.*
(ii) *S is an absolute retract in $Si(V)$.*
(iii) *S is an absolute retract in V.*

Then (i) ⇒ (iii) ⇒ (ii), and they are all equivalent if V is semisimple.

Remark 1.7. (*ii*) ⇒ (*i*) in Thm.1.6 is not true in general. As an example we can take the reduct of the variety of closure algebras where we omit $-$ and $\cdot$ (i.e. instead of the Boolean part we take any semilattice with 0 and 1). We note that in the variety of closure algebras, as well as in the variety CA_1 of cylindric algebras of dimension 1, there are no nontrivial finite retracts at all.

Theorem 1.8. *For a discriminator variety V, the following conditions are equivalent:*

(i) *Some nontrivial finite member of V is an absolute retract in V.*
(ii) *Some finite member of $Si(V)$ is maximal in $Si(V)$.*

Theorem 1.9. *Suppose V is a discriminator variety of a finite similarity type, and S is a finite maximal member of $Si(V)$. If A is a finitely generated member of V, then $A \simeq S^m \times (A/Q)$, where m is the number of non-equivalent epimorphisms from A onto S, and Q is the smallest congruence relation on A such that A/Q is in the conjugate variety of S.*

Lemma 1.10. *Suppose V is a discriminator variety of a finite similarity type, and S is a finite, maximal subdirectly irreducible member of V. If V^* is the conjugate variety of S, and if n is any positive integer, then,*

$$Fr(V,n) \simeq S^m \times Fr(V^*,n),$$

where $m = \mu(S,n)$.

Theorem 1.11. *Suppose the variety V has the congruence extension property, and suppose there exists a k–generated non–trivial finite absolute retract in V. Then, for every integer $n \geq k-1$, $Fr(V,n+1)$ is not embeddable in $Fr(V,n)$.*

PROOF: Let V, k and n be as in the statement of the theorem and let S be a non–trivial finite k–generated absolute retract in V. Assume $g : Fr(V,n+1) \rightarrowtail Fr(V,n)$ is an embedding.

First we show that g is injective over S in the sense that every surjective homomorphism $h : Fr(V, n+1) \twoheadrightarrow S$ factors through g.

Let $h : Fr(V, n+1) \twoheadrightarrow S$. Then by the congruence extension property of V, there are $S' \in V$ and homomorphisms $h' : Fr(V, n) \twoheadrightarrow S'$, $g' : S \rightarrowtail S'$ such that g' is one–one and $h'g = g'h$. Then $h''g'$ is the identity on S for some $h'' : S' \to S$ because S is an absolute retract in V. Now, $h = h^{\#}g$ where $h^{\#} = h''h'$.

Clearly, to different $h's$ different $h^{\#'}s$ belong. Therefore, to derive a contradiction, it is enough to show that there are more homomorphisms from $Fr(V, n+1)$ onto S than from $Fr(V, n)$ onto S. Let X, Y be the sets of free generators in $Fr(V, n), Fr(V, n+1)$ respectively. Then the set of homomorphisms from $Fr(V, n+1)$ onto S and the set of mappings from Y onto a generator set of S are in a one–one correspondence, and similarly for $Fr(V, n)$ and X. Let G be a generator set of S. For any set H, let $|H|$ denote the number of elements of H. If $2 \leq |G| \leq n+1$, then clearly there are more mappings from Y onto G than from X onto G, since $n = |X| < |Y|$. For other generator sets, i.e. if $|G| \leq 1$ or $|G| > n+1$, the number of surjective mappings from Y and X are the same. Since S is n–generated, non–trivial and finite, the desired conclusion follows.

Thus $Fr(V, n+1)$ is not embeddable in $Fr(V, n)$.

We now turn to atoms in free algebras. Let $\leq$ be a partial ordering on a set A with least element 0. We say that $a \in A$ is an atom with respect to $\leq$ if $0 < a$ and there is no b for which $0 < b < a$ holds. We say that $(A, \leq)$ is atomic iff below each $a \neq 0$ there is an atom.

Theorem 1.12. *Let V be a discriminator variety of a finite similarity type. Let the partial ordering $\leq$ be defined in V by a set of equations (in the language of V) such that $\leq$ has a least element in every member of V. Let n be a positive integer. Then the following hold.*

(i) *$Fr(V, n)$ is atomic with respect to $\leq$ if V is generated by finite algebras.*

(ii) *$Fr(V, n)$ has an atom if V has a nontrivial finite algebra.*

(iii) *$Fr(V, n)$ has at least k atoms if V has at least k non–isomorphic, n–generated finite maximal subdirectly irreducible members.*

1.3 Applications to relation algebras

The variety RA of all relation algebras is a discriminator variety, and hence so is every subvariety V of RA. Our results therefore apply to such a variety V provided $Si(V)$ has a finite maximal member. In Jónsson[3], Lemma 7.4, it was shown that $R(n)$, the full algebra of relations on an n–element set, is maximal in $Si(RA)$, so the results apply whenever V has some $R(n)$ as a member. Also, $R(n)$ is one–generated for every integer n. Actually all these statements apply with RA replaced by the larger variety

SA of all semiassociative relation algebras. This concept was introduced and investigated in Maddux [6–8]. We recall the relevant facts.

A nonassociative relation algebra is an algebra $A = (A_0, ;, \breve{}, 1')$ consisting of a Boolean algebra A_0, a binary operation ;, a unary operation $\breve{}$, and a distinguished element $1'$, such that $1'$ is an identity element for the operation ;, and, for a fixed element a, the operations

$$x \longrightarrow a; x, \qquad x \longrightarrow a^{\smile}; x$$

are conjugate, as well as the operations

$$x \longrightarrow x; a, \qquad x \longrightarrow x; a^{\smile}.$$

This means that, for all elements a, x, y,

$$\begin{aligned} &1'; x = x = x; 1', \\ &(a; x)y = 0 \qquad \text{iff} \qquad (a^{\smile}; y)x = 0, \\ &(x; a)y = 0 \qquad \text{iff} \qquad (y; a^{\smile})x = 0. \end{aligned}$$

If, in addition, the identity $(x; 1); 1 = x; 1$ (or $(x1'; 1); 1 = x1'; 1$ resp.) holds, then A is called a semiassociative (or weakly associative resp.) relation algebra. We denote by NA the class of all nonassociative relation algebras, and by SA (WA) the class of all semiassociative (weakly associative) relation algebras.

The class NA is a variety, and hence so are SA and WA. WA is a natural variety of algebras of binary relations. Namely, it is proved in [7] that, up to isomorphism, the elements of a WA are binary relations such that the greatest element is a reflexive and symmetric relation. Equivalently, weakly associative relation algebras are subalgebras of representable relation algebras relativized to reflexive and symmetric relations. The identity $1; 1 = 1$ holds in NA, and the extra axiom defining SA is therefore a very special case of the associative law. As was shown in Maddux [6], [8] §4, it implies that the associative law holds for the relative product of any elements $a_0, a_1, \ldots, a_n$, as long as at least one of the elements equals 1. I.e., for any such sequence, the relative product is independent of how the product is bracketed. For this reason, the theory of homomorphisms for RA carries over to SA. This is true, in particular, of the characterization of subdirect irreducibility.

Theorem 1.13. *(Maddux [6]) A nontrivial algebra A in SA is subdirectly irreducible iff $1; x; 1 = 1$ for every nonzero element x in A.*

Corollary 1.14. *SA is a discriminator variety.*

Theorem 1.15. *For every positive integer n, $R(n)$ is a maximal member of $Si(SA)$.*

Theorem 1.16. *Suppose V is a subvariety of SA, and n is a positive integer. If, for some positive integer k, $R(k)$ belongs to V, then $Fr(V, n+1)$ is not embeddable in $Fr(V, n)$.*

Below we will show that there is a subvariety V of RA for which the opposite is true, i.e. $Fr(V,m)$ is embeddable in $Fr(V,n)$ for any integers $m \geq n$. (Moreover, the subvariety ${}_\infty RRA$ of all representable relation algebras with infinite bases is such.) This will also show that in Thm.1.11, the condition asserting the existence of a finite absolute retract cannot be omitted. Also, in Thm.1.18 the conditions are necessary.

Let ${}_\infty RRA$ denote the subvariety of RA generated by $\{R(U) : U$ is an infinite set $\}$, where $R(U)$ denotes the relation set algebra of all binary relations over U. Let ω denote the smallest infinite ordinal.

Theorem 1.17. *$Fr({}_\infty RRA, \omega)$ is embeddable in $Fr({}_\infty RRA, 1)$.*

We turn to subvarieties of NA. The following lemma implies that for $n > 1$, $R(n)$ is not a retract (hence is neither maximal among the subdirectly irreducibles) in NA or WA.

Lemma 1.18. *A finite algebra is a retract in WA (in NA) iff it is a Boolean relation algebra.*

It is easy to check that every variety of Boolean algebras with operators has the congruence extension property, see e.g. Sain[12]15.6, where also more general sufficient conditions are given for a class of algebras to have the congruence extension property. Hence, every subvariety of NA has the congruence extension property, too.

Corollary 1.19. *Suppose V is a subvariety of NA and n is a positive integer. If V has a Boolean relation algebra as a member, then $Fr(V, n+1)$ is not embeddable in $Fr(V,n)$. In particular, $Fr(NA, n+1)$ is not embeddable in $Fr(NA, n)$, and $Fr(WA, n+1)$ is not embeddable in $Fr(WA, n)$.*

We turn now to applications to cylindric algebras. While relation algebras serve as algebras of binary relations, cylindric algebras serve as algebras of relations of higher rank. The basic reference on cylindric algebras is the two volume treatise Henkin–Monk–Tarski[2]. We recall here the definition and a few basic facts.

A cylindric algebra of dimension n, where n is an ordinal, is a Boolean algebra A with unary operators c_i and distinguished elements d_{ij}, for $i, j < n$, such that for all $i, j, k < n$ and $x, y \in A$, the following conditions hold:

(I) The c_i's are self–conjugate closure operations commuting with each other, i.e.

(1) $x \leq c_i x = c_i c_i x$

(2) $(c_i x) \cdot y = 0$ iff $x \cdot c_i y = 0$

(3) $c_i c_j x = c_j c_i x$

(II) The constants d_{ij} satisfy the following:

(4) $d_{ij} \cdot d_{jk} \leq d_{ik}, \quad d_{ij} = d_{ji}, \quad d_{ii} = 1$

(5) $c_i d_{ij} = 1, \quad c_k d_{ij} = d_{ij}$ if $k \neq i, j$

(6) $(c_i x) \cdot d_{ij} = x$ if $x \leq d_{ij},\ i \neq j$.

An equivalent form of saying that c_i is self–conjugate is to say that the complement of a closed element is closed. Thus, in the above definition, (2) can be replaced with

(2)′ $c_i - c_i x = -c_i x$.

We note that (6) expresses that the closure operator c_i is "discrete", or is the identity, when relativized to $d_{ij},\ i \neq j$. The class of all cylindric algebras of dimension n is written CA_n.

The primary models for the above axioms are threefold: algebras of n–ary relations, subsets of an n–dimensional geometric space, and algebras of formulas in first–order theories. Here we consider only algebras of the first kind.

Let $R_n(U)$ denote the set of all n–ary relations on U, i.e. if U^n denotes the set of all n–termed sequences of elements of U, then

$$R_n(U) = \{R : R \subseteq U^n\}.$$

Now the usual set theoretic Boolean operations are meaningful in $R_n(U)$. The unary operations c_i denote "erasing the i–th column" in a relation and the constants d_{ij} denote the identity relations: if R is an n–ary relation on U and $i, j < n$ then

$$c_i R = \{\langle s_0, \ldots, s_{i-1}, u, s_{i+1}, \ldots \rangle : s \in R,\ u \in U\}, \quad \text{and}$$
$$d_{ij} = \{s \in U^n : s_i = s_j\}.$$

$R_n(U)$ will denote also the full algebra of n–ary relations on U with the above operations. By a cylindric set algebra of dimension n and with base U we mean a subalgebra of $R_n(U)$. By a representation of a cylindric algebra A we mean an embedding of A into a subdirect product of cylindric set algebras. This is equivalent to the following. If V is a disjoint union of Cartesian spaces $U_i{}^n$, then the set $\mathcal{R}(V)$ of subrelations of V form a natural cylindric algebra. Now, a representation of A is an embedding $A \rightarrowtail \mathcal{R}(V)$ for some appropriate V. RCA_n denotes the class of all representable cylindric algebras of dimension n. RCA_n is a variety (a theorem due to Tarski), see Thm.s 3.1.103,108 in [2], Part II, and pp.43,46 and perhaps p.402 of Part I.

Remark 1.20. (Application to cylindric algebras) (i) As an application of Thm.1.11, we now show that the corresponding theorem is true for cylindric algebras in general. In more detail, the following two statements $(*)-(**)$ are true.

- $(*)$ If $V \subseteq CA_m$, m is an integer, and a full cylindric set algebra with a base of size smaller than m is in V, then $Fr(V, n+1)$ is not embeddable in $Fr(V, n)$, for any integer n.
- $(**)$ If $V \subseteq CA_m$, m is an infinite ordinal, and there is an algebra in V with finite characteristic, then $Fr(V, n+1)$ is not embeddable in $Fr(V, n)$, for any integer n.

One can obtain (*) for $m \geq 2$ as an immediate corollary of Thm.1.11, because if m is an integer then CA_m is a discriminator variety, and if $m \geq 2$ then any finite full cylindric set algebra of dimension m and with base smaller than m is a one–generated, maximal member of $Si(CA_m)$, see Comer[1]Thm.3.8(1). For $m < 2$, (*) is true because both CA_0 and CA_1 are locally finite varieties (see Remark 2.3 at the end of this paper). However, (**) can be proved as an application of Thm.1.11, too, despite the fact that if m is infinite then only discrete CA_m's are finite.

(ii) We can apply Thm.1.11 to subvarieties of the broader class $SCr_m \supset CA_m$, too. If m is an ordinal, then SCr_m and $ICrs_m$ denote the varieties of all subalgebras of cylindric–relativized algebras and of all isomorphic copies of cylindric–relativized set algebras of dimension m, respectively, cf. [2]. Then $CA_m \subset SCr_m$ and $CA_m \cap ICrs_m \subset ICrs_m \subset SCr_m$ for $m \geq 3$. Now SCr_m has the congruence extension property because every member of SCr_m is a Boolean algebra with operators. Also, it is easy to see that any finite discrete CA_m is an absolute retract in SCr_m. Hence we obtain the following (* * *) as an immediate corollary of Thm.1.11.

(* * *) Let $V \subseteq SCr_m$ be any variety such that V contains a discrete CA_m. Then for any positive integer $n, Fr(V, n+1)$ is not embeddable in $Fr(V, n)$. In particular, $Fr(SCr_m, n+1)$ is not embeddable in $Fr(SCr_m, n)$, and $Fr(ICrs_m, n+1)$ is not embeddable in $Fr(ICrs_m, n)$.

(iii) We note that in all the above applications we use the existence of the diagonal constants when showing that some particular algebras are absolute retracts. It seems that if we omit these constants then no finite absolute retracts remain in the class, e.g. there are no absolute retracts in the class Df_m of diagonal–free cylindric algebras, or in the class PA_m of polyadic algebras. The same applies to relation algebras if we omit the identity constant $1'$ from the basic operations.

We turn to applications of Thm.1.12 and Lemma 1.10. An atom of a Boolean algebra with operators is meant to be an atom of the Boolean part of the algebra. This corresponds to the case when the partial ordering $\leq$ is defined by the equation $x \leq y \overset{df}{\Longleftrightarrow} x + y = y$. RRA denotes the class of representable relation algebras.

Corollary 1.21. *Let $V \subseteq SA$ be a variety and let n be a positive integer. Then (i)–(iv) below hold.*

(i) *$Fr(V, n)$ is atomic if V is generated by finite algebras.*
(ii) *$Fr(V, n)$ is not atomless if V has a nontrivial algebra.*
(iii) *$Fr(V, n)$ has infinitely many atoms if $RRA \subseteq V$.*
(iv) *$Fr(V, n)$ has exactly 2^n ideal atoms (i.e. atoms that are ideal elements) if V contains a Boolean RA.*

Remark 1.22. (i) WA, NA are not discriminator varieties and SA, RA,

RRA are not generated by their finite members (see Jónsson [3]). Indeed, it is proved in Németi[9] that if $V \subseteq SA$ is a variety such that $R(\omega) \in V$ then $Fr(V,n)$ is not atomic. (This proof goes via showing that first-order logic can be built up in SA, and then using Gödel's incompleteness theorem.) Németi also proved that there is an atom of the ideal elements of $Fr(RA,n)$ below which there is no atom of $Fr(RA,n)$. It is also proved in [9] that $Fr(V,n)$ is not atomic if $m \geq 3$ is any ordinal, $V \subseteq CA_m$, and the full cylindric set algebra with an infinite base is in V.

(ii) Many results in [2] Chapter 2.5 follow from our theorems. Let n,m be positive integers. Then $Fr(V,n)$ is atomic if $V \subseteq CA_m$ is generated by finite algebras (this is [2] 2.5.7(i),(ii)), $Fr(CA_m,n)$ has infinitely many atoms ([2] 2.5.9), moreover if $V \subseteq CA_m$ is any variety such that all representable finite cylindric algebras are contained in V, then $Fr(V,n)$ contains infinitely many atoms. Also, the following generalization of [2] 2.5.11 (for finite dimensions) is a corollary of our results.

(*) Let n,m be positive integers, and assume that $V \subseteq CA_m$ contains a discrete CA_m. Then $Fr(V,n)$ contains exactly 2^n zero-dimensional atoms.

Finally, we give some applications of Thm.1.3. Let B_1, B_2, B_3 be the subvarieties of RA defined by the equations $0';0' = 0$, $0';0' = 1'$ and $0';0' = 1$ respectively (cf.[3]). Let n be any positive integer. It is easy to see that

$$Fr(RA,n) \simeq Fr(B_1,n) \times Fr(B_2,n) \times Fr(B_3,n).$$

Now, B_1 and B_2 are locally finite discriminator varieties, and $Si(B_1)$ is $\{R(1)\}$, $Si(B_2)$ is $\{\mathcal{E}_2, R(2)\}$ up to isomorphisms, where $\mathcal{E}_2$ is the four-element relation algebra with $0';0' = 1' \neq 1$. All these three algebras are one-generated. It is not difficult to compute the following:

$$\mu(R(1),n) = 2^n,$$
$$\mu(\mathcal{E}_2,n) = 4^n, \text{and}$$
$$\mu(R(2),n) = \frac{1}{2}(16^n - 4^n).$$

Therefore by Thm.1.3 we have

$$Fr(B_1,n) \simeq R(1)^{2^n},$$
$$Fr(B_2,n) \simeq \mathcal{E}_2^{4^n} \times R(2)^{\frac{1}{2}(16^n - 4^n)},$$

e.g $Fr(B_2,1) \simeq \mathcal{E}_2^4 \times R(2)^6$, $Fr(B_2,2) \simeq \mathcal{E}_2^{16} \times R(2)^{20}$, etc. Also, one can compute

$$|Fr(B_2,n)| = 4^{(16^n)}.$$

This result is analogous to [2] 2.5.62.

Let $RA(k)$ denote the variety generated by $R(k)$, for all integers k as in [3]. Then, as above, Thm.1.3 gives a complete description of $Fr\,(RA(k),n)$.

As an application of Thm.1.3 to cylindric algebras, now we give a complete structural description of $Fr(CA_1, n)$ for positive integers n, and show that [2] 2.5.62 is a corollary of this description. For any integer n, let $\mathcal{B}_n$ denote the simple CA_1 with n atoms, i.e. the Boolean part of $\mathcal{B}_n$ is the simple Boolean algebra with n atoms and $C_1 b = 1$ for all nonzero elements of $\mathcal{B}_n, C_1 0 = 0$.

Theorem 1.23. *Let n be any positive integer. Then*

$$Fr(CA_1, n) \simeq \prod_{m=1}^{2^n} \mathcal{B}_m{}^{\frac{2^n!}{m!(2^n-m)!}}$$

Corollary 1.24. *([2] 2.5.62) Let n be a positive integer. Then $Fr(CA_1, n)$ has $2^{2^{(n+2^n-1)}}$ elements.*

2 Non–free Generators in Free Algebras

Theorem 2.1. *For any variety $V \subseteq SA$ with $R(U) \in V$ for some infinite U, the V–free algebra on one generator also contains a single element that generates it non–freely.*

The basic construction in the proof of Theorem 2.1. We want to construct unary terms μ and ν such that

$$SA \models \nu\mu(x) = x, \qquad R(U) \not\models \mu\nu(x) = x.$$

If F is the V–free algebra on a single generator a, then the element $b = \mu(a)$ will also generate F, since $a = \nu(b)$, but b will not generate F freely, since $\mu\nu(b) = b$.

The construction of μ and ν will be motivated by describing the intended values of $\mu(R)$ and $\nu(R)$ when R is a member of the full algebra of binary relations on a set U. We want $\mu(R)$ and $\nu(R)$ to be equal to R except for certain special relations R. Since SA is a discriminator variety, such patching is possible provided the special relations are defined by a finite set of universal formulas.

The special relations that we consider are relations R with the following properties:

R totally orders U.
U has a first member (relative to R).
Every member of U except the last one (if it exists) has a successor.

For a special relation R we take $\mu(R)$ to be the total ordering of U obtained by transferring the first member of U to the last place. In defining $\nu(R)$ we consider two cases: If U has a last element, then this element is moved to the first position, but if no such element exists, then we take $\nu(R) = R$.

It is now clear how the special elements should be defined abstractly. Call an element x of a relation algebra a total order if

$$x;x \leq x, \qquad xx^{\smile} = 1', \qquad x + x^{\smile} = 1.$$

Define

$$x^{c} = x0'(x0';x0')^{-}.$$

For a total order x, x^c will serve as the successor relation. We say that x is special if x is a total order and

$$1;x0' < 1, \qquad x0';1 = x^c;1.$$

For a special element x of a simple relation algebra we define

$$\mu(x) = x(x^{\smile}0';1) + (1;x0')^{-},$$
$$\nu(x) = x(1;x^{\smile}0') + (x0';1)^{-},$$

while for all other elements $\mu(x) = \nu(x) = x$.

One readily checks that when applied to a member R of the full algebra of binary relations on a set U, these constructions yield the intended results, and hence that $\nu\mu(R) = R$. On the other hand, if R is special, and if U does not have a last element, then $\mu\nu(R) = \mu(R) \neq R$. Thus

$$RRA \models \nu\mu(x) = x, \qquad R(U) \not\models \mu\nu(x) = x.$$

This establishes the theorem for the special case $V = RRA$. To complete the proof in the general case we have to show that the identity $\nu\mu(x) = x$ holds in every semiassociative relation algebra.

Remark 2.2. (On the possibilities of generalizing Thm. 2.1.) In Jónsson–Tarski [5] it is proved that if a variety is generated by finite algebras then any n–element generator set of the n–generated free algebra generates it freely. Therefore, the assumption $R(U) \in V$ for some *infinite* U cannot be omitted from Thm.2.1.

By the same theorem in [5], the variety SA in Thm.2.1 cannot be replaced by either of the larger varieties WA or NA, for in Németi[10] is shown that these two varieties are generated by their finite members.

The cylindric algebraic counterpart of Thm.2.1, i.e. that $Fr(CA_m, 1)$ can be non–freely generated by a single element, is true for every ordinal $m \geq 3$. This is the solution of Problem 2.7 in [2]. For $m \geq 4$ this is implied by Thm.2.1, as follows. In the proof of Thm.2.1, we gave two relation algebraic terms, μ and ν, such that $SA \models \nu\mu x = x$ while $R(U) \not\models \mu\nu x = x$ for some U. Let μ', ν' be the corresponding CA_3–terms as defined in [2]§5.3. (i.e. we replace $x;y$ by $c_2(s_2^1 x \cdot s_2^0 y)$ etc). Then from $SA \models \nu\mu x = x$ we conclude that $CA_4 \models \nu'\mu' x = x$, therefore $CA_m \models \nu'\mu' x = x$ for all $m \geq 4$. Similarly, from $R(U) \not\models \mu\nu x = x$ we conclude that $\mu'\nu' x = x$ does not hold in the full cylindric set algebra of dimension m and with base U. This settles the case $m \geq 4$. The analog of Thm.2.1 for CA_3 is also true and is proved in [9]. (For $m = 3$ we note that $SA \models \nu\mu x = x$ does not imply $CA_3 \models \nu'\mu'x = x$ because in the proof of $SA \models \nu\mu x = x$ we used the Peircean law several times and in CA_3 the corresponding law does not hold.) For $m < 3$, the analog of Thm.2.1 is not true for CA_m. This is proved in [2] 2.5.23.

Remark 2.3. (On the general case, concerning both questions (a) and (b)).

Let us recall from the introduction that questions (a) and (b) are the following: (a) can a free algebra on n generators contain as a subalgebra a free algebra on $n+1$ generators? (b) can a free algebra on n generators be non–freely generated by an n–element set?

Let V be a variety in which no nontrivial equation holds, i.e. let V be a similarity class. Then the answer to (b) is clearly no, since the only irredundant generator set of any absolutely free algebra is the free one, while the answer to (a) is negative if we have at most one unary operation beside constants, positive otherwise.

Let V be any locally finite variety. Then the answers to both questions are negative. For (b) this follows from [5], since V is generated by its finite members. For (a) the answer is negative because of the following. Both $Fr(V, n+1)$ and $Fr(V, n)$ are finite. $Fr(V, n)$ is a subalgebra of $Fr(V, n+1)$, and it is a proper subalgebra because $Fr(V, n+1)$ cannot be generated by n elements, since the answer to (b) is negative. Thus $Fr(V, n+1)$ has more elements than $Fr(V, n)$, hence cannot be embedded in it. This implies e.g. that for the variety of distributive lattices the answers to both questions are negative.

The two questions (a) and (b) are completely independent of each other in the sense that for all 4 combinations of the answers there are varieties with that combination of answers. The following table demonstrates this.

$Fr(V, n+1)$ embeddable in $Fr(V, n)$	$Fr(V, 1)$ non–freely generated	examples
YES	YES	$_\infty RRA$
NO	YES	RA, SA
YES	NO	groups, lattices, similarity class of one binary operation
NO	NO	locally finite varieties, similarity class of one unary operation.

Acknowledgments: Research supported by Hungarian National Foundation for Scientific Research grant No. 1810 and by NSF grant DMS–8800290.

References

1. Comer, S.D., *Galois-theory of cylindric algebras and its applications.*, Trans. Amer. Math. Soc. **286** (1984), 771–785.
2. Henkin, L., Monk, J.D. and Tarski, A., *Cylindric Algebras, Parts I–II*, North–Holland, Amsterdam.
3. Jónsson, B., *Varieties of relation algebras*, Algebra Universalis **15** (1982), 273–298.
4. Jónsson, B. and Tarski, A., *Boolean algebras with operators. Part II*, Amer. J. Math. **74** (1952), 127–162.
5. Jónsson, B. and Tarski, A., *On two properties of free algebras*, Math. Scand. **9** (1961), 95–101.
6. Maddux, R., "Topics in relation algebras," Ph.D Thesis, University of California, Berkeley, 1978.
7. Maddux, R., *Some varieties containing relation algebras*, Trans. Amer. Math. Soc **272** (1982), 501–526.
8. Maddux, R., *Pair-dense relation algebras*, Trans. Amer. Math. Soc. (to appear).
9. Németi, I., "Free algebras and decidability in algebraic logic," Dissertation for D.Sc. with Hung. Academy of Sciences (In Hungarian.), Budapest, 1986.
10. Németi, I., *Decidability of relation algebras with weakened associativity*, Proc. Amer. Math. Soc. 100 **2** (1987), 340–344.
11. Németi, I., *On varieties of cylindric algebras with applications to logic*, Annals of Pure and Applied Logic **36** (1987), 235–277.
12. Sain, I., *Strong amalgamation and epimorphisms of cylindric algebras and Boolean algebras with operators.*, Partially abstracted in: Sain,I., *Amalgamation, Epimorphisms, and Definability Properties in Algebraic Logic*, Bulletin of the Section of Logic, Vol 18, No 2 (Warsaw–Lodz 1989), 72–78., Studia Logica (to appear).
13. Tarski, A. and Givant, S., "A formalization of set theory without variables," Colloquium publications **41**, AMer. Math. Soc., Providence, Rhode Island, 1987.
14. Werner, H., "Discriminator algebras," Akademie Verlag, Berlin, 1978.

Keywords. Discriminator varieties, relation algebras, cylindric algebras
1980 *Mathematics subject classifications:* 06, 08

The Value of Free Algebras

Joel Berman[1]

In this paper I will present, without proofs, some examples and results that illustrate how free algebraic systems arise in various areas of computer science. The algebras I consider are finite and homogeneous and have only a finite number of fundamental operations.

The following notation will be used throughout. An *algebra* $\mathbf{A} = \langle A, F \rangle$ is a set A, called the *universe* of $\mathbf{A}$, together with an indexed family F of finitary operations on A. The *equational class* generated by $\mathbf{A}$, denoted $V(\mathbf{A})$, consists of all algebras of the same similarity type as $\mathbf{A}$ that satisfy every identity that holds in $\mathbf{A}$. If W is an equational class, then $\mathbf{F}_W(X)$ is the free W algebra generated by a set X of free generators. For finite n, $\mathbf{F}_W(n)$ is the free n-generated algebra for W and it is this finitely generated free algebra for various equational classes W that is the concern of this paper.

There are three facts about free algebras that are worth reviewing.

The first property is the usual definition of $\mathbf{F}_W(X)$: For every $\mathbf{B} \in W$ and every function $f : X \to B$ there is a homomorphism $h : \mathbf{F}_W(X) \to \mathbf{B}$ such that $f(x) = h(x)$ for all $x \in X$. I mention this property primarily to call attention to the fact that I do not explicitly use it in the paper.

The second fact about free algebras, and the one which is used throughout this paper, is in the case that $W = V(\mathbf{A})$ for a given algebra $\mathbf{A}$. If $W = V(\mathbf{A})$ and if $p(x_1, \ldots x_n)$ and $q(x_1, \ldots, x_n)$ are terms in the language of $\mathbf{A}$, then by definition of W, $\mathbf{A} \models p \approx q$ if and only if $W \models p \approx q$. Here $\mathbf{A} \models p \approx q$ means that for every $v : \{x_1, \ldots, x_n\} \to A$, $p^{\mathbf{A}}(v(x_1), \ldots, v(x_n)) = q^{\mathbf{A}}(v(x_1), \ldots, v(x_n))$. The connection to the free algebra $\mathbf{F} = \mathbf{F}_W(x_1, \ldots, x_n)$ is that $W \models p \approx q$ if and only if $p^{\mathbf{F}}(x_1, \ldots, x_n) = q^{\mathbf{F}}(x_1, \ldots, x_n)$. Thus an identity that holds in W corresponds to an equality between elements of $\mathbf{F}_W(n)$ and to an equality between two term operations of $\mathbf{A}$. This interplay between identities for W, term operations on $\mathbf{A}$, and elements of $F_W(n)$ repeatedly comes up in applications of free algebras.

The third fact about free algebras that I wish to mention is an old result of Birkhoff's which for $\mathbf{A}$ finite represents $\mathbf{F}_W(n)$ as a subalgebra of $\mathbf{A}^{|A|^n}$ and gives the bound $|F_W(n)| \leq k^{k^n}$ where $|A| = k$. This provides a concrete representation of $\mathbf{F}_W(n)$ as a canonically generated subalgebra of a direct product of copies of $\mathbf{A}$.

---

[1]Department of Mathematics, Statistics, and Computer Science University of Illinois at Chicago, Box 4348, Chicago, Il 60680

This paper is in final form and no version of it will be submitted for publication elsewhere.

As an example of these notions, consider the three-element algebra

$$\mathbf{A} = \langle \{T, F, U\}, \wedge, \vee, \neg \rangle$$

in which $\wedge$ and $\vee$ are binary operations and $\neg$ is unary given by the following tables:

$\wedge$	T	F	U
T	T	F	U
F	F	F	F
U	U	U	U

$\vee$	T	F	U
T	T	T	T
F	T	F	U
U	U	U	U

$\neg$	T	F	U
	F	T	U

As the notation for the operations suggests the algebra corresponds to truth values with conjunction, disjunction and negation in a nonclassical logic. Note that $\wedge$ and $\vee$ are not commutative. McCarthy [1963], [1963a] considers this system, in which U means undefined, for a mathematical model of computation. The noncommutativity of say, $x \wedge y$, arises when doing efficient evaluation, for if x is F there is no need to evaluate y.

An automated theorem prover for this logic has been developed by Nagata, Nakanishi, and Nishimura [1975]. An investigation of the expressive power of this system, i.e. a characterization of the class of polynomial operations $f : \{T, F, U\}^n \to \{T, F, U\}$ that can be built using $\wedge, \vee$, and $\neg$ and the constants, is presented in Zaslavskii [1979]. D. Gries, in his book, "The Science of Programming" [1981] calls this system conditional logic and uses it throughout the text. More recently, Guzman and Squier [1989] provide a careful description of the free algebra for the equational class generated by this algebra and in so doing they clearly delineate the algebraic structure of this class.

The case study described in the preceeding two paragraphs is a common one in that an algebra is described for some specific purpose and various people investigate the algebra in this given context. Eventually research focuses on the free algebra in the equational class generated by the algebra and a careful analysis of the free algebra illuminates not only the structure of the free algebra but of the algebraic properties of the class as a whole.

The next example is also a three-valued non-classical logic. Let $\mathbf{A}$ be the algebra $\langle \{T, F, U\}, \wedge, \vee, \neg, T, F \rangle$ in which $\wedge$ and $\vee$ are binary operations that behave as min and max in the ordered set $F < U < T$; $\neg$ is a unary operation with $\neg F = T$, $\neg T = F$, $\neg U = U$; and T and F are nullary constant operations. $\mathbf{A}$ is called a Kleene algebra and appears in Kleene's "Introduction to Metamathematics" as a way of dealing with partial recursive predicates. Berman and Mukaidono [1984] and Blamey [1986] contain a discussion of how this algebra appears in other contexts as well.

Let $K = V(\mathbf{A})$ be the equational class of Kleene algebras and let t be an n-ary term for K. So $t^{\mathbf{A}} : \{T, F, U\}^n \to \{T, F, U\}$ and a natural question is which n-ary operations arise as term operations in this way. It is known that an operation $f : \{T, F, U\}^n \to \{T, F, U\}$ is the term operation of a term t for K if and only if f preserves the partial order

and f preserves the set $\{T, F\}$. (An n-ary operation f is said to ***preserve*** a partial order R if $(x_i, y_i) \in R$, $i = 1, \ldots, n$ imply $(f(x_1, \ldots, x_n), f(y_1, \ldots, y_n)) \in R$. In such a case f is sometimes said to be ***monotonic*** with respect to the partial order. The operation f preserves a set S if $f(x_1, \ldots, x_n) \in S$ whenever all the x_i are in S.)

This characterization of the term operations of K leads to the theory of clones and their connection to free algebras. A ***clone*** C on a set A is a set of operations on A that contains all the projections on A and is closed under composition of operations. Thus if $\mathbf{A}$ is an algebra, then the set of all term operations of $\mathbf{A}$ forms a clone on the set A. The n-ary operations in this clone correspond to the distinct elements of $\mathbf{F}_W(n)$, where W is the equational class generated by $\mathbf{A}$. For finite A it is known that every clone C on A is the set of all operations on A that preserve a particular set of relations on A. Further background on clone theory is in McKenzie, McNulty and Taylor [1987] and Szendrei [1986] and specific information on the clone of monotonic operations on the partially ordered set  is in Mukaidono and Rosenberg [1986], Berman and Mukaidono [1984], and Blamey [1986].

The class of operations on $\{T, F, U\}$ that are monotonic on the partial order 

and that preserve the set $\{T, F\}$ form a clone on $\{T, F, U\}$. This clone of course is the intersection of two clones, the clone C_1 of operations that preserve the partial order and the clone C_2 of operations that preserve $\{T, F\}$. Both of these clones have received a good deal of attention. If $\mathbf{A}_1 = \langle \{T, F, U\}, \wedge, \vee, \neg, T, F, U \rangle$ is the Kleene algebra $\mathbf{A}$ with the nullary constant U adjoined as a fundamental operation and if K_1 is $HSP(\mathbf{A}_1)$ then the set of term operations on $\mathbf{A}_1$ is known to be C_1 and the n-ary operations in C_1 correspond to the elements of $\mathbf{F}_{K_1}(n)$. The clone C_2 is the same as the set of term operations of the 3-element Łukasiewicz algebra.

I present one more example of a non-classical logic to illustrate the interplay between clones and algebras. Let

$$\mathbf{A}_4 = \langle \{T, F, N, B\}, \wedge, \vee, \neg, T, F, B, N \rangle$$

in which $\wedge$ and $\vee$ are lattice operations in the lattice $\mathbf{L}_4 =$ 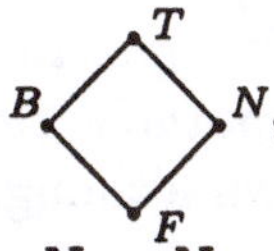

, $\neg$ is an unary operation given by $\neg T = F$, $\neg F = T$, $\neg B = B$, $\neg N = N$, and T, F, B and N are all nullary constants. So $\mathbf{A}_4$ is a De Morgan algebra in which all elements are nullary constants. Balbes and Dwinger [1974] contains material on De Morgan algebras as well as Kleene and Łukasiewicz algebras. What is a description of the clone of term operations of $\mathbf{A}_4$ using relations preserved by operations? W.J. Blok and I have shown that this clone is precisely the set of operations that preserve the partial order $R_4 =$ 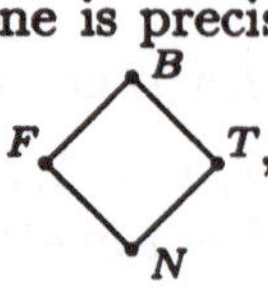

, which is obtained from $\mathbf{L_4}$ by rotating clockwise through 90°.

A semantic interpretation for for $\mathbf{A_4}$, $\mathbf{L_4}$ and R_4 is the following. View T as true, F as false and B and N as truth values denoting "both" and "neither". B indicates an overdetermined or contradictory situation while N signals that there is not enough information to assign T or F. In $\mathbf{A_4}$ the subsets $\{T, F, N\}$ and $\{T, F, B\}$ behave like the three-element Kleene algebra with respect to $\wedge, \vee, \neg, T$ and F. In R_4, points higher up in the partial order correspond to situations in which more information is known than in points below. The surprising thing is that a 90° rotation converts the lattice to the partial order. Some papers exploring this which are relevant to the themes of this conference are Scott [1973], Belnap [1976], Ginsburg [1987], and Romanowska and Trakul [1989].

For a finite algebra $\mathbf{A}$ with $W = V(\mathbf{A})$, the *free spectrum* of $\mathbf{A}$ (or of W) is the sequence $|F_w(n)|$, $n = 1, 2, \ldots$. Recent work has shown that some properties of $\mathbf{A}$ or of W can be related to properties involving the size and rate of growth of the free spectrum of $\mathbf{A}$. For example, in Hobby and McKenzie [1988] is the theorem that if W is congruence distributive than for every $0 < c < 1$ and n sufficiently large, $|F_W(n)| \geq 2^{2^{cn}}$.

Knowledge of the free spectrum of $\mathbf{A}$ may provide knowledge about $V(A)$. In some cases the free spectrum can be determined exactly, e.g. the three-element Łukasiewicz algebra mentioned earlier in the paper is easily seen to have free spectrum $2^{2^n} 3^{3^n - 2^n}$. In other cases a recursive formula or only an asymptotic formula is known, e.g. Guzman and Squier [1989] provides a recursive formula for the size of the free conditional algebra. However for the free spectrum of the algebra $\mathbf{A_4}$ (the four-element De Morgan algebra with all constants adjoined) an asymptotic estimate is all that present knowledge allows. For it can be shown by considering the join irreducibles in the lattice reduct of $\mathbf{F}_{\mathbf{A_4}}(n)$ that $\mathbf{F}_{\mathbf{A_4}}(n)$ is the direct product of two copies of the free De Morgan algebra on n free generators and it is known (e.g. Balbes and Dwinger [1974]) that the free De Morgan on n free generators has the same size as the free bounded distributive lattice on $2n$ free generators and, despite serious work on the problem, no efficiently computable closed form for the free spectrum of distributive lattices is known although a sharp asymptotic estimate is given by Korshunov [1981].

Although the theme of this paper is how free algebras can aid in our understanding of topics pertaining to computer science it is worth noting that in investigations of free spectra the computer is often used to compute the first few terms of the free spectra of small algebras. Although these computations do not provide general formulas they often give clues to possible patterns or to hidden structure that allow for the formulation and verification of conjectures concerning the free spectrum of the algebra under consideration. This is a fairly elementary yet useful application of computers to algebra and often leads to more insights about an algebra **A** especially when one goes beyond the initial representation of $\mathbf{F}_W(n)$ as a subalgebra of $\mathbf{A}^{|A|^n}$ and one uses particular structural properties of **A** to compute larger values in the free spectrum of **A**.

The next example illustrates how results on free spectra may be utilized in the study of a rather interesting class of algebras introduced by P. Winkler [1980], [1983]. These are "one-pass" algebras and for the purposes of this paper I will consider only finite algebras in this class and only present an informal definition. Let $\mathbf{A} = \langle\{0,1,\ldots,k-1\}, f_1,\ldots,f_m\rangle$ be an algebra with m fundamental operations $f_1,\ldots,f_m$. An **A**-calculator is a calculator that has a fixed finite number, say s, of registers accessed by buttons labelled $r_1,\ldots,r_s$. It also has m buttons labelled $f_1,\ldots,f_m$ and k buttons labelled $0,1,2,\ldots,k-1$ and some auxillary buttons (,), and $\leftarrow$. The **A**-calculator evaluates term operations of **A**. For example if f_1 is binary and f_2 is unary then $r_4 \leftarrow f_2(3, f_1(r_3))$ is the command: store in r_4 the result of applying f_2 to the element 3 and the element obtained by applying f_1 to the contents of r_3. **A** is called one-pass if there is an **A**-calculator with s registers such that for every term $t(x_1,\ldots,x_n)$ and every choice of $a_1,\ldots,a_n$ in $\{0,1,\ldots,k-1\}, t^{\mathbf{A}}(a_1,\ldots,a_n)$ can be evaluated on this **A**-calculator by entering each a_i exactly once. Winkler shows that commutative groups and more generally commutative semigroups are one-pass. He also points out that some of the operations that arise in statistics are one-pass, e.g. the mean $\overline{x} = (x_1+\cdots+x_n)/n$ and even the standard deviation $S(x_1,\ldots,x_n)$ provided one uses $S(x_1,\ldots,x_n) = (\frac{1}{n}(\Sigma x_i^2) - \overline{x})^{1/2}$. Here the fundamental operations of the algebra are the usual ones of addition, multiplication, division and exponentiation.

Winkler shows that every finite algebra in $W = V(\mathbf{A})$ is one-pass if there exists an integer k such that for any term $t(x_1,\ldots,x_n)$ there is an $(n-1)$-ary term q and k-ary operations $q_1,\ldots,q_{k-1}$ such that

$$\mathbf{A} \models t(x_1,\ldots,x_n) \approx q(q_1(x_1,\ldots,x_k),\ldots,q_{k-1}(x_1,\ldots,x_k),x_{k+1},\ldots,x_n).$$

Now, if Winkler's result is applied to the free algebra it gives a bound on the size of $|F_W(n)|$ in terms of $|F_W(n-1)|$ and $|F_W(k)|$. This bound shows there is constant c such that for all n, $|F_W(n)| \leq c^{n-k}$ and this is a fairly restrictive bound on the size of the free spectrum of **A**. For example, by the Hobby and McKenzie result it follows that no one-pass algebra generates a congruence distributive variety.

The next example illustrates how free algebras may be used in the theory of relational databases. It is based on joint work with W.J. Blok, (Berman and Blok [1989]).

A *relation* R on a set of attributes $\mathcal{A} = \{A_1,\ldots,A_n\}$ is a set of functions with domain $\mathcal{A}$. Elements of R are called tuples. The class of all such relations is denoted $\mathcal{R}(\mathcal{A})$.

For example suppose $n = 4$ with attributes Name, Zipcode, Area Code, and Phone Number. The relation R is a given list of tuples, say,

Name	Zipcode	Area Code	Phone Number
A. Smith	60601	312	234–5612
B. Jones	50011	515	234–5612
B. Jones	60602	312	432-0000
⋮	⋮	⋮	⋮

One primary concern of relational database theory is to develop ways for describing constraints or dependencies that hold among the tuples of a relation. These are dependencies that may arise by design or by empirical evidence. For example, does Zipcode determine Area Code in the example above? It does in the three tuples listed in the example. If there is a policy of the telphone company and the post office requiring the Zipcode to uniquely determine the Area Code, then this would be a dependency that arises by design. If, say, the relation R in the example is a list of postmasters in specific postal regions then presumably each zipcode would appear at most once and hence uniquely determine the area code. This would be an empirically determined dependency for the relation R.

Regardless of how dependencies arise, the study of dependenceis in relations is important for relational database theory. In particular, what dependences are consequences of a given set of dependencies? For example, if in the example Zipcode does determine Area Code and if Area Code with Phone Number determines Name, then by transitivity we get the dependency Zipcode with Phone Number to determine Name as well.

I will present an algebraic approach for describing those dependencies that arise from the patterns of equalities and inequalities among the values of the tuples in a relation. These are called *equational dependencies*.

For the remainder of this paper $\mathcal{A} = \{A_1, \ldots, A_n\}$ is a fixed finite set of attributes. A *valuation* on $\mathcal{A}$ is any function $v : \mathcal{A} \to \{T, F\}$. For $R \in \mathcal{R}(\mathcal{A})$ and for tuples $s, t \in R$, the valuation v_{st} on $\mathcal{A}$ is defined by $v_{st}(A_i) = T$ if $s(A_i) = t(A_i)$ and $v_{st}(A_i) = F$ is $s(A_i) \neq t(A_i)$. Let $\mathbf{B} = \langle \{T, F\}, C \rangle$ be an algebra in which the set C of fundamental operations is a subset of the familiar logical connectives of propositional logic, e.g. $C \subseteq \{\wedge, \vee, \neg, \Rightarrow, \equiv, \ldots\}$. If $p(A_1, \ldots, A_n)$ is a term for $\mathbf{B}$ built using $A_1, \ldots, A_n$ as variables and if v is a valuation on $\mathcal{A}$, then $\overline{v}(p(A_1, \ldots, A_n)) = p^{\mathbf{B}}(v(A_1), \ldots, v(A_n))$. So $\overline{v}$ maps the set of all terms for $\mathbf{B}$ with variables in $A_1, \ldots, A_n$ to $\{T, F\}$.

For a relation $R \in \mathcal{R}(\mathcal{A})$ and terms $p(A_1, \ldots, A_n), q(A_1, \ldots, a_n)$ of $\mathbf{B}$, R is said to satisfy the *equational dependency* $p \sim q$ if for all tuples $s, t \in R$, $\overline{v}_{st}(p) = \overline{v}_{st}(q)$. This is denoted $R \models p \sim q$. If Δ is a set of equational dependencies, $\Delta = \{p_1 \sim q_1, \ldots, p_k \sim q_k\}$, the equational dependency $p \sim q$ is said to be a *consequence* of Δ, denoted $\Delta \models_{\mathcal{R}} p \sim q$, if for all

$R \in \mathcal{R}(\mathcal{A})$ if $R \models p_i \sim q_i$ for all $1 \le i \le k$, then $R \models p \sim q$.

The class of equational dependencies includes various dependency conditions considered by others: Sagiv, Délobel, Parker and Fagin [1981], Czédli [1981], Demetrovics and Gyepesi [1983], Thalheim [1985] and Berman and Blok [1988].

For example, the most basic dependency considered in the theory of relational databases is that of a functional dependency. For $X, Y \subseteq \mathcal{A}$, a relation $R \in \mathcal{R}(\mathcal{A})$ satisfies the functional dependency $X \to Y$ if for all $s, t \in R$, if $s(A_i) = t(A_i)$ for all $A_i \in X$, then $s(A_j) = t(A_j)$ for all $A_j \in Y$. Thus in the example {Zipcode} $\to$ {Area Code} and {Area Code, Phone Number } $\to$ {Name}. In order to express $X \to Y$ as an equational dependency, the set C of connectives can be chosen to be $\{\wedge\}$ and $X \to Y$ holds for R precisely if $R \models \bigwedge X \sim (\bigvee X \wedge \bigvee Y)$ since for all $s, t \in R$, $\overline{v}_{st}(\bigwedge X) = \bigwedge_{A_i \in X} v_{st}(A_i) \doteq T$ if and only if $s(A_i) = t(A_i)$ for all $A_i \in X$. So a functional dependency is an example of an equational dependency.

As another example, let $C = \{\wedge, \vee\}$. Then $R \models (A_1 \wedge A_2) \sim (A_1 \vee A_2)$ means that if $s, t \in R$ agree on attribute A_1 or A_2 then they agree on both A_1 and A_2. It is easily verified that this is equivalent to $R \models A_1 \sim A_2$. Moreover, $(A_1 \wedge A_2) \sim (A_1 \vee A_2) \models_{\mathcal{R}} A_1 \sim (A_1 \vee A_2)$ and $A_1 \sim (A_1 \vee A_2) \models_{\mathcal{R}} A_2 \sim (A_1 \wedge A_2)$ follow by simple computations.

Combinatorial properties involving familiar configurations can be expressed using equational dependencies. In Berman and Blok [1988], [1989] examples involving mutually non-attacking queens, incidence structures in abstract geometries, and statements about error-correcting codes are all phrased using the language of equational dependencies.

It turns out that the basic situation of $\Delta \models_{\mathcal{R}} p \sim q$ can be described using free algebras. First some notation is needed. For an algebra $\mathbf{A}$ and $a, b \in A$, the smallest congruence relation on $\mathbf{A}$ that identifies a and b is denoted $\theta(a, b)$. The *quasivariety* generated by $\mathbf{A}$, $Q(\mathbf{A})$, is the class of all algebras of the same similarity type as $\mathbf{A}$ that satisfy every quasi-identity that holds in $\mathbf{A}$. In general $Q(\mathbf{A})$ is properly contained in $V(\mathbf{A})$. However if $\mathbf{B} = \langle \{T, F\}, C \rangle$ has $\{T\}$ as a subuniverse, then $V(\mathbf{B}) = Q(\mathbf{B})$. This odd fact is true because for such $\mathbf{B}$ the equational class $V(\mathbf{B})$ has only one subdirectly irreducible member (e.g. Taylor [1975]) and it accounts for the hypothesis in the following.

THEOREM. *Suppose $\mathbf{B} = \langle \{T, F\}, C \rangle$ has $\{T\}$ as a subuniverse and suppose p, q, p_j, q_j ($j \in J$) are n-ary terms for $\mathbf{B}$ with variables in $\mathcal{A} = \{A_1, \ldots, A_n\}$. Let $\mathbf{F}$ denote the free algebra for $V(\mathbf{B})$ freely generated by $\mathcal{A}$ and let Δ denote the set of equational dependencies $\{p_j \sim q_j : j \in J\}$. The following are equivalent.*

(i) $\Delta \models_{\mathcal{R}} p \sim q$.

(ii) *For all valuations $v : \mathcal{A} \to \{T, F\}$, if $\overline{v}(p_j) = \overline{v}(q_j)$ for all $j \in J$, then $\overline{v}(p) = \overline{v}(q)$.*

(iii) $p \equiv q \bigvee_{j \in J} \theta(p_j, q_j)$ *in the congruence lattice of* **F**.

The proof of this is in Berman and Blok [1989]. The equivalence of (ii) and (iii) holds for any algebra **B** in which $V(\mathbf{B}) = Q(\mathbf{B})$. The equivalence of (i) and (ii) is obtained by considering v_{st} and certain $R \in \mathcal{R}$ that have at most two tuples.

One consequence of this theorem is that the study of $\Delta \models_{\mathcal{R}} p \sim q$ reduces to the investigation of congruences on free algebras and such congruences have received a good deal of attention in the literature. It is interesting to translate the various concepts defined for functional dependencies on relational databases to their corresponding conditions on the congruence lattice of the free $\wedge$-semilattice. This can be done for most of the definitions given in, say, Chapter 5 of Maier [1983]. For example, a set D of functional dependencies is defined to be canonical if every functional dependency in D is of the form $X \to A_i$ and D is left-reduced and nonredundant. The corresponding notion using (iii) of the Theorem is that $\bigvee \theta(p_j, q_j)$ is an irredundant join of the $\theta(p_j, q_j)$ and each $\theta(p_j, q_j)$ is a maximal join irreducible congruence relation in the congruence lattice of $\mathbf{F}_W(\mathcal{A})$.

The *inference problem* for a class of (equational) dependencies is to provide an algorithm that decides for an arbitrary finite set Δ of dependences whether or not a particular dependency is a consequence of Δ. For equational dependencies in which all the operations in C preserve $\{T\}$ the Theorem may be used. By (ii) in the Theorem there is a semantic approach: Test all 2^n valuations. This shows the time complexity of the inference problem is in co-NP. It turns out that for some algebras **B** in the Theorem the inference problem is co-NP complete while for all the others the inference problem is of polynomial time complexity. Details and proofs are in Berman and Blok [1989] and the arguments used there are based on an analysis of the word problem for free algebras in $V(\mathbf{B})$.

My final example connects a problem in relational database theory with some of the non-classical logics discussed in the first part of the paper. The problem is how to deal with null values in a relation. Consider the example used when introducing relational databases and suppose the following tuples are elements of the relation R.

Name	Zipcode	Area Code	Phone Number
C. Smith	60601	312	123–456?
C. Smith	60601	312	123–?567
C. Smith	60601	312	???-????
D. Jones	60602	312	
E. Jones	60602	312	
F. Smyth	60603	342	765–4321

The question marks in the first three entries indicate that the digits are missing or otherwise not known. Are these three C. Smiths the same and more generally how do we reason about patterns of equality and inequality when data are lacking? The entries for D. Jones and E. Jones are also problematic. It is known that neither have a telephone and so what is the pattern of equality or inequality for their phone number attribute? The difficulty with F. Smyth is that the Area Code entry is clearly in error since the middle digit of an area code is always 0 or 1.

A hard problem in database research is to develop a theory that can handle missing, erroneous or inapplicable data. In the case of equational dependencies it is not at all clear how the valuation v_{st} should be defined for the pathological tuples just described.

One approach would be to expand the range of the valuations to $\{T, F, U\}$ where U means unkown or undecided. Then if s and t are tuples in a relation and A_i is an attribute, let

$$v_{st}(A_i) = \begin{cases} T & \text{if both } s(A_i) \text{ and } t(A_i) \text{ are known, correct, and equal.} \\ F & \begin{array}{l}\text{if both } s(A_i) \text{ and } t(A_i) \text{ are known and correct} \\ \text{but not equal}\end{array} \\ U & \text{otherwise.} \end{cases}$$

If this definition is made then what interpretation to the connectives $\wedge, \vee, \neg$ etc. should be used?

One possibility, and the most conservative, is what I call the Bochvar system (Bochvar [1939]) with connectives

$\wedge$	T	F	U
T	T	F	U
F	F	F	U
U	U	U	U

$\vee$	T	F	U
T	T	T	U
F	T	F	U
U	U	U	U

$\neg$	T	F	U
	F	T	U

Here U is an absorbing element in the corresponding algebra. The free algebras for the equational class generated by this algebra are known. One can then define equational dependencies for this algebra $\langle \{T, F, U\}, \wedge, \vee, \neg \rangle$. Another interpretation could be the Kleene system described earlier in the paper. However, we have not pursued the extension of equational dependencies to these non-classical logics since it is not at all clear that such an extension would help in the handling of null values for relational dependencies. Chapter 12 of Maier [1983] contains a discussion of null values and references to the literature.

References

R. BALBES AND PH. DWINGER [1974], Distributive Lattices, U. of Missouri Press.

N.D. BELNAP, JR. [1976], How a Computer Should Think, in Contemporary Aspects of Philosophy, ed. G. Ryle.

J. Berman and W.J. Blok, [1988] Positive Boolean dependencies, *Inf. Processing Letters*, **27**, 147–150.

J. Berman and W.J. Blok, [1989], Equational Dependencies, Research Report no. 32, Dept. of Math., Stat., and C.S., Univ. of Illinois at Chicago.

J. Berman and M. Mukaidono [1984], Enumerating fuzzy switching functions and free Kleene algebras, *Comp. and Math. with Applications* **10**, pp. 25–35.

S. Blamey [1986], Partial Logic, in Handbook of Philosophical Logic, Vol. **III**, ed. D. Gabbay and F. Guenthner, D. Reidel Pub, pp. 1–70.

D.A. Bochvar [1939], On a 3-valued logical calculus and its application to the analysis of contradictions, (Russian), Math. Sbornik **4**, 287–308. Eng. Trans. in Hist. Philos. Logic **2**(1981), 87–112.

S. Burris and H.P. Sankappanavar [1981], A Course in Universal Algebra, Graduate Texts in Mathematics, vol. 78, Springer-Verlag.

G. Czédli [1981], On dependencies in the relational model of data, *Elektron. Informationsverab. u.Kybernet.*, **17**, 103–112.

J. Demetrovics and Gy. Gyepesi [1983], Some generalized type functional dependencies formalized as equality set on matrices, *Discrete Applied Math.*, **6**, 35–47.

M.L. Ginsburg [1987], Multi-valued Logics, I,II,II. preprints.

D. Gries [1981], The Science of Programming, Springer-Verlag.

F. Guzman [1989], The implications in conditional logic, Proceedings of the Conference on Algebra and Computer Science, Iowa State University.

F. Guzman and C. Squier [1989], The algebra of conditional logic, *Algebra Universalis*, to appear.

D. Hobby and R. McKenzie [1988], The Structure of Finite Algebras, Contemporary Mathematics, **76**, Amer. Math. Soc. Pub.

S. Kleene [1952], Introduction to Metamathematics, pp. 332-340, Van Nostrand.

A.D. Korshunov [1981], On the number of monotonic Boolean functions, (Russian), Problemy Kibernetiki **38**, 5–108.

D. MAIER [1983], The Theory of Relational Databases, Computer Science Press.

J. MCCARTHY [1963], A basis for a mathematical theory of computation, in Computer Programming and Formal Systems, P. Braffort and D. Hirschberg eds., North-Holland Publishing Co., 33–70.

J. MCCARTHY [1963a], Predicate calculus with "undefined" as a truth value, Stanford Artificial Intelligence Project, Memo 1, March 1963.

R. MCKENZIE, G. MCNULTY AND W. TAYLOR [1987], ALgebras, Lattices, Varieties, vol. 1, Wadsworth.

M. MUKAIDONO AND I.G. ROSENBERG [1986], k-valued functions for treating ambiquities: Their clone and a normal form. 18th Internat. Symposium on Multiple-valued Logic, IEEE.

M. NAGATA, M. NAKANISHI, AND T. NISHIMURA [1975], Implementation of Łukasiewicz's, Kleene's and McCarthy's 3-valued logics, Sci. Rep. Tokyo Kyoiku Daigaku Sect A **13**, no. 347–365, 90–100. *Math. Rev.* **53** #2055.

A. ROMANOWSKA AND A. TRAKUL [1989], On the structure of some bilattices, preprint.

Y. SAGIV, C. DELOBEL, D.S. PARKER AND R. FAGIN [1981], An equivalence between relational database dependencies and a fragment of propositional logic, *J. ACM*, **28**, 435–453. Corrigendum *J. ACM*, **34**(1987), 1016–1018.

D.S. SCOTT [1973], Models of various type-free calculi, in *Logic, Methodology and Philosophy of Science* **IV**, ed. P. Suppes et al., North-Holland, 157–187.

A. SZENDREI [1986], Clones in Universal Algebra, SMS **99**, University of Montreal Press.

W. TAYLOR [1975], The fine spectrum of a variety, *Algebra Universalis*, **5**, 263–303.

B. THALHEIM, [1985], Funktionale Abhängigkeiten in relationalen Datenstrukturen, *Elektron. Informationsverarb. u. Kybernet.*, **21**, 23–33.

P.M. WINKLER [1980], Classification of algebraic structures by work space, *Algebra Universalis* **11**, 320–33.

P.M. WINKLER [1983], Polynomial Hyperforms, *Algebra Universalis* **17**, 101-109.

I.D. ZASLAVSKII [1979], Realization of three-valued logical functions by means of recursive and Turing operators. (Russian), Studies in the theory of algorithms and mathematical logic, Nauka, 52–61. *Math. Rev.* **81i:03031**.

Continuations of Logic Programs

H. Peter Gumm[1]

0 Background

In the realm of functional programming a wealth of techniques have been explored to transform a program into another equivalent program with the transformed program exhibiting certain computational advantages over the original. Often the transformation involves a "generalization" of the original task, where this generalization requires the addition of further parameters, called "accumulating" parameters. This technique is particularly useful in transforming functional programs into tailrecursive form. A typical example of such a transformation is the generalization of a linearly recursive function such as "factorial" into a tailrecursive function $fact'(n,m) = fact(n) * m$.

On first sight, such generalizations appear to involve quite an insight into the particular problem at hand, but they turn out to be instances of a very general method of transformations based on "continuations". The transformation always succeeds on linearly recursive programs, insight is only requested to further simplify the resulting program.

A ***continuation*** is a function of one parameter, representing some remaining computation necessary to transform an intermediate value into a final outcome. In the example of the standard definition of the factorial function, after having finished the inner recursive call to $f(n-1)$ the resulting value still has to be multiplied with n, so the continuation would be $\lambda w.n * w$. A representation of this continuation is all that has to be stored in the accumulating parameter. In the preceding case it suffices to simply represent $\lambda w.n * w$ by n. Further optimizations are possible, if the space of representations can be endowed with a monoid structure so the abstraction function maps this monoid homomorphically to the monoid of continuations with composition (denoted by $\circ$) as operation. In the "factorial" example, the monoid is the multiplication monoid on the natural numbers, so $\boldsymbol{abs}(n * m) = \lambda w.(n * m) * w = \lambda w.n * (m * w) = \lambda w.n * w \circ \lambda w.m * w = \boldsymbol{abs}(n) \circ \boldsymbol{abs}(m)$. An excellent account of the technique is given in [W].

The general method of transforming a linearly recursive functional program into tailrecursive form can then be sketched briefly as follows. Let

$$f(x) = \textbf{ if } g(x) \textbf{ then } h(x) \textbf{ else } \Phi(x, f(r(x)))$$

[1]Dept. of Mathematics and Computer Science, SUNY College at New Paltz, New Paltz, N.Y.,12561, `gummp@snynewba.bitnet`

This paper is in final form and no version of it will be submitted for publication elsewhere.

be a linearly recursive program. Generalize it to a function $cf(x, \gamma)$ by introducing a further parameter γ to represent a continuation with the intention

$$cf(x, \gamma) := (\gamma \circ f)(x).$$

Then

$$\begin{aligned} cf(x, \gamma) &= \textbf{if } g(x) \textbf{ then } (\gamma \circ h)(x) \textbf{ else } \gamma(\Phi(x, f(r(x)))) \\ &= \textbf{if } g(x) \textbf{ then } (\gamma \circ h)(x) \textbf{ else } (\gamma \circ \lambda w.\Phi(x, w) \circ f)r(x) \\ &= \textbf{if } g(x) \textbf{ then } (\gamma \circ h)(x) \textbf{ else } cf(r(x), \gamma \circ \lambda w.\Phi(x, w)), \end{aligned}$$

a function which is now tailrecursive. The original function f can be recreated using the identity function id in $f(x) = cf(x, id)$.

The second argument to cf, which is a function, can be represented (encoded) by the pieces of data γ and x as $p(x, \gamma)$ with some constructor p, and with a constant id serving as the representation of the identity function. As long as p is a free constructor, i.e. it essentially pushes values on a stack beginning with the empty stack id, we can uniquely decode the function it represents. The decoding map is defined on the space of representations by

$$decode(id) = \lambda x.x$$

and

$$decode(p(x, y)) = decode(y) \circ \lambda w.\Phi(x, w).$$

In many cases simpler representations can be found by means of a binary operation $*$ defined on the space of representations, so that

$$decode(p(x * y, z)) = decode(p(x, p(y, z))),$$

in particular, such a $*$ can always be chosen associative.

1 Logic Programs

In the mathematical semantics of logic programs the order of the predicates in a clause should not matter, but of course it does make a difference to the termination properties of a PROLOG program. More importantly, most practical logic programs contain nonlogical predicates, such as arithmetical predicates, predicates causing side effects or any system predicates that require certain arguments to be bound before execution. Clearly, such predicates cannot be freely permuted with other predicates of the same clause.

Thus the notion of a "tailrecursive" predicate makes sense even in logic programming and indeed most PROLOG compilers will generate more efficient code if a program is tailrecursive. Interpreters will also have to store less backtrackpoints, if a recursive call is the last call in the last applicable clause in the program. Several authors have studied how to apply the unfold/fold technique to transform logic programs into tailrecursive form [TS], [D].

Here we explore a possible way how to give a meaning to continuations in logic programming. There are various ways of doing so , and the continuation may either represent an extra goal to be solved, or a relation between intermediate values and output values, in the case of a logic program whose intended use is to transform input into output. Following this, but independent of the method, simplifications may be applied on the ensuing program, often getting rid of the auxiliary predicate.

A logic program for a predicate q is called *linearly recursive*, if there is at most one recursive call to q in each clause . q is called *tailrecursive*, if it is linearly recursive and each call to q, if any, occurs as the last goal in the clauses body. Thus a typical linearly recursive program would be of the form:

$$\mathtt{q}_i(\mathtt{r}_j) \;:-\; \mathtt{g}_i(\mathtt{s}_j). \tag{1}$$

$$\mathtt{q}_i(\mathtt{t}_j) \;:-\; \mathtt{h}_i(\mathtt{u}_j), \mathtt{q}_i(\mathtt{v}_j), \mathtt{p}_i(\mathtt{w}_j). \tag{2}$$

where we abbreviate termlists such as $\mathtt{t}_1(x_1, \ldots, x_n), \ldots, \mathtt{t}_k(x_1, \ldots, x_n)$ by $\mathtt{t}_i(\mathtt{x}_j)$. There may be at most one recursive call to q in the body of each clause, yet there may be several clauses such as (2).

The following predicate will serve as a prototype to demonstrate the transformation. The formulation for a general linear recursive program will be obvious, but tedious.

The predicate relates a list to its length and can be used either to calculate the length of a list, or to provide a list of a given length. It is obviously linearly recursive, but not tailrecursive, and reordering of the subgoals in its body is not possible, since V must be bound, when the predicate "U is V + 1" is encountered.

```
length([ ],0).
length([H|T],U) :- length(T,V),U is V + 1.
```

The idea corresponding to the continuation transformation in functional programming would be to generalize the "length" predicate so that it also incorporates the calculation ensuing after the recursive call in its body, creating a new generalized predicate that will become tailrecursive. For this we extend `length` by a further argument position that is to encode the "ensuing calculation".

In functional programs this ensuing calculation is represented as a function of one argument, the "continuation", in logic programming it ought to be a goal, representing the task yet to be solved. The new predicate, say `cLength`, is intended to have the semantics :

$$\mathtt{cLength(L,N,}\Gamma\mathtt{)} \Longleftrightarrow \mathtt{length(L,N),}\Gamma. \tag{3}$$

The relationship between `length` and `cLength` would then be defined as

$$\texttt{length(L,N)} \;:-\; \texttt{cLength(L,N,true)}. \tag{4}$$

Using (3) as a definition of `cLength` we shall have to remove the reference to the old `length`-predicate. Partial evaluation (see [V],[K]) of `length(L,N)` in (3) yields the clauses :

$$\texttt{cLength([ ],0,}\Gamma\texttt{)} \;:-\; \Gamma. \tag{5}$$

$$\texttt{cLength([H|T],U,}\Gamma\texttt{)} \;:-\; \texttt{length(T,V),U is V+1,}\Gamma. \tag{6}$$

Now, we have to fold the right hand side with (3) which is trivial by simply letting the new continuation be the conjunction of the goals "`U is V+1`" and "Γ", i.e.

$$\texttt{cLength([H|T],U,}\Gamma\texttt{)} \;:-\; \texttt{cLength(T,V,(U is V+1,}\Gamma\texttt{))}. \tag{7}$$

The new program for length consists of (4), (5), and (7) . Some PROLOG implementations would need to replace the call to Γ in the body of (5) with an explicit "`call(`Γ`)`".

It is clear that the new program for `length` is equivalent to the old version and also that the new program has become tailrecursive. Instead of creating backtrackpoints as is necessary in a call to (6), the extra argument in `cLength` is used to store the necessary information for the ensuing goals. The necessary information to recreate the "ensuing calculation" is completely provided by the variables `U`,`V` and Γ, so we choose a function symbol `p` to encode the continuation as `p(U,V,`Γ`)`, with a constant `done` representing the goal `true`. Of course, we need a decoding predicate `run` now, which is given by

```
run(done).
run(p(U,V,G))  :- U is V+1,run(G).
```

The new cLength then becomes :

```
cLength'([ ],0,G)  :- run(G).
cLength'([H|T],U,G)  :- cLength'(T,V,p(U,V,G)).
```

with initial call :

```
length(L,N)  :- cLength'(L,N,done).
```

Thus, not surprisingly, forming `p(U,V,G)` amounts to pushing `U` and `V` onto the stack `G`, with `done` representing the empty stack. Clearly, also, there are some savings possible, since not both `U` and `V` need to be pushed onto the stack, in particular, since it is obvious that `V` is local to the body of (6), but we shall see a refined version of the transformation and with it an improved version of `length` in the next section.

2 List Recursion

There may be several reasons why a linear recursive program cannot be turned into a tailrecursive program by simply switching the order of the subgoals in the clauses of (2). In the majority of cases though, there will be some value computed in the recursive call to q which is subsequently needed by p. (If q and p do not have any variables in common there is no reason why they could not be interchanged, unless they create sideeffects.) Taking this fact into account, we can improve upon the previous transformation. In this chapter we shall demonstrate this for programs recursing over lists, and in the following chapter we give an example of the same transformation in the context of graphs.

List recursion seems to be a rather typical case where linear recursive programs arise. (Stretching this point somewhat, recursion over natural numbers can be seen as a special case of list recursion). The general form of a program recursing over lists can be written as

$$\mathtt{q([\ ],c).} \tag{8}$$
$$\mathtt{q([H|T],M)\ :-\ q(T,K),r(H,K,M).} \tag{9}$$

(Additional nonrecursive clauses or additional goals in (8) and (9) would only complicate notation). The body of (9) can be viewed as a relational product (join) of the relations q(-,-) and r(H,-,-). Thus continuations should become relations and they should be composed using relational composition $\circ$. Augmenting q with a further argument to hold the representation of a continuation and introducing a ternary relation abs(-,-,-) that decodes the representation of a continuation, so that $\mathtt{abs(rep(C),-,-) = C}$, we introduce the generalization cq(-,-,-) of q(-,-) with the intention

$$\mathtt{cq(L,M,R)} \iff \mathtt{q(L,U),abs(R,U,M).}$$

To recover the original predicate, we set :

$$\mathtt{q(L,M)\ :-\ cq(L,M,id).}$$

and

$$\mathtt{abs(id,X,X).}$$

Next we use the defining clauses for q to partially evaluate the definition of cq :

$$\mathtt{cq([\ ],M,R)\ :-\ abs(R,c,M).}$$
$$\mathtt{cq([H|T],M,R)\ :-\ q(T,K),r(H,K,U),abs(R,U,M).}$$

We need the right hand side to be of the form q(T,K), abs(Ω,K,M), so Ω must encode H and R. Hence we choose a binary function symbol p and the clause

$$\mathtt{abs(p(H,R),K,M)\ :-\ r(H,K,U),abs(R,U,M).}$$

This gives us the final program

$$\mathtt{q(L,M)\ :-\ cq(L,M,id).}$$

$$\begin{aligned}\texttt{cq([ ],M,R)} &\ :-\ \texttt{abs(R,c,M).}\\ \texttt{cq([H|T],M,R)} &\ :-\ \texttt{cq(T,M,p(H,R)).}\end{aligned}$$

$$\begin{aligned}&\texttt{abs(id,X,X).}\\ &\texttt{abs(p(H,R),K,M)}\ :-\ \texttt{r(H,K,U),abs(R,U,M).}\end{aligned}$$

Here all predicates are tailrecursive. The functor p is free, that is, the data structure built as representation of the continuation is isomorphic to a stack, with id corresponding to the empty stack. If there were several clauses in the original program containing a call to q, we would need a constructor p_i for each of them, together with a corresponding clause for abs.

Suppose that q is called with its first argument bound to a list l, then the role of cq is merely to push the elements of l so they can be retrieved by abs and processed in reverse order.

In special cases various optimizations are possible. If, for example, r(H,K,M) does not depend on H, such as in the length predicate of the previous chapter, then p becomes essentially unary and the continuations can be represented by natural numbers, id, p(id), p(p(id)),... . Another important case is when a binary operation $\diamond$ can be defined such that $\texttt{r}(x,-,-)\circ\texttt{r}(y,-,-) = \texttt{r}(x\diamond y,-,-)$. W.l.o.g. we can assume a right unit e with r(e,X,X). Then the transformed program simplifies to

$$\begin{aligned}\texttt{q(L,M)} &\ :-\ \texttt{cq(L,M,e).}\\ \texttt{cq([ ],M,R)} &\ :-\ \texttt{r(R,c,M).}\\ \texttt{cq([H|T],M,R)} &\ :-\ \texttt{cq(T,M,H}\diamond\texttt{R).}\end{aligned}$$

and abs becomes superfluous. (Since any call to the program will be made through a call to q, the last argument of cq will always be bound.) Examples of programs amenable to the latter simplification are e.g. programs combining the elements of a list by an associative operation. The length program, again, is a special case here, setting $\texttt{e} = 0$ and $\texttt{H}\diamond\texttt{R} := \texttt{R}+1$.

3 Modifying a Search Program

The previous transformation is taylored to, but not limited to programs recurring over lists. As an example, suppose a graph is given by a relation edge(_,_) relating pairs of nodes. Reachability can then be defined as the transitive hull of the edge relation :

$$\begin{aligned}&\texttt{reach(X,X).}\\ &\texttt{reach(X,Y)}\ :-\ \texttt{reach(X,Z),edge(Z,Y).}\end{aligned}$$

The predicate cReach will be introduced again, with the intention:

$$\texttt{cReach(X,Y,G)} \iff \texttt{reach(X,U),abs(G,U,Y).}$$

This leads to

```
reach(X,Y)  :- cReach(X,Y,done).

            abs(done,X,X).
```

Partially evaluating the body of this definition we get

```
   cReach(X,Y,G)  :- abs(G,X,Y).
   cReach(X,Y,G)  :- cReach(X,Y,p(G)).
abs(p(G),Z,Y)  :- edge(Z,U),abs(G,U,Y).
```

The domain of continuation representations, again, is isomorphic to the natural numbers, and it seems that renaming **abs** into **distance** is more appropriate, we get :

```
reach(X,Y)  :- cReach(X,Y,0).


cReach(X,Y,N)  :- distance(X,Y,N).
cReach(X,Y,N)  :- cReach(X,Y,succ(N1)).

distance(X,X,0).
distance(X,Y,succ(N))  :- edge(X,U),distance(U,Y,N).
```

Thus, whereas in the original program a call such as `reach(a,b)` results in a depth first search backwards from `b`, the transformed program will do an exhaustive search, increasing the boundaries of the search space with each call to **cReach**. The original program, by contrast, is likely to be caught in infinite loops. Logically, though, the two programs are equivalent.

4 Difference Lists

Difference lists are a representation of the list data structure, that is particularly efficient for the "append" operation, in that appending of two difference lists can be achieved totally by unification(see [ZG]). A difference list `d(A,B)` represents a list that satisfies $\mathtt{d(A,B)} \oplus \mathtt{B} = \mathtt{A}$, where `A` and `B` are lists and $\mathtt{A} \oplus \mathtt{B}$ denotes the list obtained by appending `A` to `B`. The program

```
append(d(X,Y),d(Y,Z),d(X,Z)).
```

appends two difference lists and, obviously, leaves all the work to the unification routine. Difference lists are particularly useful in parsing, where the append program is used to split a list of incoming tokens into pieces. Each piece is then parsed by parsers responsible for the individual nonterminals of the grammar. Applying the continuation transformation onto a simple minded version of a parser we shall see that difference lists quite naturally come about as representations of continuations. Let us take a typical clause of a grammar such as

< sentence >::=< nounPhrase >< verbPhrase >< nounPhrase >

and a corresponding parser that works on a list of tokens to construct an abstract syntax tree:

```
pSent(In,mkSent(N,V,M)) :- pNP(A,N),
                           append(A,R1,In),
                           pVP(B,V),
                           append(B,R2,R1),
                           pNP(C,M),
                           append(C,[],R2).
```

together with some simple definitions of pNP and pVP such as e.g.

```
pNP([the,X],subj(the,X)) :- noun(X).
pNP([Y],person(Y)) :- name(Y).
pVP([eats],verb).
pVP([likes],verb).
```

The last call to append, in pSent could, of course, be dispensed with, but it serves to show the regular structure of the parser. Every subparser is now paired with its own continuation. Since those all have the same structure, we need only one representation, resp. decoding predicate to work for all. This yields:

```
cpNP(N,p(R1,In)) :- pNP(A,N),append(A,R1,In).
cpVP(V,p(R2,R1)) :- pVP(B,V),append(B,R2,R1).
pSent(In,mkSent(N,V,M)) :- cpNP(N,p(R1,In)),
                           cpVP(V,p(R2,R1)),
                           cpNP(M,p([ ],R2)).
```

partial evaluation of cpNP and cpVP gives

```
cpNP(subj(the,X),p(R1,In)) :- noun(X),append([the,X],R1,In).
cpNP(person(Y),p(R1,In)) :- name(Y),append([Y],R1,In).
cpVP(verb,p(R2,R1)) :- append([eats],R2,R1).
cpVP(verb,p(R2,R1)) :- append([likes],R2,R1).
```

Next, we can partially evaluate the append subgoals, obtaining:

```
cpNP(subj(the,X),p(R1,[the,X|R1])) :- noun(X).
cpNP(person(Y),p(R1,[Y|R1])) :- name(Y).
```

```
cpVP(verb, p(R2, [eats|R2])).
cpVP(verb, p(R2, [likes|R2])).
```

Note now, that the previous continuation representation has turned into a difference list, since `p(B,A)` can be interpreted as the difference list `d(A,B)`. Actually, one would probably want to relinquish the constructor p altogether, and simply list the components in separate argument positions resulting in the final program, that is a substantial improvement over the initial program, since the calls to append have disappeared.

```
pSent(In, mkSent(N, V, M)) :- cpNP(N, R1, In),
                              cpVP(V, R2, R1),
                              cpNP(M, [ ], R2).

cpNP(subj(the, X), R1, [the, X|R1]) :- noun(X).
cpNP(person(Y), R1, [Y|R1]) :- name(Y).

cpVP(verb, R2, [eats|R2]).
cpVP(verb, R2, [likes|R2]).
```

5 Left Recursion

Left recursion is a problem frequently encountered in constructions of recursive descent parsers. Its solution is well known, we will nevertheless derive it here again, to show that it may as well be considered an instance of a continuation based transformation. Let the grammar be given as $A \Leftarrow \alpha \mid A\beta$, and let pA, p$\alpha$ and pβ be the associated PROLOG predicates. We disregard as inessential here the fact that pA, pα, pβ usually would have some arguments. The PROLOG program for the grammar,

$$\text{pA} :- \text{p}\alpha.$$

$$\text{pA} :- \text{pA}, \text{p}\beta.$$

would suffer from left recursion. Introducing a continuation parameter and a decoding predicate `abs(_)`, so that

$$\text{cpA(G)} \Longleftrightarrow \text{pA}, \text{abs(G)}$$

we get

$$\text{pA} :- \text{cpA(id)}.$$

$$\text{cpA(G)} :- \text{p}\alpha, \text{abs(G)}.$$

$$\text{cpA(G)} :- \text{cpA(s(G))}.$$

where

$$\text{abs(id)}.$$

$$\text{abs(s(G))} :- \text{p}\beta, \text{abs(G)}.$$

Once again, the continuations can be represented by the natural numbers, moreover, the last clause for cpA together with the fact that the original call to cpA is with argument `id`, indicate that the argument is really superfluous. We replace it by "_", and it turns out that `abs(_)` will also be called with argument "_", thus we can eliminate the continuation parameter from `cpA` and from `abs`, obtaining the final program

$$\begin{aligned} \texttt{pA} &:- \texttt{cpA}. \\ \texttt{cpA} &:- \texttt{p}\alpha, \texttt{abs}. \\ \texttt{abs}&. \\ \texttt{abs} &:- \texttt{p}\beta, \texttt{abs}. \end{aligned}$$

which is the familiar transformation for leftrecursive grammars.

6 Conclusion

We have demonstrated that the concept of continuation based transformations can be successfully carried over from functional programming to logic programming. Various known techniques of logic programming, such as removal of linear recursion, parsing by difference lists and removal of left recursion in grammars can be considered as special instances of the technique.

7 References

[D] S. Debray "Optimizing Almost-Tail-Recursive Prolog Programs," *Proc. IFIP International Conference on Functional Programming Languages and Computer Architecture.* Nancy, France, 1985.

[K] H. J. Komorowski "Partial evaluation as a means for inferencing data structures in an applicative language : A theory and implementation in the case of Prolog," *Proceedings of the 9th ACM Symposium on Principles of Programming Languages*, Albuquerque, New Mexico, 255–267 (1982).

[TS] S. Tamaki and T. Sato "Unfold/Fold Transformations of Logic Programs," *Proc. 2nd. Logic Programming Conference*, Uppsala, Sweden, 1984.

[V] R. Venken "A Prolog meta-interpreter for partial evaluation and its application to source to source transformation and query optimization," in *T.O'Shea (ed.): ECAI-84. Advances in Artificial Intelligence*, Pisa, Italy, 91–100. North-Holland, 1984.

[W] M. Wand "Continuation based program transformation strategies," *Journal of the ACM,* **27**(1980)164–180.

[ZG] J. Zhang and P. W. Grant "An Automatic Difference-list Transformation Algorithm for Prolog," in *Proceedings of ECAI-88. European Conf. on Artificial Intelligence*, Munich 1988.

On Cylindric Algebraic Model Theory

István Németi[1]

Contents

1 Introduction

If algebraic logic is somehow based on the algebraization of logic (cf. §4.3 of [8]), then, by model theory being a part of logic, we can ask ourselves, what the result of this algebraization process is when it is applied to model theory. It is this particular "subset" of algebraic logic that is called *algebraic model theory*. §4.3 entitled "Connections between logic and cylindric algebras" of the Henkin–Monk–Tarski monograph [8] on algebraic logic provides a consize introduction to algebraic model theory too. In particular, the "Concluding remarks" (item 4.3.68 pp. 176–178) of §4.3 of [8] gives a brief survey of the more advanced results and research directions in this field. The main purpose of the present paper is, on the one hand, to prove some of the results announced (but not proved) in 4.3.68 of [8], and on the other hand, to present some of the theory and results that were mentioned (but not presented in detail) there.

It turns out that the algebraic counterparts of models are cylindric set algebras (i.e. algebras of relations of some fixed rank). Properties of models (like being saturated, or universal) translate to algebraic properties of

---

[1]Mathematical Institute of the Hungarian Academy of Sciences, Budapest, PF.127, H–1364, Hungary
This paper is in final form and no version of it will be submitted for publication elsewhere.

cylindric set algebras. Further, connections between models (like elementary embeddability, or one being an ultrapower of the other) translate to special kinds of isomorphisms between cylindric set algebras.

The algebraic counterparts of models of classical first-order logic ($L_{\omega\omega}$) form a narrow subclass of that of all cylindric set algebras. Therefore the algebraic counterparts of the model theoretic theorems do not always generalize (from this narrow subclass) to all cylindric set algebras. The answer to the question of whether and to what extent they do, translates back to logic and gives rise to logical results of an abstract model theoretic flavour. The reason for the latter is that the logical counterparts of cylindric set algebras not in the above mentioned "narrow subclass" are models of logics other than $L_{\omega\omega}$.

The structure of this paper is the following.

In the second half of the present §1 and in §2 we introduce cylindric set algebras and related machinery, together with some of the most essential algebraic concepts corresponding to model theoretic ones.

In §3.1 we prove the equivalence of certain algebraic concepts with certain model theoretic ones. This gives rise to natural algebraic counterparts for some well known model theoretic results.

As outlined in the above paragraph mentioning $L_{\omega\omega}$, in §3.2 we look into the algebraic generalizability of these algebraic counterpart theorems and in §3.3 we draw some (abstract) model theoretic conclusions concerning e.g. an infinitary logic originating with Henkin and Tarski. The methods can be extended to Keisler's logic with infinitary predicates, cf. [13].

In §3.2 we obtain a solution for Open Problem 3.2 of Henkin–Monk–Tarski [8]. We note that Theorem 7 in §3.2, besides solving the mentioned problem from [8], also solves Problems II.6–7 on p. 310 of [9]. Our solution implies that I.7.30(g) of [9] does not extend from cylindric set algebras to regular cylindric set algebras.

*****

Next we recall some basic concepts and symbols from [8]. *BA* denotes the class of all Boolen algebras. From now on, throughout the paper, α is an arbitrary but fixed ordinal. Recall from [8] p. 430, Def. 2.7.1 that an *α-dimensional BA with operators*, a Bo_α, is an algebra $\mathfrak{A} = \langle A, +, -, c_i, d_{ij}\rangle_{i,j\in\alpha}$ such that $\mathfrak{Bl}\,\mathfrak{A} \stackrel{\text{def}}{=} \langle A, +, -\rangle$ is a *BA* (with "+" denoting join), the d_{ij}'s are constants, and the operations c_i ($i \in \alpha$) are additive, i.e. $\mathfrak{A} \models c_i(x+y) = c_i x + c_i y$ for all $i \in \alpha$. If $\mathfrak{A} \in Bo_\alpha$ then $\mathfrak{Bl}\,\mathfrak{A}$ is called the *Boolean reduct* of $\mathfrak{A}$. Note that $BA = Bo_0$. Bo_α's were introduced and extensively investigated in Jónsson–Tarski [12].

The step of moving from propositional to first-order logic can be visualized as moving from the *BA* of unary relations to *BA*'s of arbitrary (say α-ary) relations. This *unary* $\longmapsto$ α-ary move is often called 1-dimensional$\longmapsto$ α-dimensional in [8,9]. The most natural way of generalizing Boolean *set* algebras to higher dimensions (or arities) is to replace the "points" of the Boolean *set* algebra with sequences. So the greatest element $1^{\mathfrak{A}}$ of the new BA $\mathfrak{A}$ is not only a set but a set of sequences (i.e. $1^{\mathfrak{A}} \subseteq {}^{\alpha}U$ for some set U and ordinal α). If we fix the length of these sequences to be α then the resulting class of BA's is called Crs_{α} (cylindric-relativized set algebras). Of course, to make all this meaningful, we have to derive some new operations (c_i, d_{ij}; $i, j \in \alpha$) from the structure of these sequences.

We shall need the classes $Cs_{\alpha} \subset Crs_{\alpha} \subset Bo_{\alpha}$. For any $x \in \mathfrak{A} \in Bo_{\alpha}$, $\Delta x \stackrel{\text{def}}{=} \{i \in \alpha : c_i x \neq x\}$. Recall from [8] or [9] that a Crs_{α} is a Boolean set algebra $\mathfrak{A}$ with unit (greatest element) $1^{\mathfrak{A}} = V \subseteq {}^{\alpha}U$ for some U. The non BA operations c_i and d_{ij} are defined from the structure of the Cartesian space ${}^{\alpha}U$, as follows:

$$d_{ij}^{\mathfrak{A}} = D_{ij}^{[V]} \ = \{f \in V : f_i = f_j\} \quad \text{and}$$
$$c_i^{\mathfrak{A}}(x) = C_i^{[V]}x = \{f \in V : f \restriction (\alpha \smallsetminus \{i\}) \subseteq g \in x \text{ for some } g\}$$

for all $x \subseteq V$ and $i, j \in \alpha$. See Figure 1. $\mathfrak{A}$ is said to be a Cs_{α} (a cylindric set algebra) iff $V = {}^{\alpha}U$. U is called the base of $\mathfrak{A}$ and is denoted by $base(\mathfrak{A})$. In general, the base of a Crs_{α} $\mathfrak{A}$ with unit V is $base(\mathfrak{A}) \stackrel{\text{def}}{=} \bigcup\{Rg(s) : s \in V\}$.

Let $\mathfrak{A} \in Cs_{\alpha}$ with base U. Then, by definition,

(1) $\mathfrak{A} \in {}_{\kappa}Cs_{\alpha}$ iff $|U| = \kappa$,

(2) ${}_{\infty}Cs_{\alpha} = \bigcup\{{}_{\kappa}Cs_{\alpha} : \kappa \geq \omega\}$,

(3) $x \in A$ is *regular* iff $(\forall f, g \in {}^{\alpha}U)[f \restriction \Delta x \subseteq g \in x \Rightarrow f \in x]$.
$\mathfrak{A} \in Cs_{\alpha}^{reg}$ iff $(\forall x \in A)(x$ is regular$)$.

Cylindric algebras (CA_{α}'s) are those Bo_{α}'s which satisfy equations (C1)–(C7) in [8]. We do not need to recall these equations for the purposes of this paper (since they are more relevant to the proof theoretic aspects of algebraic logic than to the model theoretic ones, cf. §4.3 of [8]).

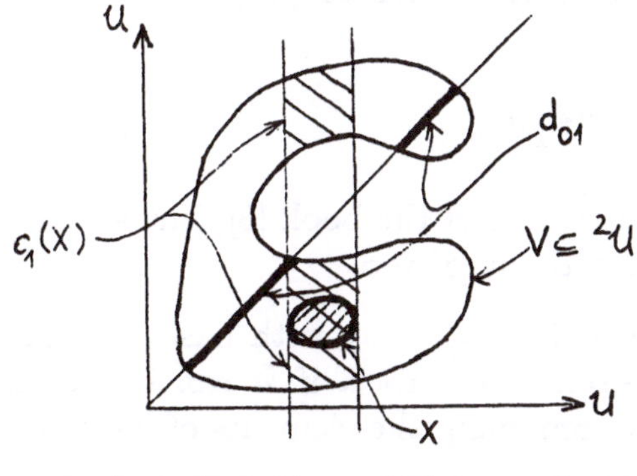

Figure 1: A typical Crs_2. (Observe that $c_1c_0c_1x = c_0c_1x$ is not valid in this Crs_2, while it *is* valid in Cs_2.)

REMARK 1. One thing that makes Crs_α more important, in many respects, than Cs_α is the following result of D.Resek and R.J.Thompson (see Andréka–Thompson [4], Resek–Thompson [17], and/or Remark 3.2.88 in Part II of [8] p. 101). Let $s^i_j \stackrel{\text{def}}{=} c_i(d_{ij} \cdot x)$ for $i \neq j$ and $s^i_i x = x$. Consider the equation scheme

(C8) $\{s^e_i s^i_j s^j_k s^k_e c_e x = s^e_j s^j_k s^k_i s^i_e c_e x : i, j, k, e \in \alpha\}$.

Its intuitive meaning is that if we want to permute the i–th, j–th, and k–th member of a sequence $(i \to j, j \to k, k \to i)$ using the e–th place as "auxiliary store" then we can do this two equivalent ways:

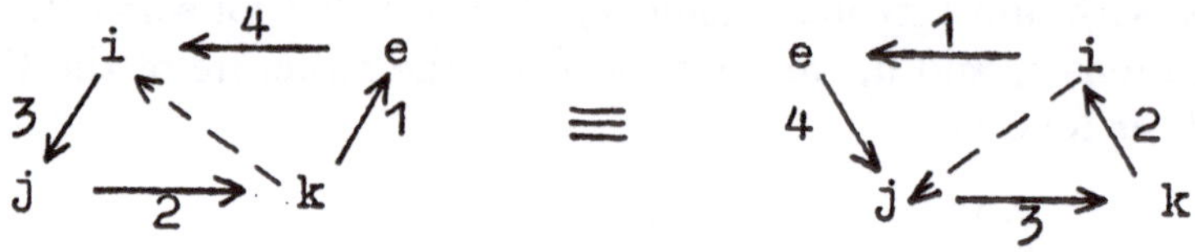

Figure 2

Let for example $\langle i, j, k, e \rangle = \langle 0, 1, 2, 3 \rangle$. Then $s^3_0 s^0_1 s^1_2 s^2_3 c_3\{\langle a, b, c, \ldots \rangle\} = c_3\{\langle c, a, b, \ldots \rangle\}$, which clearly corresponds to permuting the first 3 members of the sequence $\langle a, b, c, \ldots \rangle$. Now if we add to the equations (C0)–(C7) in [8] defining CA's the single new scheme (C8) then *all* new CA's become *representable* as Crs_α's. That is,

$$Mod(C0 - C8) \subseteq \text{Isomorphic copies of } Crs_\alpha\text{'s}.$$

Summing up, by adding a single schema (C8) to the axioms defining CA's, the much discussed negative representation property of CA's (see [8] for example) becomes positive! We should perhaps note that $Cs_\alpha \models (C8)$. ■

Concerning the class Crs_α we note that, up to isomorphisms, it is a variety with a decidable but not finitely axiomatizable equational theory. Further, Crs_α has the strong amalgamation property, cf. §5.5 of [8], [14], [15].

2 Algebraic Concepts

We use the notational system of the book [8], but we will try to keep the paper understandable without [8] too.

DEFINITION 2. Let $\mathfrak{A}, \mathfrak{B} \in Crs_\alpha$ with bases U and W respectively. Then $\mathfrak{B}$ is *ext–base–isomorphic* to $\mathfrak{A}$ if $\mathfrak{A}$ can be obtained from $\mathfrak{B}$ by restricting its unit and renaming the elements of its base. In more detail: Assume that there is a $\mathfrak{C}$ with unit $Z \subseteq 1^{\mathfrak{B}}$ such that $rl^B_Z \stackrel{\text{def}}{=} rl_Z \stackrel{\text{def}}{=} rl(Z) \stackrel{\text{def}}{=} \langle x \cap Z : x \in B \rangle : \mathfrak{B} \rightarrowtail\!\!\!\!\rightarrow \mathfrak{C}$ is an isomorphism. Then $\mathfrak{B}$ is said to be *ext–isomorphic* to $\mathfrak{C}$. Further, $\mathfrak{A}$ and $\mathfrak{C}$ are *base–isomorphic* if they are "the same Crs_α except for renaming the elements of the bases" (i.e. $Z \subseteq {}^\alpha H$ for some H and there is a bijection $f : H \rightarrowtail\!\!\!\!\twoheadrightarrow U$ such that $\mathfrak{A}$ is obtained from $\mathfrak{C}$ by consequently replacing the elements of the base H of $\mathfrak{C}$ according to f). Now, $\mathfrak{B}$ is ext–base–isomorphic to $\mathfrak{A}$ iff there is a $\mathfrak{C}$ as above (i.e. $\mathfrak{B}$ is ext–isomorphic to $\mathfrak{C}$ and $\mathfrak{C}$ is base–isomorphic to $\mathfrak{A}$).

Finally, $\mathfrak{A}$ is *sub–base–isomorphic* or *sub–isomorphic* to $\mathfrak{B}$ iff $\mathfrak{B}$ is ext–base–isomorphic or ext–isomorphic resp. to $\mathfrak{A}$.

For a more detailed formulation of this definition see Def.3.1.41 of [8] or [9] p. 156 Def.3.1 and p. 181 Def.3.12. ■

DEFINITION 3. Let $\mathfrak{A} \in BA$. Then Fu $\mathfrak{A}$ denotes the set of all ultrafilters on $\mathfrak{A}$ and Fl $\mathfrak{A}$ denotes the set of all (nonempty) filters on $\mathfrak{A}$. ■

Recall from [8] that for a subset F of a Bo_α , the c_i–image of F is

$$c_i{}^*F = \{c_i(x) : x \in F\}.$$

DEFINITION 4. Let $\mathfrak{A} \in Crs_\alpha$ with unit V . Then

(i) $\mathfrak{A}$ is defined to be *saturating* if

$$(\forall i \in \alpha)(\forall F \in Fl\mathfrak{B}\mathfrak{l}\,\mathfrak{A})C_i^{[V]}(\bigcap F) = \bigcap c_i^* F.$$

(ii) $\mathfrak{A}$ is *compact* iff $(\forall F \in \mathrm{Fu}\,\mathfrak{B}\mathfrak{l}\,\mathfrak{A})\bigcap F \neq 0$. ■

We note that the properties introduced so far are not abstract in the sense that they are not preserved under isomorphisms—rather, they are properties of a "concrete algebra" or, in other words, of a given "representation" of an algebra. To illustrate this, we quote from [14] the following:

LEMMA 5. *Every Crs_α is isomorphic to a compact and saturating Crs_α*

This lemma was proved in [14] as Lemma 5.1. ■

We have defined saturating and compact Crs_α's as well as base-isomorphisms and sub-isomorphisms. These algebraic notions correspond to notions in model theory (saturated model, universal model, definitional equivalence of models, elementary submodel respectively), see §4.3 of [8]. The algebraic counterparts are more general than the original logical notions. Thus the logical theorems do not carry always over (or not unchanged) to the algebraic forms. We shall look into these matters in the next section. The results do have some abstract model theoretic consequences (in the sense that they go beyond first–order logic).

3 Algebraic Model Theory

3.1 Connections between some model theoretic and cylindric algebraic concepts

We will use the concepts and notation of §4.3 ("Connections between logic and CA's") of [8]. In particular, a model may have infinitary relations too. That is, a model is $\mathfrak{M} = \langle M, R_i \rangle_{i \in I}$ such that $M \neq \emptyset$ and to each $i \in I$ there is an ordinal $\rho(i)$ with $R_i \subseteq {}^{\rho(i)}M$. The language or similarity type of $\mathfrak{M}$ is of the form $\Lambda = \langle \alpha, I, \rho \rangle$ with $\alpha \geq \rho(i)$ for all $i \in I$.

The set denoted by $\Phi\mu_r^\Lambda$ in [8] p. 153 of formulas of Λ (i.e. of $\mathfrak{M}$) is built up from the atomic formulas $R_i(v_0, \ldots, v_j, \ldots)_{j<\rho(i)}$ and $v_j = v_k$ for $i \in I$ and $j, k \in \alpha$ the usual (finitary) way (i.e. via finitely many applications of the finitary connectives "$\wedge$", "$\neg$", and "$\exists v_j$"). Here $\{v_j : j \in \alpha\}$ is the set of variables occurring in $\Phi\mu_r^\Lambda$. If $\mathfrak{M}$ is given, the choice of α is not quite arbitrary since $\alpha \geq \rho(i)$ has to be satisfied. Unless otherwise specified, we will always assume that $\alpha \geq \omega$. The rest of Λ is determined by $\mathfrak{M}$. Therefore we use $\mathcal{L}(\mathfrak{M})$ to denote $\Phi\mu_r^\Lambda$ i.e. the set of formulas in the language of $\mathfrak{M}$ if α is recoverable from context. This finitary logic of infinitary structures was introduced originally by Henkin and Tarski [10] (cf. Henkin [7], Jaskowski [11]), and was investigated later in section VI of [1], where it was denoted by ${}_cL_F^t$. Recall from [8] that $\Lambda = \langle \alpha, I, \rho \rangle$ and its models $\mathfrak{M}$ are called *ordinary* iff $Rg(\rho) \subseteq \omega \leq \alpha$ (i.e. iff all relations are finitary and there are infinitely many variables just as in "ordinary" $L_{\omega\omega}$).

We mention that if Λ is ordinary, then for each formula φ of Λ in the usual (e.g. Chang– Keisler's) sense there is $\varphi' \in \Phi\mu_r^\Lambda$ such that φ and φ' are equivalent (i.e. are satisfied in each model under the same evaluations of variables). This is proved in [8] 4.3.6.

The languages $\Lambda = \langle \alpha, I, \rho \rangle$ are generalizations of the ordinary one in two directions: If $\alpha < \omega$ then we have ordinary models but we can use only finitely many variables in our language. If $\alpha \geq \omega$ then we may have infinitary relations in our models, but the language is very similar to the ordinary one. Cylindric algebraic theorems have model theoretic consequences of both kinds. In the present paper we concentrate on the case $\alpha \geq \omega$.

We use the words *κ-saturated* and *κ-universal* for ordinary models as defined in Chang–Keisler [5]. Throughout $\kappa > 1$ is a cardinal, α, β, and γ are ordinals. Cardinals are special ordinals and κ^+ is the successor cardinal of κ. $H \subseteq_\kappa G$ means that $H \subseteq G$ and $|H| < \kappa$.

For a model $\mathfrak{M}$, the corresponding natural cylindric set algebra of dimension α , $\mathfrak{Cs}_\alpha^{\mathfrak{M}} = \mathfrak{Cs}_\alpha^{(\mathfrak{M})} = \mathfrak{Cs}^{\mathfrak{M}}$ was defined on p. 154 of §4.3 of [8] as follows. Let

$$Cs_\alpha^{(\mathfrak{M})} = \{\{q \in {}^\alpha M : \mathfrak{M} \models \varphi[q]\} : \varphi \in \Phi\mu_r^\Lambda\}.$$

Then $\mathfrak{Cs}^{\mathfrak{M}}$ is the cylindric set algebra with universe $Cs_\alpha^{(\mathfrak{M})}$. (Actually $\mathfrak{Cs}^{\mathfrak{M}}$ was denoted by ${}^r\mathfrak{Cs}^{\mathfrak{M}}$ in [8] but for simplicity we omit the superscript "r".)

THEOREM 1. *Let $\mathfrak{M}$ be a model of an ordinary language. Then*

$$\mathfrak{M} \textit{ is } |\alpha|^+\textit{-saturated iff } \mathfrak{Cs}_\alpha^{(\mathfrak{M})} \textit{ is saturating.}$$

Proof: Notation: If k is a function and $i \in \mathrm{Do}(k)$ then $k(i/a)$ denotes the function we obtain from k by replacing its i–th value with a. Assume $\mathfrak{Cs}_\alpha^{(\mathfrak{M})}$ is saturating. We show that $\mathfrak{M}$ is $|\alpha|^+$-saturated.

Let $\langle m_i\rangle_{i<\xi} \in {}^\xi M$ with $|\xi| \leq |\alpha|$. Let $\mathfrak{M}^+ \stackrel{\text{def}}{=} \langle \mathfrak{M}, m_i\rangle_{i<\xi}$. Let $\Sigma \subseteq \mathcal{L}(\mathfrak{M}^+)$ be a set of formulas such that each $\varphi \in \Sigma$ has one free variable v_0. Assume that

(∗) every finite subset of Σ is satisfiable in $\mathfrak{M}^+$ (i.e. $(\forall\Sigma_0 \subseteq_\omega \Sigma)\mathfrak{M}^+ \models \exists v_0 \bigwedge\Sigma_0$).

We have to show that the whole Σ is satisfiable in $\mathfrak{M}^+$ too. Let $\eta\colon \xi \rightarrowtail \alpha \smallsetminus 1$ be arbitrary. (Note that $\alpha \smallsetminus 1 = \alpha \smallsetminus \{0\} = \{i : 0 < i < \alpha\}$.) For every $\varphi \in \mathcal{L}(\mathfrak{M}^+)$ with one free variable v_0 define φ' to be the formula obtained from φ by replacing m_i with v_{η_i} for every $i < \xi$. Then $\varphi' \in \mathcal{L}(\mathfrak{M})$ (since $\mathfrak{M}$ is ordinary) and

(⁑) $\mathfrak{M} \models \varphi'[k]$ iff $\mathfrak{M}^+ \models \varphi[k_0]$, for every $k \in {}^\alpha M$ such that $(\forall i < \xi)k(\eta_i) = m_i$.

Let $F \stackrel{\text{def}}{=} \{((\bigwedge\Sigma_0)')^{\mathfrak{M}} : \Sigma_0 \subseteq_\omega \Sigma\}$. Then $F \subseteq Cs^{(\mathfrak{M})}$ and F is directed downward. Let $k \in {}^\alpha M$ be such that $(\forall i < \xi)k(\eta_i) = m_i$. By (∗) and (⁑) we have $k \in c_0\sigma$ for every $\sigma \in F$. By saturatingness of $\mathfrak{Cs}_\alpha^{(\mathfrak{M})}$ we have $k \in c_0(\bigcap F)$. Let $k' \stackrel{\text{def}}{=} k(0/a)$ be such that $k' \in \bigcap F$. Then $(\forall\varphi \in \Sigma)k' \in (\varphi')^{\mathfrak{M}}$ i.e. $\mathfrak{M}^+ \models \varphi[a]$. Thus $a \in M$ satisfies the whole Σ in $\mathfrak{M}^+$.

Conversely, assume $\mathfrak{M}$ is $|\alpha|^+$–saturated. We show that $\mathfrak{Cs}_\alpha^{(\mathfrak{M})}$ is saturating. Let $F \subseteq Cs^{(\mathfrak{M})}$ be downward directed, $i \in \alpha$, and $k \in \bigcap c_i^* F$. We have to show $k \in c_i \bigcap F$. For every $x \in F$ let $\varphi_x \in \mathcal{L}(\mathfrak{M})$ be such that $x = \varphi_x^{\mathfrak{M}}$. Let φ'_x denote the formula obtained from φ_x by replacing v_j with k_j (as a constant symbol denoting itself) for every $j \in \alpha \smallsetminus \{i\}$. Let $\Sigma \stackrel{\text{def}}{=} \{\varphi'_x : x \in F\}$ and $\mathfrak{M}^+ \stackrel{\text{def}}{=} \langle \mathfrak{M}, k_j\rangle_{j\in\alpha}$. Then $\Sigma \subseteq \mathcal{L}(\mathfrak{M}^+)$ such that every $\varphi \in \Sigma$ has only one free variable, v_i. By $k \in \bigcap c_i^* F$ and directedness of F we have that every finite subset of Σ is satisfiable in $\mathfrak{M}^+$ (because: Let $\Sigma_0 \subseteq_\omega \Sigma$. By directedness of F, there is $x \in F$ such that $x \leq \varphi^{\mathfrak{M}}$ for every $\varphi' \in \Sigma_0$. Then, by $k \in c_i(\varphi'_x)^{\mathfrak{M}}$ we have $\mathfrak{M}^+ \models \exists v_i \varphi'_x$, hence $\mathfrak{M}^+ \models \exists v_i \bigwedge\Sigma_0$).

By $|\alpha|^+$–saturatedness of $\mathfrak{M}$ then Σ is satisfiable in $\mathfrak{M}^+$. Let $a \in M$ be such that $\mathfrak{M}^+ \models \Sigma(v_i/a)$. Then $k(i/a) \in \bigcap F$. ■

As usual, by the cardinality of a language (or similarity type) $\Lambda = \langle\alpha, I, \rho\rangle$ we understand $|I|$ i.e. the number of relation symbols in Λ. Note that in particular this cardinality of Λ may be much smaller than α.

To be able to appreciate the (algebraic) generality of Thm. 1 above (and the theorems below), we recall from [8] the following.

FACT 6.

(i) *Every* $\mathfrak{A} \in Cs_\alpha^{reg} \cap Lf_\alpha$ *is of the form*

$$\mathfrak{A} = \mathfrak{Cs}_\alpha^{\mathfrak{M}}$$

for some ordinary model $\mathfrak{M}$.

(ii) *Further, if* $\mathfrak{A}$ *is generated by* $\leq \kappa$ *elements, then the language of* $\mathfrak{M}$ *is of cardinality* $\leq \kappa$. ∎

THEOREM 2. *(G. Serény) Let* $\mathfrak{M}$ *be a model of an ordinary language. Then (i)–(ii) below hold.*

(i) $\mathfrak{Cs}_\alpha^{(\mathfrak{M})}$ *is compact* $\Longrightarrow$ $\mathfrak{M}$ *is* $|\alpha|^+$*–universal.*

(ii) *Assume that the language or similarity type of* $\mathfrak{M}$ *is of cardinality* $\leq |\alpha|$. *Then*

$$\mathfrak{Cs}_\alpha^{(\mathfrak{M})} \textit{is compact} \iff \mathfrak{M} \textit{ is } |\alpha|^+\textit{-universal.}$$

Proof: Throughout, $\equiv_e$, $Rg(f) = Rgf$, and $\mathrm{Do}(f) = \mathrm{Do}\,f$ denote elementary equivalence, range, and domain of f respectively.

Assume that $\mathfrak{Cs}_\alpha^{(\mathfrak{M})}$ is compact. We show that $\mathfrak{M}$ is $|\alpha|^+$–universal. Let $\mathfrak{N} \equiv_e \mathfrak{M}$, $|N| \leq |\alpha|$. We have to show that $\mathfrak{N}$ is elementarily embeddable into $\mathfrak{M}$. Let $k \in {}^\alpha N$ be such that $Rg(k) = N$. Let $F \stackrel{\text{def}}{=} \{\varphi^{\mathfrak{M}} : \mathfrak{N} \models \varphi[k] \text{ and } \varphi \in \mathcal{L}(\mathfrak{M})\}$. Then it is easy to check that F is a proper filter of $\mathfrak{Bl}\ \mathfrak{Cs}_\alpha^{(\mathfrak{M})}$. Thus $\bigcap F \neq 0$. Let $h \in \bigcap F$. Now $(\forall\varphi \in \mathcal{L}(\mathfrak{M}))(\mathfrak{M} \models \varphi[h] \text{ iff } \mathfrak{N} \models \varphi[k])$. Let $n \stackrel{\text{def}}{=} \{\langle k_i, h_i\rangle : i \in \alpha\}$. Then it is not hard to check that $n : \mathfrak{N} \rightarrowtail \mathfrak{M}$ is an elementary embedding of $\mathfrak{N}$ into $\mathfrak{M}$.

Assume the cardinality conditions of (ii). Then $|Cs_\alpha^{\mathfrak{M}}| \leq |\alpha|$. Assume that $\mathfrak{M}$ is $|\alpha|^+$–universal. We show that $\mathfrak{Cs}_\alpha^{(\mathfrak{M})}$ is compact. Let F be a proper filter of $\mathfrak{Bl}\ \mathfrak{Cs}_\alpha^{(\mathfrak{M})}$. For every $x \in F$ let $\varphi_x \in \mathcal{L}(\mathfrak{M})$ be such that $\varphi_x^{\mathfrak{M}} = x$. For every $i < \alpha$ let k_i be a new constant symbol and for every $\varphi \in \mathcal{L}(\mathfrak{M})$ let φ' denote the formula we obtain from φ by replacing each variable v_i with the constant symbol k_i. Let $\Sigma \stackrel{\text{def}}{=} \{\varphi'_x : x \in F\}$. Let $\mathfrak{M} = \langle M, R\rangle = \langle M, R_i\rangle_{i\in\mathrm{Do}(R)}$. By $|Cs^{(\mathfrak{M})}| \leq |\alpha|$ we have $|Rg(R)| \leq |\alpha|$. Let $R' \subseteq R$ be such that $Rg(R)' = Rg(R)$ and $|R'| \leq |\alpha|$. Let $\mathfrak{M}' \stackrel{\text{def}}{=} \langle M, R'\rangle$. Then $\mathfrak{M}'$ is definitionally equivalent to $\mathfrak{M}$ and $|\mathcal{L}(\mathfrak{M}')| \leq |\alpha|$. Let $Th(\mathfrak{M}') \stackrel{\text{def}}{=} \{\varphi \in \mathcal{L}(\mathfrak{M}') : \mathfrak{M}' \models \varphi\}$. Then $|Th(\mathfrak{M}') \cup \Sigma| \leq |\alpha|$. Let Λ be an (ordinary) language of cardinality $\leq |\alpha|$ such that $Th \stackrel{\text{def}}{=} Th(\mathfrak{M}') \cup \Sigma \subseteq \Phi\mu_\tau^\Lambda$. Every finite subset of Th has a model (namely, a reduct of $\mathfrak{M}$ enriched with appropriate constants). Thus by the compactness and the downward Löwenheim–Skolem theorems, there is a model $\mathfrak{N} = \langle N, P\rangle$ of Th such that $|N| \leq |\alpha|$. Let $\mathfrak{N}'$ be the model similar to $\mathfrak{M}$ and obtainable from $\mathfrak{N}$ the natural way, i.e. let $\mathfrak{N}' \stackrel{\text{def}}{=} \langle N, P'\rangle$ where $\text{Do}(P') = \text{Do}(R)$, $P' \restriction \text{Do}(R') \subseteq P$, and $(\forall \xi, \eta \in \text{Do}(R))[P'_\xi = P'_\eta \text{ iff } R_\xi = R_\eta]$. (This is not contradictory by $\mathfrak{N} \models Th(\mathfrak{M}')$.) Now $\mathfrak{N}' \equiv_e \mathfrak{M}$ by $\mathfrak{N} \models Th(\mathfrak{M}')$ and by the construction of $\mathfrak{N}', \mathfrak{M}'$; hence by $|N| \leq |\alpha|$ and $|\alpha|^+$–universality of $\mathfrak{M}$ there is an elementary embedding $n : \mathfrak{N}' \rightarrowtail \mathfrak{M}$ of $\mathfrak{N}'$ into $\mathfrak{M}$. Define $h \in {}^\alpha M$ as $h_i \stackrel{\text{def}}{=} n(k_i^{\mathfrak{N}})$ for every $i < \alpha$. Let $x \in F$. Then $h \in \varphi_x^{\mathfrak{M}}$ by $\mathfrak{N} \models \varphi'_x$ and since n is an elementary embedding. Thus $h \in \bigcap F$. ■

To formulate the algebraic counterpart of κ–saturated models for an arbitrary cardinal κ, we need the following.

Let $\mathfrak{A} \in Bo_\alpha$. Recall from [8] that $Nr_\beta \mathfrak{A} = \{x \in A : (\forall i \in \alpha)[i \notin \beta \Rightarrow c_i x = x]\}$ is called the neat β–reduct of $\mathfrak{A}$. We will think of $Nr_\beta \mathfrak{A}$ as a Boolean algebra i.e. as a subalgebra of $\mathfrak{Bl}\,\mathfrak{A}$. Intuitively, $\mathfrak{A}$ is κ–saturating or κ–compact iff $(\forall \beta < \kappa)(Nr_\beta \mathfrak{A}$ viewed as a Crs_β is saturating or compact respectively). In more detail:

DEFINITION 7. Let $\kappa \leq |\alpha|^+$ be a cardinal and $\mathfrak{A} \in Crs_\alpha$. Then:

(i) $\mathfrak{A}$ is *κ-saturating* iff $(\forall \beta < \kappa)(\forall F \in \text{Fl}\, Nr_\beta \mathfrak{A})(\forall i \in \alpha) C_i(\bigcap F) = \bigcap(c_i{}^* F)$.

(ii) $\mathfrak{A}$ is *κ–compact* iff $(\forall \beta < \kappa)(\forall F \in \text{Fu}\, Nr_\beta \mathfrak{A}) \bigcap F \neq 0$. ■

THEOREM 3. *Let $\mathfrak{M}$ be a finitary (i.e. ordinary) model and $\kappa \leq |\alpha|^+$. Then*

(i) *$\mathfrak{M}$ is κ–saturated iff $\mathfrak{Cs}_\alpha^{\mathfrak{M}}$ is κ-saturating.*

(ii) *Assume that the language or similarity type of $\mathfrak{M}$ is of cardinality $< \kappa$, and $\kappa > \omega$. Then $\mathfrak{M}$ is κ-universal iff $\mathfrak{Cs}_\alpha^{\mathfrak{M}}$ is κ-compact.*

Proof: First we note that (i) makes sense for finite κ too. In the literature, however, κ–saturatedness is studied primarily for $\kappa \geq \omega$. Therefore the various equivalent definitions of κ–saturatedness slightly disagree for finite κ (cf. e.g. Prop. 5.1.1 in [5]). Indeed, if in (i) we want to use, for finite κ, the definition as worded above Prop. 5.1.1 in Chang–Keisler [5] then we have to write

$$\mathfrak{M} \text{ is } \kappa\text{-saturated} \iff \mathfrak{Cs}^{\mathfrak{M}} \text{ is } \kappa + 1\text{-saturating.}$$

On the other hand, if we use the wording of the definition in §3.3 herein, then the formulation in Theorem 3(i) above is correct. (The two definitions of κ–saturated are equivalent for infinite κ.)

Apart from this, the proof is similar to those of Theorems 1, 2 above. The condition $\kappa > (\omega +$ the number of relation symbols in $\mathfrak{M})$ of (ii) is needed only in the "κ–universal $\Rightarrow$ κ–compact" direction. Namely these are needed to make the downward Löwenheim–Skolem theorem applicable.

(Indeed, if one of these two fail then $\mathfrak{M}$ can be vacuously κ–universal without $\mathfrak{Cs}^{\mathfrak{M}}$ being κ–compact.) The condition $\kappa \leq |\alpha|^+$ is assumed only to make Definition 7 applicable.   ■

The following is not difficult to check, see [8] 4.3.45 and 4.3.68(7), (10). Definitional equivalence of two models with the same universe is introduced in [8] Part I p. 56, and in [8] 4.3.45, but it can be found also in Def. 17 in §3.3 herein.

PROPOSITION 8. *Let* $\mathfrak{M}, \mathfrak{N}$ *be ordinary models,* $\alpha \geq \omega$. *Then (i)–(v) below hold.*

  (i) $\mathfrak{Cs}_{\alpha}^{\mathfrak{M}} = \mathfrak{Cs}_{\alpha}^{\mathfrak{N}}$ *iff* $\mathfrak{M}$ *is definitionally equivalent to* $\mathfrak{N}$.
  (ii) $\mathfrak{Cs}_{\alpha}^{\mathfrak{M}} \cong \mathfrak{Cs}_{\alpha}^{\mathfrak{N}}$ *iff* $\mathfrak{M}$ *is elementarily equivalent to a model definitionally equivalent to* $\mathfrak{N}$.
  (iii) $\mathfrak{Cs}_{\alpha}^{\mathfrak{M}}$ *is base–isomorphic to* $\mathfrak{Cs}_{\alpha}^{\mathfrak{N}}$ *iff* $\mathfrak{M}$ *is isomorphic to a model definitionally equivalent to* $\mathfrak{N}$.
  (iv) $\mathfrak{Cs}_{\alpha}^{\mathfrak{M}}$ *is sub–isomorphic to* $\mathfrak{Cs}_{\alpha}^{\mathfrak{N}}$ *iff* $\mathfrak{M}$ *is an elementary submodel of a model definitionally equivalent to* $\mathfrak{N}$.
  (v) $\mathfrak{Cs}_{\alpha}^{\mathfrak{M}}$ *is sub–base–isomorphic to* $\mathfrak{Cs}_{\alpha}^{\mathfrak{N}}$ *iff* $\mathfrak{M}$ *is isomorphic to an elementary submodel of a model definitionally equivalent to* $\mathfrak{N}$.   ■

3.2 INVESTIGATING THE ALGEBRAIC CONCEPTS (DERIVED FROM THE MODEL THEORETIC ONES)

Let $\mathfrak{A}$ be a Crs_{α} with base U and unit V. Let $X \subseteq U$. Recall from [8] 4.3.68(10), that $\mathfrak{A}_X$ is the subalgebra of the full $Crs_{\alpha}\mathfrak{Sb}(V)$ with unit V generated by $A \cup \{\{f \in V : f_0 = u\} : u \in X\}$. Also recall from [8] that Lf_{α} is the class of *locally finite* dimensional CA_{α}'s i.e. of those CA's every element x of which is such that Δx is finite. Cs_{α}^{reg} was recalled from [8] in the Introduction. Intuitively, $Cs_{\omega}^{reg} \cap Lf_{\omega}$ is the class of algebras of finitary relations (because each element x of a Cs_{α}^{reg} is basically a Δx–ary relation).

Next we state an immediate algebraic corollary of Theorems 1,2 above. (The proof uses well known properties of saturated– and universal models from e.g. Chang–Keisler [5] together with Fact 6 and Prop. 8 above. However, Corollary 9 has a natural algebraic proof, too, see Serény [21], [23].)

COROLLARY 9. *(Serény) Let* $\mathfrak{A} \in Cs_{\alpha}^{reg} \cap Lf_{\alpha}$, $\alpha \geq \omega$. *Then (i)–(x) below hold.*

  (i) $\mathfrak{A}$ *is saturating* $\not\Longleftarrow\!\!\!\Longrightarrow$ $\mathfrak{A}$ *is compact.*
  (ii) $|base(\mathfrak{A})| < \omega \Rightarrow \mathfrak{A}$ *is saturating.*
  (iii) $\mathfrak{A}$ *is saturating* $\Rightarrow (|base(\mathfrak{A})| < \omega$ *or* $|base(\mathfrak{A})| > |\alpha|)$.
  (iv) $\mathfrak{A}$ *is saturating,* $X \subseteq base(\mathfrak{A})$, $|X| \leq |\alpha| \Rightarrow \mathfrak{A}_X$ *is saturating.*
  (v) $\mathfrak{A}$ *is saturating iff* $(\forall X \subseteq base(\mathfrak{A}))[|X| \leq \alpha \Rightarrow \mathfrak{A}_X$ *is 2-compact].*

(vi) *If* $\mathfrak{A} \cong \mathfrak{B} \in Cs_\alpha^{reg}$, *both* $\mathfrak{A}$ *and* $\mathfrak{B}$ *are saturating and* $|base(\mathfrak{A})| = |base(\mathfrak{B})| \leq |\alpha|^+$ *then* $\mathfrak{A}$ *and* $\mathfrak{B}$ *are base–isomorphic.*

(vii) $\mathfrak{A}$ *is sub–isomorphic to a saturating* Cs_α^{reg}.

(viii) $\mathfrak{A}$ *is sub–isomorphic to a compact* Cs_α^{reg}.

(ix) *If* $\mathfrak{A} \cong \mathfrak{B} \in Cs_\alpha^{reg}$ *then there is a* $\mathfrak{C} \in Cs_\alpha^{reg}$ *ext–base–isomorphic to both* $\mathfrak{A}$ *and* $\mathfrak{B}$.

(x) $[\mathfrak{A} \cong \mathfrak{B} \in {}_\kappa Cs_\alpha^{reg}$, $\kappa \leq \alpha$, *and* $\mathfrak{A}$ *is compact*$] \Longrightarrow \mathfrak{A}$ *is ext-base-isomorphic to* $\mathfrak{B}$. ■

Let us see to what extent (i)–(x) above generalize from $Cs_\alpha^{reg} \cap Lf_\alpha$ to other classes.

Weak cylindric set algebras (Ws_α's) and *generalized cylindric set algebras* (Gs_α's) are defined in [8], [9]. Let $\mathfrak{A} \in Crs_\alpha$ with unit V. $\mathfrak{A}$ is said to be a Ws_α iff $V = {}^\alpha U^{(p)} \stackrel{\text{def}}{=} \{s \in {}^\alpha U : |\{i \in \alpha : s_i \neq p_i\}| < \omega\}$ for some U and p. $\mathfrak{A}$ is said to be a Gs_α iff V is a disjoint union of Cartesian spaces, i.e. if $V = \bigcup\{{}^\alpha U_i : i \in I\}$ for some system $\langle U_i : i \in I\rangle$ of disjoint sets. In this case, U_i, $(i \in I)$ is called a *subbase* of $\mathfrak{A}$. $\mathfrak{A}$ is called *minimal* if $\mathfrak{A}$ is generated by $\{d_{ij}^{\mathfrak{A}} : i,j \in \alpha\}$ and $\mathfrak{A}$ is called *discrete* if $d_{ij}^{\mathfrak{A}} = V$ for all $i,j \in \alpha$. $\mathfrak{A}$ is said to be *dimension complemented*, a Dc_α, iff $(\forall x \in A)\Delta x \neq \alpha$.

We shall see below in Theorems 4–7 that in (i) both regularity and local finiteness are needed, (ii) generalizes to Crs_α, (iii) generalizes to Gs_α but not to Ws_α, in (vi) both regularity and "$\cap Lf$" are needed, (vii) generalizes to all interesting classes, (viii) generalizes to Crs_α, ${}_\infty Cs_\alpha$ but not to ${}_\infty Cs^{reg}$, ${}_\kappa Cs$ $(\kappa < \omega)$, (ix) generalizes to ${}_\infty Cs$ but not to $Cs_\alpha^{reg} \cap Dc_\alpha$. By Thm. 4(v) we have that (v) does not generalize to Cs_α^{reg} or $Cs_\alpha \cap Lf_\alpha$.

In passing we note that Shelah [24] proved recently that in (vi) saturatingness cannot be replaced with the condition that [$\mathfrak{A}$ and $\mathfrak{B}$ have no ext-isomorphic images (except themselves) and $Nr_n\mathfrak{A}$ is atomic for all $n < \omega$].

In connection with Corollary 9(iv) see Cor.27 in §3.3, in connection with (v) we mention the following. Serény showed the existence of a saturating $\mathfrak{A} \in Cs_\alpha \cap Lf_\alpha$ with finite base such that $\mathfrak{A}$ is not 1–compact. (His proof goes as follows: Let $\kappa < \omega$. Let $\mathfrak{A}$ be any ${}_\kappa Cs_\alpha$ generated by a downward directed set X with empty intersection such that $(\forall x \in X)\Delta x = 0$. The existence of such an X is easy to see. Then clearly $\mathfrak{A}$ is not 1–compact, and $\mathfrak{A}$ is saturating by Thm. 4(ii) below.) Therefore direction $\Longrightarrow$ of (v) does not generalize to $Cs_\alpha \cap Lf_\alpha$. Serény has a candidate for a counterexample showing that direction $\Longleftarrow$ of (v) does not generalize to $Cs_\alpha^{reg} \cap Dc_\alpha$.

We state Theorems 4–7 together, their proofs will come after stating Theorem 7. In the next two theorems we investigate the behaviour of saturating Crs_α's.

Theorem 4(v) below is due to G. Serény, and its proof can be found in Serény [21], [23]. We note that by combining the proofs in the two papers (the present one and [23]), (iv) and (v) below can be improved to state that the isomorphic but not (ext–base–) isomorphic members of K claimed to exist in (iv) and (v) are *both* saturating and compact. $\mathbf{I}K \stackrel{\text{def}}{=} \{\mathfrak{A} : \mathfrak{A} \cong \mathfrak{B}$ for some $\mathfrak{B} \in K\}$. Recall that we have a convention that throughout this paper $\alpha \geq \omega$ unless otherwise specified. Therefore we note that (ii) and (iii) below hold for finite α, too.

THEOREM 4.

(i) *Let $\alpha \geq \omega$ and $1 < \kappa < \omega$ or $\kappa > \alpha$. Let $K = {}_\kappa Cs_\alpha^{reg} \cap Dc_\alpha$ or $K = {}_\kappa Cs_\alpha \cap Lf_\alpha$. Then there are elements of K which are saturating but not compact.*

(ii) *Every Crs_α with finite base is saturating.*

(iii) *Let $\mathfrak{A} \in Gs_\alpha$ be saturating. Then each subbase of $\mathfrak{A}$ is of cardinality $> |\alpha|$ or is finite. Every minimal Ws_α with unit ${}^\alpha U^{(p)}$, $|Rg(p)| < \omega$ is saturating.*

(iv) *Let α, κ and K be as in (i). Then there are isomorphic saturating elements of K such that they are not base–isomorphic.*

(v) *Let α and κ be as in (i). Let $K = {}_\kappa Cs_\alpha^{reg}$ or $K = {}_\kappa Cs_\alpha \cap Lf_\alpha \cap \mathbf{I}Cs_\alpha^{reg}$. Then there are isomorphic compact elements of K such that they are not ext–base–isomorphic (i.e. there is no ext–base–isomorphism between them).*

The class $Gws_\alpha (\supset Gs_\alpha \cup Ws_\alpha)$ was defined in [8]. We do not repeat the definition.

COROLLARY 10. *Every minimal Gws_α with unit $\bigcup\{{}^\alpha U_i{}^{(p_i)} : i \in I\}$ is saturating iff $(\forall i \in I)$ either $|U_i| < \omega$ or $|U_i \smallsetminus Rg(p_i)| \geq \omega$. The "if" direction holds for $\alpha < \omega$ too.*

The "only if" direction of this corollary is easy, cf. Lemma 15 later, while the "if" direction is immediate by the second part of Thm. 4(iii) because of the following:

FACT 11. *Let $\mathfrak{A} \in Crs_\alpha$ and $X \subseteq Nr_0\mathfrak{A}$ a partition of $1^\mathfrak{A}$. Then*

$$\mathfrak{A} \text{ is saturating} \iff (\forall x \in X)\mathfrak{Rl}_x\mathfrak{A} \text{ is saturating.}$$

For K as in Theorem 5 below, ${}_nK$ was defined both in [8] and on p. 134 of [9] with the exception that ${}_nCrs_\alpha$ is available only in [9]. The reader with no access to [9] may safely replace ${}_nCrs$ with Crs's of base of cardinality n. These definitions are natural generalizations of that of ${}_nCs$ in the present paper's Introduction, e.g. $\mathfrak{A} \in {}_nGs_\alpha$ if all subbases of $\mathfrak{A}$ have cardinality n. Gws, Gws^{wd}, Gws^{norm}, Gws^{comp} were defined in [8], [9]—we do not repeat their definitions. Strong sub–isomorphism is a strengthened version of the concept introduced in Definition 2 herein. Namely, let $\mathfrak{A}, \mathfrak{B} \in Crs_\alpha$. We say that $\mathfrak{A}$ is strongly sub–isomorphic to $\mathfrak{B}$ if $rl({}^\alpha base(\mathfrak{A})) : \mathfrak{B} \rightarrowtail\!\!\!\!\rightarrow \mathfrak{A}$ is an isomorphism.

The next theorem holds for arbitrary (including finite) α. The same applies to Lemmas 14,15 below.

THEOREM 5.

(i) *Let K be one of Crs_α, Cs_α, Cs_α^{reg}, Gs_α, Ws_α, Gws_α, Gws_α^{wd}, Gws_α^{norm}, Gws_α^{comp}. Assume $\mathfrak{A} \in K$. Then $\mathfrak{A}$ is strongly sub-isomorphic to a saturating $\mathfrak{B} \in K$. If $\mathfrak{A} \in {}_nK$ then $\mathfrak{B} \in {}_nK$, for all $n \in \omega$. If $\mathfrak{A} \in K^{reg}$ then $\mathfrak{B} \in K^{reg}$.*

(ii) *Let $K = Crs_\alpha$ or $K = {}_\infty Cs_\alpha$. Then every $\mathfrak{A} \in K$ is strongly sub-isomorphic to a compact and saturating $\mathfrak{B} \in K$.*

Next we investigate compact Crs_α's. In connection with Thm. 5(ii) above we note that Serény [22] proved that every $\mathfrak{A} \in K$ is isomorphic to a compact $\mathfrak{B} \in K$ where $K \in \{Lf_\alpha, Gws_\alpha, Gs_\alpha, Gs_\alpha^{reg} \cap Lf_\alpha, Cs_\alpha^{reg} \cap Lf_\alpha\}$, and where α is any, possibly finite, ordinal. The next theorem as well as its proof are due to G. Serény [21]. We include them with his permission. If we compare the next theorem with Theorems 4,5 then we will see that there are fewer compact algebras than saturating ones. We recall from [8] that an algebra is called weakly subdirectly indecomposable if it is not isomorphic to a nontrivial subdirect product of two factors. In connection with Thm. 6(ii) below we mention that each Ws_α is weakly subdirectly irreducible.

THEOREM 6. *(G. Serény)*

(i) *There is $\mathfrak{A} \in {}_\kappa Cs_\alpha^{reg} \cap Dc_\alpha$ which is isomorphic to no compact Cs_α^{reg}.*

(ii) *Let $\kappa < \omega$ and $K = {}_\kappa Cs_\alpha^{reg} \cap Dc_\alpha \cap \mathbf{I}Ws_\alpha$ or $K = {}_\kappa Cs_\alpha \cap Lf_\alpha$. Then there is $\mathfrak{A} \in K$ which is isomorphic to no compact Cs_α.*

(iii) *No nondiscrete Ws_α is compact.*

We note that in [23] Serény improves the $\mathbf{I}Ws_\alpha$ part of (ii) above by proving that there are simple $Cs_\alpha^{reg} \cap Dc_\alpha$'s which are isomorphic to no compact Cs_α.

The next theorem solves Problem 3.2 of [8] as well as Problem II.3.11 of [9] negatively. It is proved in [9] II.3.10 that Corollary 9(ix) herein generalizes to ${}_\infty Cs$, Crs, Gs, Crs^{reg} but does not generalize to $Ws \cap Lf$. The next theorem states that Corollary 9(ix) does not generalize to ${}_\infty Cs^{reg} \cap Dc$, either.

THEOREM 7. *Let $\alpha \geq \omega$ and $1 < \kappa$. Then*

$$(\exists \mathfrak{A}, \mathfrak{B} \in {}_\kappa Cs_\alpha^{reg} \cap Dc_\alpha)[\mathfrak{A} \cong \mathfrak{B} \text{ but no } Cs_\alpha^{reg} \text{ is ext-base-isomorphic to both } \mathfrak{A} \text{ and } \mathfrak{B}].$$

Moreover, $\mathfrak{A}$ can be chosen to be locally countable dimensional, i.e. $(\forall x \in A)|\Delta x| \leq \omega$.

Proof of Theorem 4: (ii): Let $\mathfrak{A} \in Crs_\alpha$ with base U. Assume $|U| < \omega$. We show that $\mathfrak{A}$ is saturating. Let $F \in \mathrm{Fl}\,\mathfrak{Bl}\,\mathfrak{A}$ and assume $s \in \bigcap\{c_j b : b \in F\}$. We show $s \in C_j \bigcap F$. Let $X \stackrel{\text{def}}{=} \{p \in 1^{\mathfrak{A}} : p \restriction (\alpha \smallsetminus \{j\}) \subseteq s\}$. Then $|X| \leq |U| < \omega$ and $(\forall y)[s \in C_j y \text{ iff } y \cap X \neq 0]$. Assume $X \cap \bigcap F = 0$.

Then $(\forall p \in X)(\exists Y_p \in F) p \notin Y_p$. But then $Y \stackrel{\text{def}}{=} \bigcap\{Y_p : p \in X\} \in F$ and $X \cap Y = 0$ contradicting $s \in c_j Y$.

(iii): Assume $\mathfrak{A} \in Gs_\alpha$ and U is a subbase of $\mathfrak{A}$ with $\omega \leq |U| \leq \alpha$. Let $k \in {}^\alpha U$ be such that $Rg(k \restriction (\alpha \smallsetminus 1)) = U$ and let $F = \{\prod\{-d_{0j} : j \in J\} : J \subseteq_\omega \alpha \smallsetminus 1\}$. Then $k \in \bigcap c_0^* F$ but $k \notin c_0 \bigcap F$.

Let $\mathfrak{A}$ be a minimal Ws_α with unit ${}^\alpha U^{(p)}$, $|Rg(p)| < \omega$. By Thm. 4(ii), we may assume that $|U| \geq \omega$. Let $F \subseteq A$ be a downward directed subset of A, and $k \in {}^\alpha U^{(p)}$. Then $|Rg(k)| < \omega$ by $|Rg(p)| < \omega$. Let $i \in \omega$. Denote $U_x \stackrel{\text{def}}{=} \{u \in U : k(i/u) \in x\}$. Then the following holds:

$$(\star) \qquad U_x \cap U_y = U_{x \cdot y}, \quad U_{-x} = U \smallsetminus U_x.$$

Now by minimality of $\mathfrak{A}$, and by [8] 2.2.24 we have that $\mathfrak{A}$ is Boolean generated by $G = \{d_{hj} : h, j \in \alpha\} \cup \{a_h : h < \alpha\}$, where $\Delta a_h = 0$ for all h. Let $H \stackrel{\text{def}}{=} U \smallsetminus Rg(k)$. By $|U \smallsetminus Rg(p)| \geq \omega$, $H \neq 0$. Let $B \stackrel{\text{def}}{=} \{U_x : x \in A\}$. Since $\mathfrak{A}$ is Boolean generated by G, $(\star)$ implies that B is Boolean generated by $\{\{k_j\} : j \in \alpha\}$. Clearly, $(\forall Y \in B)[|Y| < \omega$ or $H \subseteq Y]$. If $(\exists x \in F)|U_x| < \omega$, then $\bigcap\{U_x : x \in F\} \neq 0$ hence we are done. Else $\bigcap\{U_x : x \in F\} \supseteq H \neq 0$ proving saturatedness of $\mathfrak{A}$.

(i)(a): Assume $\alpha \geq \omega$ and $1 < \kappa < \omega$ or $\kappa > |\alpha|$. Let $H \subseteq \alpha$ be infinite and coinfinite. Let $R \stackrel{\text{def}}{=} \{s \in {}^\alpha\kappa : (\forall i, j \in H) s_i = s_j\}$. Let $\mathfrak{A} \stackrel{\text{def}}{=} \mathfrak{Sg}^{(\mathfrak{Sb}\,{}^\alpha\kappa)}\{R\}$. Then $\mathfrak{A} \in {}_\kappa Cs_\alpha^{reg} \cap Dc_\alpha$ by [9] II.1.4(ii)b because R is an atom of $\mathfrak{Cl}_{(\alpha \smallsetminus H)}\mathfrak{A} = \{a \in A : \Delta(a) \subseteq H\}$. $\mathfrak{A}$ is not compact since $F \stackrel{\text{def}}{=} \{-R, d_{ij} : i, j \in H\}$ generates a proper filter with empty intersection. It remains to show that $\mathfrak{A}$ is saturating. If $\kappa < \omega$ then we are done by (ii). Assume $\kappa > |\alpha|$. Now we use the following lemma:

LEMMA 12. *Let $\mathfrak{A} \in Cs_\alpha$. Assume*[2] *$|A| < |base(\mathfrak{A})|$ and $U(x, k, i) \stackrel{\text{def}}{=} \{u : k(i/u) \in x\}$ is finite or cofinite in $base(\mathfrak{A})$ for every $x \in A$, $k \in 1^{\mathfrak{A}}$, and $i \in \alpha$. Then $\mathfrak{A}$ is saturating.*

Proof of Lemma 12: Assume the conditions. Let $F \subseteq A$ be downward directed and assume $k \in \bigcap c_i^* F$. Consider $S \stackrel{\text{def}}{=} \{U(x, k, i) : x \in F\}$. Then S is a downward directed set of subsets of $U \stackrel{\text{def}}{=} U(1, k, i)$. If S has a finite member or S is finite then we are done. Assume therefore $|U| > |A| \geq |S| \geq \omega$ and $(\forall x \in F)|U \smallsetminus U(x, k, i)| < \omega$. Thus $|\bigcup\{U \smallsetminus V : V \in S\}| = |S| < |U|$, showing $\bigcap S \neq 0$. Then $k \in c_i \bigcap F$. **QED (Lemma 12)**

Let us return to our $\mathfrak{A} \in {}_\kappa Cs_\alpha^{reg} \cap Dc_\alpha$. Now $|A| = |\alpha| < \kappa = base(\mathfrak{A})$. Next we show, by an elimination of quantifiers argument, that A is Boolean generated by $G = \{d_{ij}, c_{(\Gamma)}R : \Gamma \subseteq_\omega \alpha, i, j \in \alpha\}$. In the following, $d_\Omega \stackrel{\text{def}}{=} \prod\{d_{ij} : i, j \in \Omega\}$, $\bar{d}_\Sigma \stackrel{\text{def}}{=} \prod\{-d_{ij} : i, j \in \Sigma, i \neq j\}$. It is enough to show

[2]In general form the lemma reads: Let $\mathfrak{A} \in Crs_\alpha$. Assume $U(x, k, i)$ is finite or co–finite in $U(1, k, i)$ and $|U(1, k, i)| > |A|$, for every ....

that

$$\sigma \stackrel{\text{def}}{=} c_i(d_\Omega \cdot \bar{d}_\Sigma \cdot \prod_{j\in J} c_{(\Gamma_j)}R \cdot \prod_{k\in K} -c_{(\Delta_k)}R)$$

is a Boolean combination of elements of G, where Ω, Σ, Γ_j, Δ_k are finite subsets of α, and J, K are finite sets. We may assume that $i \in \Omega \cap \Sigma \cap H$, $i \notin \Gamma_j$, $i \notin \Delta_k$. Let $h \in H \setminus (\Omega \cup \Sigma \cup \Gamma_j \cup \bigcup\{\Delta_k : k \in K\})$.

Case 1 $\Omega \setminus \{i\} \neq 0$ or $J \neq 0$.

By $c_{(\Gamma_j)}R \subseteq d_{ih}$ we may assume $\Omega \setminus \{i\} \neq 0$. Let $p \in \Omega$, $p \neq i$. Let $\Omega' \stackrel{\text{def}}{=} \Omega \setminus \{i\}$, $\Sigma' \stackrel{\text{def}}{=} (\Sigma \setminus \{i\}) \cup \{p\}$, $\Gamma_j' \stackrel{\text{def}}{=} (\Gamma_j \setminus \{p\}) \cup \{i\}$, $\Delta_k' \stackrel{\text{def}}{=} (\Delta \setminus \{p\}) \cup \{i\}$. Then

$$\sigma = d_{\Omega'} \cdot \bar{d}_{\Sigma'} \cdot \prod_{j\in J} d_{ph} \cdot c_{(\Gamma_j')}R \cdot \prod_{k\in K} d_{ph} \cdot -c_{(\Delta_k')}R.$$

This can be seen as follows. By $p \in \Omega$, $p \neq i$ we have $\sigma = c_i(d_\Omega \cdots) = c_i(d_{ip} \cdot d_\Omega \cdots) = s^i_p(d_\Omega \cdots)$. Now $s^i_p d_\Omega = d_{\Omega'}$, $s^i_p \bar{d}_\Sigma = \bar{d}_{\Sigma'}$, $s^i_p c_{(\Gamma)}R = d_{ph} \cdot c_{(\Gamma')}R$ where $\Gamma' = (\Gamma \setminus \{p\}) \cup \{i\}$, and s^i_p is a Boolean homomorphism (i.e. $s^i_p(x \cdot y) = s^i_p x \cdot s^i_p y$, and $s^i_p - x = -s^i_p x$).

Case 2 $\Omega = J = 0$. (Recall that $\kappa \geq \omega$.) Then $\sigma = \bar{d}_{(\Sigma \setminus \{i\})}$.

Now it is not hard to check that $U(x,k,i)$ is finite or cofinite in κ for every $x \in A, k \in {}^\alpha\kappa, andi \in \alpha$ (by ($\star$) above, it is enough to check this for $x \in G$). Thus $\mathfrak{A}$ is saturating by Lemma 12.

(i)(b): Let $R \stackrel{\text{def}}{=} \bigcup\{{}^\alpha\kappa^{(\bar{u})} : u \in \kappa\}$ where $\bar{u} = \langle u : i \in \alpha\rangle$. Let $\mathfrak{A} \stackrel{\text{def}}{=} \mathfrak{Sg}^{(\mathfrak{Sb}^\alpha\kappa)}\{R\}$. Then $\mathfrak{A} \in {}_\kappa Cs_\alpha \cap Lf_\alpha$. We will use the following general lemma.

LEMMA 13. *Let $\mathfrak{A} \in Cs_\alpha$ with base U be generated by some $X \subseteq Nr_0\mathfrak{A}$, $|X| \leq |\alpha|$, $\alpha \geq \omega$. Then (i)–(ii) below hold.*

(i) *$\mathfrak{A}$ is saturating iff $|U| < \omega$ or $|U| > |\alpha|$.*

(ii) *Assume that X is a finite partition of $1^\mathfrak{A}$. Then $\mathfrak{A}$ is compact iff for each equivalence relation e on α ,with $\leq |U|$ blocks in case $|U| < \omega$, and for each $x \in X$ there is $f \in x$ with kernel e.*

Proof: By [8] 2.2.24 we have that each element of A is a Boolean combination of elements from $G = \{d_{ij}, x : i,j \in \alpha, x \in X\}$.

(i): Direction $\Longrightarrow$ follows from Thm. 4(iii). To prove the other direction, assume $|U| < \omega$ or $|U| > |\alpha|$. If $|U| < \omega$ then $\mathfrak{A}$ is saturating by Thm. 4(ii). Assume therefore $|U| > |\alpha| = |A|$. Now $U(a,k,i)$ is finite or cofinite in U for all $a \in G$, $k \in {}^\alpha U$ and $i \in \alpha$. Then the same holds for all $a \in A$ by ($\star$) above, because A is Boolean generated by G. By Lemma 12 then $\mathfrak{A}$ is saturating.

(ii): Let $\{x_0,...,x_n\} \stackrel{\text{def}}{=} X$. Let $F \in \mathrm{Fu}\,\mathfrak{Bl}\,\mathfrak{A}$ be arbitrary. Then $e \stackrel{\text{def}}{=} \{(i,j) : d_{ij} \in F\}$ is an equivalence relation on α and there is a unique $k \leq n$ with $x_k \in F$. We will show that $\bigcap F \neq 0$ iff there is $f \in x_k$ with kernel e. Assume $f \in x_k$ with kernel e. Let $a \in F$ be arbitrary. Then there is

$g \leq a, g \in F$ such that g is a product of elements from $G \cup \{-d_{ij}, -x_m : i,j \in \alpha, m \leq n\}$. Then, since F is downward directed, g must be a product of elements from $\{d_{ij} : (i,j) \in e\} \cup \{-d_{ij} : (i,j) \notin e\} \cup \{-x_m : k \neq m \leq n\}$. But then $f \in g \leq a$, showing that $f \in \bigcap F$. The other direction is trivial. Now let e be an equivalence relation on α and $k \leq n$. Then there is $F \in \text{Fu}\,\mathfrak{Bl}\,\mathfrak{A}$ with $e = \{(i,j) : d_{ij} \in F\}$ and $x_k \in F$ iff e has $\leq |U|$ blocks if $|U| < \omega$. **QED(Lemma 13)**

We note that Lemma 13(ii) above is used implicitly in Serény [21].

Returning to the proof of (i)(b): $\mathfrak{A}$ is saturating by $\Delta(R) = 0$ and $\mathfrak{A}$ is not compact because there is no $f \in -R$ with kernel $\alpha \times \alpha$.

Notation: If $h : \mathfrak{A} \longrightarrow \mathfrak{B}$ is a homomorphism then $h^*\mathfrak{A}$ denotes the h–image of $\mathfrak{A}$, i.e. $h^*\mathfrak{A}$ is the unique algebra for which $h : \mathfrak{A} \twoheadrightarrow h^*\mathfrak{A}$ is surjective.

(iv): Let $\kappa \geq 2, \alpha \geq \omega, H \subseteq \alpha, |H| \geq \omega, |\alpha \smallsetminus H| \geq \omega, R \overset{\text{def}}{=} \{s \in {}^\alpha\kappa : (\forall i \in H) s_i = 0\}, S \overset{\text{def}}{=} \{s \in {}^\alpha\kappa : (\forall i \in H) s_i = 0 or (\forall i \in H) s_i = 1\}, \mathfrak{A} = \mathfrak{Sg}^{(\mathfrak{Sb}\,{}^\alpha\kappa)}\{R\}, \mathfrak{B} \overset{\text{def}}{=} \mathfrak{Sg}^{(\mathfrak{Sb}\,{}^\alpha\kappa)}\{S\}$. Then $\mathfrak{A}, \mathfrak{B} \in Cs_\alpha^{reg} \cap Dc_\alpha$ by [9] II.1.3 or by [8] 3.1.63.

It can be checked that a similar (to the above) elimination of quantifiers argument works for both $\mathfrak{A}$ and $\mathfrak{B}$. (In fact, Case 1 goes through without any change. Case 2 will change as follows: $c_i(\bar{d}_\Sigma \cdots) = c_i(-d_{ih} \cdot \bar{d}_\Sigma \cdots) + c_i(d_{ih} \cdot \bar{d}_\Sigma \cdots)$. The second summand falls in Case 1, hence we have to care for the first summand only. But by $\{i,h\} \cap (\Delta_k) = \emptyset$ we have $-d_{ih} \subseteq -c_{(\Delta_k)}R$, hence $c_i(-d_{ih} \cdot \bar{d}_\Sigma \cdot \prod_{k \in K} -c_{(\Delta_k)}R) = c_i(-d_{ih} \cdot \bar{d}_\Sigma)$.) Therefore

$$(\star\star) \qquad A = \{\Sigma X : X \subseteq_\omega G^*\}, \text{ where}$$

$$G^* = \left\{ \prod_{i \in I} d_{\Omega_i} \cdot \bar{d_\Sigma} \cdot \prod_{j \in J} c_{(\Gamma_j)}R \cdot \prod_{k \in K} -c_{(\Delta_k)}R \ : \ \Omega_i, \Sigma, \Gamma_j, \Delta_k \subseteq_\omega \alpha \right\}.$$

And similarly for $\mathfrak{B}$.

From this, by an argument similar to the one in the proof of (i), one can show that $\mathfrak{A}$ and $\mathfrak{B}$ are saturating, if $\kappa < \omega$ or $\kappa > |\alpha|$.

Next we show $\mathfrak{A} \cong \mathfrak{B}$. Let $V = {}^\alpha\kappa^{(p)}$ where $p \in {}^\alpha\kappa$, $(\forall i \in \alpha) p_i = 0$. Define $h(x) \overset{\text{def}}{=} x \cap V$ for every element x of A or B. Then h is a homomorphism on both $\mathfrak{A}$ and $\mathfrak{B}$ because $\Delta(V) = 0$ (see [8] 2.3.26(i), p. 291). Also $h^*\mathfrak{A} = h^*\mathfrak{B}$ since $h(R) = h(S)$ and $\mathfrak{A}, \mathfrak{B}$ are generated by R, S respectively. Thus it is enough to show that $\mathfrak{A} \cong h^*\mathfrak{A}, \mathfrak{B} \cong h^*\mathfrak{B}$. To show $\mathfrak{A} \cong h^*\mathfrak{A}$, it is enough to show that $(\forall a \in A)[a \neq 0 \Rightarrow h(a) \neq 0]$. Let $a \in A, a \neq 0$. By $(\star\star)$, we may assume that $a = \prod d_{\Omega_i} \cdot \bar{d_\Sigma} \cdot \prod c_{(\Gamma_j)}R \cdot \prod -c_{(\Delta_k)}R$ for some $\Omega, \Sigma, \Gamma_j, \Delta_k$. Now it is not difficult to check that $a \cap V \neq 0$. To show $\mathfrak{B} \cong h^*\mathfrak{B}$ we use a similar argument to the above. Clearly, $\mathfrak{A}$ and $\mathfrak{B}$ are not base–isomorphic (to show this, observe that $(\forall b \in B)(\forall z \in b)(\exists u, v \in \kappa)[z \restriction H \neq [u,v] \circ z \restriction H$ and $[u,v] \circ z \in b]$, while this is not true for A).

To construct suitable counterexamples $\mathfrak{A}', \mathfrak{B}' \in Cs_\alpha \cap Lf_\alpha$, let $R' \overset{\text{def}}{=}$

$\bigcup\{{}^{\alpha}\kappa^{(s)} : s \in R\}$, $S' \stackrel{\text{def}}{=} \bigcup\{{}^{\alpha}\kappa^{(s)} : s \in S\}$, $\mathfrak{A}' \stackrel{\text{def}}{=} \mathfrak{Sg}\{R'\}$, $\mathfrak{B}' \stackrel{\text{def}}{=} \mathfrak{Sg}\{S'\}$. The proof for $\mathfrak{A}'$ and $\mathfrak{B}'$ is similar to the above one.

As was already indicated, the proof of **(v)** is in Serény [23]. **QED (Theorem 4)**

For the proof of Theorem 5 we have to recall the following. On p. 181 of [9] as well as in [8] 3.1.89 the algebraic counterpart of ultraproducts of models modulo an ultrafilter F was introduced. This algebraic counterpart is a sub–base–isomorphism denoted by ud_{cF} or ud^{A}_{cF} in [9]. In particular, if $\mathfrak{M}$ is an ordinary model and $\mathfrak{N}$ is the usual ultrapower of $\mathfrak{M}$ modulo F then $ud_{cF} : \mathfrak{Cs}^{(\mathfrak{M})} \rightarrowtail\!\!\!\rightarrow \mathfrak{Cs}^{(\mathfrak{N})}$ is the sub–base–isomorphism induced by the usual elementary embedding of $\mathfrak{M}$ into $\mathfrak{N}$. If $\mathfrak{A} = \mathfrak{Cs}^{(\mathfrak{M})}$ then we write ud^{A}_{cF} to indicate the domain of our isomorphism. Using the notation in [8], if $d : \mathfrak{A} \rightarrowtail {}^{I}\mathfrak{A}/F$ is the diagonal embedding, then $ud^{A}_{cF} = Rep(F, A, c) \circ d$. We should note that ud_{cF}, besides the ultrafilter F, has another parameter c which is called in [8], [9] an $(F, base\mathfrak{A}, \alpha)$–choice function. In passing, we note that the parameter c is needed when extending ultraproducts from ordinary models to infinitary ones, i.e. to ones having infinitary relations too. We need this step since there are Cs's which are obtainable in the form $\mathfrak{Cs}^{(\mathfrak{M})}$ only if $\mathfrak{M}$ is infinitary. We will use the following:

LEMMA 14. *Every Crs_α is sub–isomorphic to a saturating Crs_α and is isomorphic to a compact one. Moreover: To every $\mathfrak{A} \in Crs_\alpha$ there are an ultrafilter D and a $(D, base\mathfrak{A}, \alpha)$–choice function c such that ud^{A}_{cD} is a sub–base–isomorphism onto a saturating Crs_α $ud^{A}_{cD}{}^{*}\mathfrak{A}$.*

The proof of this lemma is on p. 695 of [14] i.e. in the proof of Lemma 7 there. In particular the present lemma is implied by statements (i), (ii) in the quoted proof there. ■

Recall that rl^{A}_{Q} is the function $\langle x \cap Q : x \in A\rangle$. Some notation: $\mathfrak{Rl}_Q\mathfrak{A}$ is the subalgebra of $\mathfrak{Sb}Q$ generated by $Rg(rl^{A}_{Q})$, $\text{Is}(\mathfrak{A}, \mathfrak{B})$, $\text{Ism}(\mathfrak{A}, \mathfrak{B})$ denote the sets of all isomorphisms from $\mathfrak{A}$ onto, and into $\mathfrak{B}$, respectively. $\text{Is}(\mathfrak{A}) \stackrel{\text{def}}{=} \{f : (\exists\mathfrak{B})f \in \text{Is}(\mathfrak{A}, \mathfrak{B})\}$. We will need the following lemma.

LEMMA 15. *Let $\mathfrak{B}$ be a saturating Crs_α with unit V. Let $Q \subseteq V$ with $\Delta^{[V]}Q = 0$. Then $\mathfrak{Rl}_Q\mathfrak{B}$ is saturating.*

Proof: Let $p \in Q$ and $F \in \text{Fl}\,\mathfrak{Bl}\ \mathfrak{Rl}_Q\mathfrak{B}$, $i \in \alpha$. Assume $(\forall x \in F)p \in c^{(Q)}_i x$. There is $G \in \text{Fl}\,\mathfrak{Bl}\ \mathfrak{B}$ with $rl_Q{}^{*}G = F$, namely $G = \{y \in B : y \cap Q \in F\}$ is such. Hence $(\forall x \in G)p \in c^{V}_i x$. Since $\mathfrak{B}$ is saturating $(\exists q \in \bigcap G)p \in c^{V}_i\{q\}$. Thus $q \in c^{V}_i\{p\} \subseteq c^{V}_i Q = Q$. Hence $(\forall x \in G)q \in rl_Q x$ thus $q \in \bigcap F$. Thus $p \in c^{V}_i\{q\} = c^{Q}_i\{q\} \subseteq c^{Q}_i(\bigcap F)$. We proved $\bigcap\{c_i x : x \in F\} = c_i(\bigcap F)$. **QED (Lemma 15)**

Proof of Theorem 5: (i): Let $\mathfrak{A} \in Crs_\alpha$ with unit V. Let F and c be such that $ud_{cF} \in \text{Is}\,\mathfrak{A}$ is a sub–base–isomorphism and $ud_{cF}{}^{*}\mathfrak{A}$ is saturating. Such F and c exist by Lemma 14. Then $ud_{cF}{}^{*}\mathfrak{A}$ is base–isomorphic to some

$\mathfrak{N}$ ext–isomorphic to $\mathfrak{A}$. Hence $\mathfrak{N}$ is saturating with $rl_V \in \mathrm{Is}(\mathfrak{N},\mathfrak{A})$. If $\mathfrak{A}$ is regular then so is $\mathfrak{N}$ by [9] I.7.6. If $K = Cs_\alpha$ then rl_V is a strong ext–isomorphism, and $\mathfrak{N} \in Cs_\alpha$ by [9] I.7.4(i); therefore we are done.

Let $Q \stackrel{\text{def}}{=} 1^{\mathfrak{N}} \cap \left(\bigcup\{{}^\alpha(base\mathfrak{N})^{(q)} : q \in V\}\right)$ and $\mathfrak{R} \stackrel{\text{def}}{=} \mathfrak{Rl}_Q\mathfrak{N}$. By Lemma 15 we have that $\mathfrak{R}$ is saturating. Clearly, $V \subseteq Q \subseteq 1^{\mathfrak{N}}$ and thus $rl_Q \in \mathrm{Is}\,\mathfrak{N}$. This implies that if $\mathfrak{N}$ is regular then so is $\mathfrak{R}$.

Let $\tilde{t} : (ud_{cF}{}^*\mathfrak{A}) \rightarrowtail\!\!\!\!\twoheadrightarrow \mathfrak{N}$ be the base–isomorphism mentioned above. Let $U = base(V)(= base\mathfrak{A})$. Then $t : {}^IU/F \rightarrowtail\!\!\!\!\twoheadrightarrow (base\mathfrak{N}) \supseteq U$, with $t(\bar{u}) = u$ for $u \in U$ and $\bar{u} = \langle u : i \in I\rangle$.

We will first show that $Q \cap {}^\alpha U = V$. Let $Q^+ = \tilde{t}^{-1}(Q)$, $V^+ = \tilde{t}^{-1}(V)$, $U^+ = \tilde{t}^{-1*}U$ etc. In particular $u^+ = \bar{u}$ and $U^+ = \{\bar{u} : u \in U\}$. Further $Q^+ \subseteq ud_{cF}(V)$. Let $q \in Q^+ \cap {}^\alpha(U^+) = (Q \cap {}^\alpha U)^+$. Then by $q \in ud_{cF}(V)$ and using Def. 3.1.89 of [8],

$$(\text{☆☆☆}) \qquad (c^+q)_i \in V \text{ for some } i \in I.$$

By $q \in {}^\alpha U^+$, $q = \langle \bar{u}_k : k < \alpha\rangle$ for some $u \in {}^\alpha U$. By the choice of c in the proof in [14] quoted in the proof of our present Lemma 14, $c(k, q_k) = \bar{u}_k$ for all $k < \alpha$. Therefore $(c^+q)_i = u$ for all $i \in I$, hence by (☆☆☆) $u \in V$. Then $t \circ q = u \in V$ proving $q \in V^+$. Thus $Q^+ \cap {}^\alpha U^+ = V^+$ proving $Q \cap {}^\alpha U = V$.

Now we have that $rl_V^R = rl_{({}^\alpha U)}^R \in \mathrm{Is}(\mathfrak{R},\mathfrak{A})$ is a strong sub–isomorphism by the definition of Q. Let $K \in \{Crs_\alpha, Gws_\alpha, Gws_\alpha^{norm}, Gws_\alpha^{comp}\}$. Then $\mathfrak{A} \in K$ implies $\mathfrak{N} \in K$ by [9] I.7.4(i)+II.7.10(ii). By the definition of Q we have $\mathfrak{N} \in K$ implies $\mathfrak{R} \in K$. If $\mathfrak{A} \in Ws_\alpha$ then $\mathfrak{N} \in Gws_\alpha^{comp}$ and therefore $\mathfrak{R} \in Ws_\alpha$. We are done with the case $K \in \{Crs_\alpha, Gws_\alpha, Gws_\alpha^{comp}, Gws_\alpha^{norm}, Ws_\alpha\}$.

The case $K \in \{Gs_\alpha, Gws_\alpha^{wd}\}$ remains to be proved. Assume $\mathfrak{A} \in Gs_\alpha$. Let $I \stackrel{\text{def}}{=} \{{}^\alpha U : U \text{ is a subbase of } \mathfrak{A}\}$. Then $V = \bigcup I$ is a disjoint union of Cs_α units. Let $Y \in I$. Then $\mathfrak{Rl}_Y\mathfrak{A}$ is a Cs_α and hence is strongly sub–isomorphic to some saturating $\mathfrak{B}_Y \in Cs_\alpha$. Consider the system $\mathfrak{B} \stackrel{\text{def}}{=} \langle \mathfrak{B}_Y : Y \in I\rangle \in {}^ICs_\alpha$. We may assume $(\forall Y, W)(Y \neq W \Rightarrow base(\mathfrak{B}_Y) \cap base(\mathfrak{B}_W) = 0)$. Let $W \stackrel{\text{def}}{=} \bigcup\{{}^\alpha base(\mathfrak{B}_Y) : Y \in I\}$. Let $h_Y \in \mathrm{Is}(\mathfrak{Rl}_Y\mathfrak{A}, \mathfrak{B}_Y)$ be a sub–isomorphism for all $Y \in I$. (Exists.) Let $f \stackrel{\text{def}}{=} \langle \bigcup\{h_Y(X \cap Y) : Y \in I\} : X \in A\rangle$. Then $f \in \mathrm{Ism}(\mathfrak{A}, \mathfrak{Sb}W)$ by Section Products of [9]. Let $\mathfrak{C} \stackrel{\text{def}}{=} f^*\mathfrak{A}$. Let $x \in A$. Then $({}^\alpha U) \cap f(x) = {}^\alpha U \cap \bigcup\{h_Y(x \cap Y) : Y \in I\} = \bigcup\{{}^\alpha U \cap h_Y(x \cap Y) : Y \in I\} = \bigcup\{(x \cap Y : Y \in I)\} = x$. Thus $f : \mathfrak{A} \rightarrowtail\!\!\!\!\twoheadrightarrow \mathfrak{C}$ is a strong sub–isomorphism. Clearly $\mathfrak{C}$ is a saturating Gs_α. By II.7.10.1 of [9] p. 246, if $\mathfrak{A} \in Gs_\alpha^{reg}$ then $\mathfrak{C} \in Gs_\alpha^{reg}$, too.

The case $K = Gws_\alpha^{wd}$ is completely analogous to the above one, therefore we omit it.

(ii)(a): Let $\mathfrak{A} \in Crs_\alpha$. Then by Lemma 14 there are a compact $\mathfrak{B} \in Crs_\alpha$ and an $f \in \mathrm{Is}(\mathfrak{A}, \mathfrak{B})$. We may assume $base(\mathfrak{A}) \cap base(\mathfrak{B}) = 0$, e.g. by I.3.1 of [9]. Let $h \stackrel{\text{def}}{=} \langle x \cup fx : x \in A\rangle$, $V \stackrel{\text{def}}{=} 1^{\mathfrak{A}}$ and $W \stackrel{\text{def}}{=}$

$1^{\mathfrak{B}}$. Then $h \in \text{Ism}(\mathfrak{A}, \mathfrak{Sb}(V \cup W))$ by I.6.2 of [9]. Let $\mathfrak{C} \stackrel{\text{def}}{=} h^*\mathfrak{A}$. Then $rl_V^C = rl_{(^\alpha base\mathfrak{A})} \in \text{Is}(\mathfrak{C}, \mathfrak{A})$ showing that $\mathfrak{A}$ is strongly sub–isomorphic to $\mathfrak{C}$. Since $\mathfrak{B}$ is compact and sub–isomorphic to $\mathfrak{C}$, also $\mathfrak{C}$ is compact. We have seen that $\mathfrak{A}$ is strongly sub–isomorphic to a compact $\mathfrak{B} \in Crs_\alpha$. By Theorem 5(i), we have that $\mathfrak{B}$ is strongly sub–isomorphic to a saturating $\mathfrak{C} \in Crs_\alpha$. Now, $\mathfrak{A}$ is strongly sub–isomorphic to $\mathfrak{C}$ and $\mathfrak{C}$ is compact since $\mathfrak{B}$ is compact.

(ii)(b): Let $\mathfrak{A} \in {}_\infty Cs_\alpha$ with unit V. The above proof of (ii)(a) yields a compact $\mathfrak{C} \in {}_\infty Gws_\alpha$ with $rl_V \in \text{Is}(\mathfrak{C}, \mathfrak{A})$ (by [9] I.7.4(i)). By 7.12.1(ii) and (iii) of [9] p. 250, every ${}_\infty Gws_\alpha$ is sub-isomorphic to a ${}_\infty Cs_\alpha$. Hence $\mathfrak{C}$ is sub-isomorphic to some $\mathfrak{B} \in Cs_\alpha$. Compactness of $\mathfrak{C}$ implies that $\mathfrak{B}$ is compact. Thus $rl_V \in \text{Is}(\mathfrak{B}, \mathfrak{A})$ and $\mathfrak{B}$ is a compact ${}_\infty Cs_\alpha$. By Lemma 14 we have $ud_c \in \text{Is}(\mathfrak{B}, \mathfrak{N})$ for some saturating $\mathfrak{N}$. By I.7.4 of [9], $\mathfrak{N} \in Cs_\alpha$. Hence $\mathfrak{B}$ is sub–isomorphic to some saturating $\mathfrak{M} \in Cs_\alpha$. Compactness of $\mathfrak{B}$ implies compactness of $\mathfrak{M}$. Thus $\mathfrak{M}$ is a saturating and compact ${}_\infty Cs_\alpha$ and $rl_V \in \text{Is}(\mathfrak{M}, \mathfrak{N})$. **QED (Theorem 5)**

Proof of Theorem 6: **(i):** Let $\alpha \geq \omega$ and $2 \leq \kappa \stackrel{\text{def}}{=} U$. Let $\alpha = H \cup G$ such that $H \cap G = 0$, $0 \in H$ and $|H| \geq \omega$, $|G| \geq \omega$. Let $Y \stackrel{\text{def}}{=} \{q \in {}^\alpha U : q_0 = 0\}$, $X \stackrel{\text{def}}{=} \{q \in {}^\alpha U : (\forall i \in H) q_i = 0\}$, and $\mathfrak{A} \stackrel{\text{def}}{=} \mathfrak{Sg}^{(\mathfrak{Sb}^\alpha U)}\{X, Y\}$. $\mathfrak{A}$ is regular by [8] 3.1.63. Let $D \stackrel{\text{def}}{=} \{\prod\{d_{0i} : i \in \Gamma\} \cdot Y - X : \Gamma \subseteq H,\ |\Gamma| < \omega\}$. Then $(\exists D^+ \in \text{Fu}\,\mathfrak{B}\mathfrak{l}\ \mathfrak{A}) D \subseteq D^+$. But $\bigcap D^+ = 0$, proving that $\mathfrak{A}$ is not compact. See Figure 3.

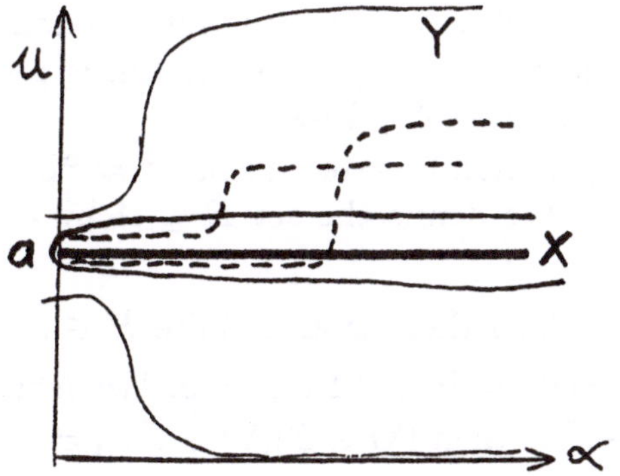

Figure 3

Let $\mathfrak{B} \in Cs_\alpha^{reg}$ and $h \in \text{Is}(\mathfrak{A}, \mathfrak{B})$. Then by regularity of $\mathfrak{B}$ there is a such that $h(Y) = \{p \in {}^\alpha(base\mathfrak{B}) : p_0 = a\}$, since $Y \cdot c_0(d_{0i} \cdot Y) \leq d_{0i}$ in $\mathfrak{A}$. By $h(X) \subseteq h(Y)$ and $(\forall i \in H) h(X) \subseteq d_{0i}$ this implies

$$h(X) = \{p \in {}^\alpha(base\mathfrak{B}) : (\forall i \in H) p_i = a\}.$$

But then $\bigcap(h^* D) = 0$, proving that $\mathfrak{B}$ is not compact.

(ii)(a): Let $0 < \kappa < \omega \leq \alpha$. Let $\mathfrak{C} = \mathfrak{Sb}({}^\alpha\kappa^{(p)})$ for some $p \in {}^\alpha\kappa$. Let $H \stackrel{\text{def}}{=} \{2 \cdot n : n \in \omega\}$, and $\mathfrak{A} \stackrel{\text{def}}{=} \mathfrak{Sg}^{(\mathfrak{C})}\{x \in C : \Delta x \cap H = 0\}$. Then $\mathfrak{A} \in {}_\kappa Ws_\alpha \cap Dc_\alpha$. By [8] 3.1.102, $\mathfrak{A} \in \mathbf{I}_\kappa Cs_\alpha^{reg}$, too. Let $G \stackrel{\text{def}}{=} \alpha \smallsetminus H$ and $\mathfrak{Q} \stackrel{\text{def}}{=} \mathfrak{Sb}({}^G\kappa^{(p \restriction G)})$. Let $rs_G(x) \stackrel{\text{def}}{=} \{s \restriction G : s \in x\}$ for all $x \in A$. Then $rs_G \in \mathrm{Is}(\mathfrak{Nr}_G\mathfrak{A}, \mathfrak{Q})$ is easy to check. Hence $\mathfrak{Bl}\ \mathfrak{Q} \cong \mathfrak{Bl}\ \mathfrak{Nr}_G\mathfrak{A}$ proving $|\mathrm{Fu}\,\mathfrak{Bl}\ \mathfrak{Q}| \leq |\mathrm{Fu}\,\mathfrak{Bl}\ \mathfrak{A}|$. Since $|\alpha| = |G| = |1^{\mathfrak{Q}}|$ and $Q = Sb1^{\mathfrak{Q}}$ we have $|\mathrm{Fu}\,\mathfrak{Bl}\ \mathfrak{Q}| = 2^{2^{|\alpha|}}$, see e.g. Ex. 6.1.5 on p. 323 of [5]. Assume $\mathfrak{A} \cong \mathfrak{B} \in Cs_\alpha$. Then $\mathfrak{B} \in {}_\kappa Cs_\alpha$ since in Sec. Subalgebras of [9] ${}_nCs_\alpha = Cs_\alpha \cap \mathbf{I}_n Cs_\alpha$ is proved for all $n \in \omega$. Thus $|1^{\mathfrak{B}}| \leq 2^{|\alpha|} < 2^{2^{|\alpha|}} \leq |\mathrm{Fu}\,\mathfrak{Bl}\ \mathfrak{A}| = |\mathrm{Fu}\,\mathfrak{Bl}\ \mathfrak{B}|$ proves that $\mathfrak{B}$ is not compact.

(ii)(b): For each $u < \kappa$ let $\overline{u} \stackrel{\text{def}}{=} \langle u : i \in \alpha\rangle$ and let $x_u \stackrel{\text{def}}{=} {}^\alpha\kappa^{(\overline{u})}$. Let $x_\kappa \stackrel{\text{def}}{=} {}^\alpha\kappa \smallsetminus \bigcup\{x_u : u < \kappa\}$ and let $\mathfrak{A}$ be the Cs_α with base κ and generated by $\{x_u : u \leq \kappa\}$. Then $\mathfrak{A} \in {}_\kappa Cs_\alpha \cap Lf_\alpha$. Let $h \in \mathrm{Is}(\mathfrak{A}, \mathfrak{B})$, $\mathfrak{B} \in Cs_\alpha$. Then $|base(\mathfrak{B})| = \kappa$ by $\kappa < \omega \leq \alpha$. Also, $\mathfrak{B}$ is generated by $\{h(x_u) : u \leq \kappa\}$ which is a partition of $1^{\mathfrak{B}}$. By $|base(\mathfrak{B})| = \kappa$ there are only κ–many sequences in $1^{\mathfrak{B}}$ with kernel $\alpha \times \alpha$. Thus one of $\{h(x_u) : u \leq \kappa\}$ does not contain any sequence with kernel $\alpha \times \alpha$, showing that $\mathfrak{B}$ is not compact.

We note that almost the same argument proves the existence of a $\mathfrak{C} \in Cs_\alpha^{reg} \cap Dc_\alpha$ isomorphic to no compact Cs_α. Namely, let $y_u \stackrel{\text{def}}{=} \{f \in {}^\alpha\kappa : (\forall n \in \omega) f(2n) = u\}$ for every $u < \kappa$, $y_\kappa \stackrel{\text{def}}{=} {}^\alpha\kappa \smallsetminus \bigcup\{y_u : u < \kappa\}$, and let $\mathfrak{C}$ be the Cs_α with base κ generated by $\{y_u : u \leq \kappa\}$. Then in any isomorphic image of $\mathfrak{C}$, the sets $\{y_u\} \cup \{d_{ij} : i, j \in \alpha\}$ have the finite intersection property for all $u \leq \kappa$, but the intersection can be nonempty only for κ–many of them if the algebra we are in is a Cs. However, this does not prove (ii)(a), because $\mathfrak{C} \notin \mathbf{I}Ws_\alpha$.

(iii): Let $\mathfrak{A} \in Ws_\alpha$ be nondiscrete. Then $1^{\mathfrak{A}} = {}^\alpha U^{(p)}$ for some U and $p \in {}^\alpha U$ with $|U| > 1$. We define the set $E \subseteq A$ by a distinction of two cases.

<u>Case 1</u> $|Rg(p)| < \omega$. Then there is an infinite $H \in \alpha/(\ker p)$. Let $|S| \geq \omega \leq |T|$ be such that $S \cup T = H$ and $S \cap T = 0$. We define $E \stackrel{\text{def}}{=} \{-d_{ij} : i \in S \text{ and } j \in T\}$. If $q \in \bigcap E$ then $(\forall i \in S)(\forall j \in T) q_i \neq q_j$ hence $q \notin {}^\alpha U^{(p)}$, thus $\bigcap E = 0$.

<u>Case 2</u> $|Rg(p)| \geq \omega$. Then $E \stackrel{\text{def}}{=} \{d_{ij} : i, j \in \alpha\}$.

By Cases 1–2 above we defined E. Clearly $(\exists D \in \mathrm{Fu}\,\mathfrak{Bl}\ \mathfrak{A}) E \subseteq D$. But $\bigcap E = 0$. **QED (Theorem 6)**

Proof of Theorem 7: Let $\alpha \geq \omega$ and $\kappa > 1$. Let $H, L \subseteq \alpha$ with $|H| = |L| = \omega$ and $|\alpha \smallsetminus (H \cup L)| \geq \omega$, $H \cap L = 0$. Notation: $\overline{n} \stackrel{\text{def}}{=} \langle n : i \in \alpha\rangle$, i.e. $\overline{n} = \alpha \times \{n\}$. Let $p \stackrel{\text{def}}{=} \overline{1}$, $q \stackrel{\text{def}}{=} (\overline{1} \restriction L) \cup (\overline{0} \restriction H)$, and $t \stackrel{\text{def}}{=} (\overline{0} \restriction L) \cup (\overline{1} \restriction H)$, $V \stackrel{\text{def}}{=} {}^\alpha\kappa$ and $K \stackrel{\text{def}}{=} L \cup H$. Let $P \stackrel{\text{def}}{=} \{f \in V : p \restriction K \subseteq f\}$, $Q \stackrel{\text{def}}{=} \{f \in V :$

$q \subseteq f\}$, $T \stackrel{\text{def}}{=} \{f \in V : t \subseteq f\}$. Let $E \stackrel{\text{def}}{=} \{f \in V : f_0 < 2\}$, $\mathfrak{C} \stackrel{\text{def}}{=} \mathfrak{Sb}V$. Let $\mathfrak{A} \stackrel{\text{def}}{=} \mathfrak{Sg}^{(\mathfrak{C})}\{E, P, Q\}$ and $\mathfrak{B} \stackrel{\text{def}}{=} \mathfrak{Sg}^{(\mathfrak{C})}\{E, P, T\}$. See Figure 4. By [8] 3.1.63, $\mathfrak{A}$ and $\mathfrak{B}$ are regular, hence $\mathfrak{A}, \mathfrak{B} \in {}_\kappa Cs_\alpha^{reg} \cap Dc_\alpha$.

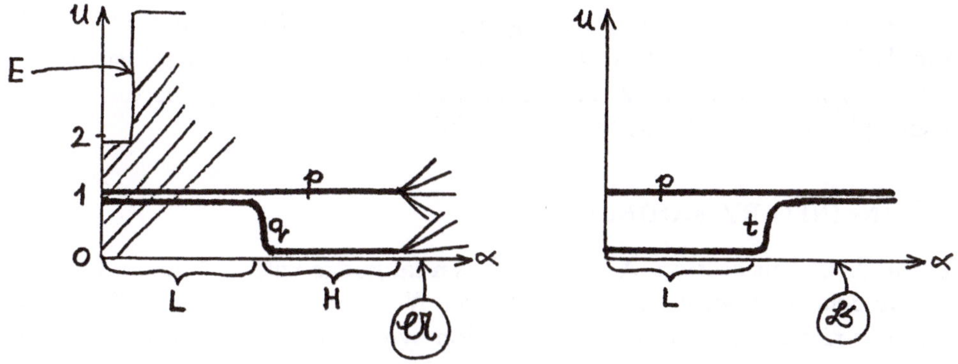

Figure 4

Next we show $\mathfrak{A} \cong \mathfrak{B}$. Let $V_q \stackrel{\text{def}}{=} \bigcup\{{}^\alpha\kappa^{(s)} : s \in Q\}$, $V_t \stackrel{\text{def}}{=} \bigcup\{{}^\alpha\kappa^{(s)} : s \in T\}$, $V_r \stackrel{\text{def}}{=} V \smallsetminus (V_q \cup V_t)$, $\mathfrak{C}_q \stackrel{\text{def}}{=} \mathfrak{Sb}V_q$, $\mathfrak{C}_t \stackrel{\text{def}}{=} \mathfrak{Sb}V_t$, $\mathfrak{C}_r \stackrel{\text{def}}{=} \mathfrak{Sb}V_r$. Then $\mathfrak{C} \cong \mathfrak{C}_q \times \mathfrak{C}_t \times \mathfrak{C}_r$ and specially,

$$h \stackrel{\text{def}}{=} \langle (x \cap V_q, x \cap V_t, x \cap V_r) : x \in C \rangle : \mathfrak{C} \rightarrowtail\!\!\!\!\rightarrow \mathfrak{C}_q \times \mathfrak{C}_t \times \mathfrak{C}_r,$$

by [8] 3.1.76. Clearly, $\mathfrak{C}_q$ is base–isomorphic to $\mathfrak{C}_t$, let $i : \mathfrak{C}_q \rightarrowtail\!\!\!\!\rightarrow \mathfrak{C}_t$ be this base–isomorphism. Then $f \stackrel{\text{def}}{=} \langle (i^{-1}(b), i(a), c) : (a, b, c) \in C_q \times C_t \times C_r \rangle$ is an automorphism of $\mathfrak{C}_q \times \mathfrak{C}_t \times \mathfrak{C}_r$, thus $k \stackrel{\text{def}}{=} h^{-1} \circ f \circ h$ is an automorphism of $\mathfrak{C}$. It can be checked that $k(E) = E$, $k(P) = P$, and $k(Q) = T$, showing that $\mathfrak{A} = \mathfrak{Sg}^{(\mathfrak{C})}\{E, P, Q\} \cong \mathfrak{Sg}^{(\mathfrak{C})}\{E, P, T\} = \mathfrak{B}$.

Assume $\mathfrak{N} \in Cs_\alpha^{reg}$ would be ext–base–isomorphic to both $\mathfrak{A}$ and $\mathfrak{B}$. Say, $f : \mathfrak{N} \rightarrowtail\!\!\!\!\rightarrow \mathfrak{A}$ and $h : \mathfrak{N} \rightarrowtail\!\!\!\!\rightarrow \mathfrak{B}$ are the two ext–base–isomorphisms. By the choice of f, there are W and $\mathfrak{M} \in Cs_\alpha^{reg}$ with $rl({}^\alpha W) : \mathfrak{N} \rightarrowtail\!\!\!\!\rightarrow \mathfrak{M}$ and $\mathfrak{M}$ base–isomorphic to $\mathfrak{A}$. We may assume $\mathfrak{M} = \mathfrak{A}$. Let $U = base(\mathfrak{N})$. Then $\kappa \subseteq U$ and $f = rl({}^\alpha\kappa) \in \mathrm{Is}(\mathfrak{N}, \mathfrak{A})$. There are $E^+, P^+, Q^+ \in N$ with $fE^+ = E$, $fP^+ = P$ and $fQ^+ = Q$. By regularity of $\mathfrak{N}$ and by $\Delta(E^+) = \Delta(E) = \{0\}$, there is $H \subseteq U$ such that $E^+ = \{g \in {}^\alpha U : g_0 \in H\}$. By $E = E^+ \cap ({}^\alpha\kappa)$ then $2 \subseteq H$, and by $E \cdot c_0(d_{01} \cdot E) \cdot c_0(d_{02} \cdot E) \cdot \overline{d}_{\{0,1,2\}} = 0$ we have $|H| \leq 2$, thus $H = 2$. Therefore $E^+ = \{g \in {}^\alpha U : g_0 \in 2\}$. By $(\forall i \in K) c_0(d_{0i} \cdot E) \supseteq P$ we have $(\forall g \in P^+) g \restriction K \in {}^K 2$, and by $P = P^+ \cap {}^\alpha\kappa$, $(\forall i, j \in K) P \leq dij$ we then have $P^+ = \{g \in {}^\alpha U : g \restriction K \subseteq p\}$. Similarly, $Q^+ = \{g \in {}^\alpha U : q \subseteq g\}$. We have seen that $P^+, Q^+ \in N$

are very much similar to $P, Q \in A$. The crucial question is "what are $h(P^+), h(Q^+) \in B$"? Can they be different from $P, T \in B$? Let $x = h(P^+)$. Then $x \in B$, $\Delta(x) = K$ and $x \leq d_{ij}$ for all $i, j \in K$. Assume $s \in x$ with $s \upharpoonright K \neq p \upharpoonright K$. Let $Q = {}^{\alpha}\kappa^{(s)}$. Then rl_Q is a homomorphism on $\mathfrak{B}$ by $\Delta(Q) = \emptyset$. By $rl_Q(P) = rl_Q(T) = 0$ then $rl^*_Q\mathfrak{B} \in Lf_\alpha$. Thus $\Delta(x \cap Q)$ is finite, contradicting $0 \neq x \cap Q \leq d_{ij}$ for all $i, j \in K$. Therefore $(\forall s \in x)s \upharpoonright K = p \upharpoonright K$ and then $x = P$ because $\mathfrak{B}$ is regular. The proof of $h(Q^+) = T$ is completely analogous, we omit it. But then $h(Q^+) = T$ and $h(P^+) = P$ prove that h cannot be an ext–base–isomorphism since $(\forall i \in L)[q_i = p_i$ but $t_i \neq p_i]$. This proves that no Cs^{reg}_α is ext–base–isomorphic to both $\mathfrak{A}$ and $\mathfrak{B}$. **QED(Theorem 7)**

3.3 Infinitary models

Throughout this section, $\Lambda = \langle \alpha, I, \rho \rangle$ is a language with $\alpha \geq \omega$, and $\mathfrak{M}, \mathfrak{N}$ are models of Λ with universes M and N respectively. We will call Λ *variable–rich* if the arity of every relation symbol is co–infinite in α, i.e. if $(\forall i \in I)|\alpha \smallsetminus \rho(i)| \geq \omega$. Λ is called *full* if every relation symbol of Λ has arity α, i.e. if $Rg(\rho) = \{\alpha\}$. For simplicity, we will write $\Phi\mu^\Lambda$ instead of $\Phi\mu^\Lambda_r$. Let $\varphi \in \Phi\mu^\Lambda$. Recall from [8] that $\mathfrak{M} \models \varphi$ iff $(\forall k \in {}^\alpha M)\mathfrak{M} \models \varphi[k]$, $\varphi^{\mathfrak{M}} = \{k \in {}^\alpha M : \mathfrak{M} \models \varphi[k]\}$, and $\Theta\rho\mathfrak{M} = \{\varphi \in \Phi\mu^\Lambda : \mathfrak{M} \models \varphi\}$. We say that $\mathfrak{M}$ is a model of φ if $\mathfrak{M} \models \varphi$; $\mathfrak{M}$ is a model of $\Gamma \subseteq \Phi\mu^\Lambda$ if $\mathfrak{M}$ is a model of each $\varphi \in \Gamma$, and $\Gamma \models \varphi$ iff every model of Γ is also a model of φ. (We note that in [8], these are defined for ordinary Λ. If Λ is ordinary, then there is no real difference between formulas and sentences, and accordingly $\Theta\rho\mathfrak{M}$ is defined in [8] as the set of sentences (i.e. formulas with no free variables) valid in $\mathfrak{M}$. However, if Λ is not ordinary, then it is important that $\Theta\rho\mathfrak{M}$ is the set of formulas, and not only sentences, valid in $\mathfrak{M}$.) A set $\Gamma \subseteq \Phi\mu^\Lambda$ is said to be *complete* iff $(\forall \varphi \in \Phi\mu^\Lambda)(\Gamma \models \varphi$ or $\Gamma \cup \{\varphi\}$ has no model). There are models $\mathfrak{M}$ such that $\Theta\rho\mathfrak{M}$ is not complete, thus $\mathfrak{M} \models \Theta\rho\mathfrak{N} \not\Rightarrow \mathfrak{N} \models \Theta\rho\mathfrak{M}$. (We shall see such models in Example 20.) We call $\mathfrak{A} \in Cs_\alpha$ trivial iff $|A| = 1$ (i.e. if $base(\mathfrak{A}) = 0$).

Basic concepts (elementary submodels, elementary equivalence etc)

FACT 16.

(i) *Every nontrivial $\mathfrak{A} \in Cs_\alpha$ is of the form $\mathfrak{A} = \mathfrak{Cs}^{\mathfrak{M}}_\alpha$ for some model $\mathfrak{M}$. If $\mathfrak{A} \in Cs^{reg}_\alpha \cap Dc_\alpha$ then $\mathfrak{M}$ can be chosen to have a variable rich language.*

(ii) *$\mathfrak{Cs}^{\mathfrak{M}}_\alpha$ is simple $\iff$ $\Theta\rho\mathfrak{M}$ is complete.*

(iii) *$\mathfrak{Cs}^{\mathfrak{M}}_\alpha$ is subdirectly irreducible $\iff (\exists \varphi \in \mathcal{L}(\mathfrak{M}) \smallsetminus \Theta\rho\mathfrak{M})(\forall \psi \in \mathcal{L}(\mathfrak{M}) \smallsetminus \Theta\rho\mathfrak{M})\Theta\rho\mathfrak{M} \cup \{\psi\} \models \varphi$.*

(iv) $\mathfrak{Cs}_\alpha^{\mathfrak{M}}$ *is weakly subdirectly irreducible* $\iff$ $(\forall\varphi,\psi \in \Phi\mu^\Lambda)$ $[(\forall$*finite sequences* $\overline{x}$ *of variables*$)\mathfrak{M} \models (\forall\overline{x}\varphi \vee \forall\overline{x}\psi) \implies (\mathfrak{M} \models \varphi \vee \mathfrak{M} \models \psi)]$.

Proof: In the proof we will use the following two observations (which are easy to check):

($\star$) Suppose that f is a homomorphism of $\mathfrak{Cs}_\alpha^{\mathfrak{M}}$ onto a nontrivial Cs_α. Then there is a model $\mathfrak{N}$ of the same language as $\mathfrak{M}$ such that $\varphi^{\mathfrak{N}} = f\varphi^{\mathfrak{M}}$ for all $\varphi \in \mathcal{L}(\mathfrak{M})$.

($\star\star$) Suppose that $\mathfrak{N} \models \Theta\rho\mathfrak{M}$. There is a homomorphism $f\colon \mathfrak{Cs}_\alpha^{\mathfrak{M}} \longrightarrow \mathfrak{Cs}_\alpha^{\mathfrak{N}}$ such that $f\varphi^{\mathfrak{M}} = \varphi^{\mathfrak{N}}$ for all $\varphi \in \mathcal{L}(\mathfrak{M})$.

The proof of **(i)** is easy, we leave it to the reader.

(ii): Assume that $\mathfrak{Cs}_\alpha^{\mathfrak{M}}$ is simple, and that $\Theta\rho\mathfrak{M} \cup \{\varphi\}$ has a model $\mathfrak{N}$. Let $f : \mathfrak{Cs}_\alpha^{\mathfrak{M}} \longrightarrow \mathfrak{Cs}_\alpha^{\mathfrak{N}}$ be the homomorphism as in ($\star\star$). Then $f\varphi^{\mathfrak{M}} = \varphi^{\mathfrak{N}} = 1$. By simplicity of $\mathfrak{Cs}_\alpha^{\mathfrak{M}}$, and since $\mathfrak{Cs}_\alpha^{\mathfrak{N}}$ is nontrivial, f is one–one, so $\varphi^{\mathfrak{M}} = 1$, and hence $\mathfrak{M} \models \varphi$. Thus $\Theta\rho\mathfrak{M}$ is complete. Assume that $\mathfrak{Cs}_\alpha^{\mathfrak{M}}$ is not simple. By $\mathbf{H}Cs_\alpha \subseteq \mathbf{SP}Cs_\alpha$ (see [8] 3.1.107–108), then there is a homomorphism $f : \mathfrak{Cs}_\alpha^{\mathfrak{M}} \longrightarrow \mathfrak{A}$ onto a nontrivial $\mathfrak{A} \in Cs_\alpha$ such that f is not one–one. By ($\star$) then there are a model $\mathfrak{N}$ and a formula φ such that $f(\varphi^{\mathfrak{M}}) = 1$, $\varphi^{\mathfrak{M}} \neq 1$ and $\psi^{\mathfrak{N}} = f(\psi^{\mathfrak{M}})$ for all $\psi \in \Phi\mu^\Lambda$. Then $\mathfrak{N} \models \Theta\rho\mathfrak{M} \cup \{\varphi\}$ but $\Theta\rho\mathfrak{M} \not\models \varphi$, showing that $\Theta\rho\mathfrak{M}$ is not complete.

(iii): Let $\mathfrak{A} \in CA_\alpha$, $y \in A$. We will denote by $Fg^{(\mathfrak{A})}\{y\}$ the filter generated by $\{y\}$ in $\mathfrak{A}$, i.e.

$$Fg^{(\mathfrak{A})}\{y\} = \{x : f(x) = 1 \text{ for all homomorphisms } f \text{ on } \mathfrak{A} \text{ such that } f(y) = 1\}.$$

To prove (iii), first we prove the following statement:

$$(\star\star\star) \qquad \Theta\rho\mathfrak{M} \cup \{\psi\} \models \varphi \text{ iff } \varphi^{\mathfrak{M}} \in Fg^{(\mathfrak{Cs}_\alpha^{\mathfrak{M}})}\{\psi^{\mathfrak{M}}\}.$$

Proof of ($\star\star\star$): Let $F = Fg^{(\mathfrak{Cs}_\alpha^{\mathfrak{M}})}\{\psi^{\mathfrak{M}}\}$. Assume $\varphi^{\mathfrak{M}} \notin F$. Let $f\colon \mathfrak{Cs}_\alpha^{\mathfrak{M}} \longrightarrow \mathfrak{B}$ be such that $f(\psi^{\mathfrak{M}}) = 1$, $f(\varphi^{\mathfrak{M}}) \neq 1$. By $\mathbf{H}Cs_\alpha \subseteq \mathbf{SP}Cs_\alpha$ we may assume $\mathfrak{B} \in Cs_\alpha$. By ($\star$) then $\Theta\rho\mathfrak{M} \cup \{\psi\} \not\models \varphi$. Assume $\varphi^{\mathfrak{M}} \in F$ and let $\mathfrak{N} \models \Theta\rho\mathfrak{M} \cup \{\psi\}$ be arbitrary. By ($\star\star$), there is a homomorphism $f : \mathfrak{Cs}_\alpha^{\mathfrak{M}} \longrightarrow \mathfrak{Cs}_\alpha^{\mathfrak{N}}$ such that $f\chi^{\mathfrak{M}} = \chi^{\mathfrak{N}}$ for all $\chi \in \mathcal{L}(\mathfrak{M})$. Then $f(\psi^{\mathfrak{M}}) = 1$ by $\mathfrak{N} \models \psi$, therefore $f(\varphi^{\mathfrak{M}}) = 1$ by $\varphi^{\mathfrak{M}} \in Fg^{(\mathfrak{Cs}_\alpha^{\mathfrak{M}})}\{\psi^{\mathfrak{M}}\}$. Thus $\mathfrak{N} \models \varphi$, showing that $\Theta\rho\mathfrak{M} \cup \{\psi\} \models \varphi$. **QED($\star\star\star$)**

We are ready to prove (iii). We recall from [8] that $\mathrm{Ig}^{(\mathfrak{A})}\{y\}$ denotes the ideal generated by $\{y\}$ in $\mathfrak{A}$. Then $Fg^{(\mathfrak{A})}\{y\} = \{x : -x \in \mathrm{Ig}^{(\mathfrak{A})}\{-y\}\}$. We will use 2.4.44 of [8] saying that (a),(b) below are equivalent.

(a) $\mathfrak{A} \in CA_\alpha$ is subdirectly irreducible;

(b) $(\exists x \in A, x \neq 0)(\forall y \in A, y \neq 0)x \in \mathrm{Ig}^{(\mathfrak{A})}\{y\}$.

Now (b) is clearly equivalent with

(c) $(\exists x \in A, x \neq 1)(\forall y \in A, y \neq 1)x \in Fg^{(\mathfrak{A})}\{y\}$.

Now substituting $\mathfrak{Cs}_\alpha^{\mathfrak{M}}$ in place of $\mathfrak{A}$ we obtain

(d) $(\exists\varphi \in \mathcal{L}(\mathfrak{M}) \setminus \Theta\rho\mathfrak{M})(\forall\psi \in \mathcal{L}(\mathfrak{M}) \setminus \Theta\rho\mathfrak{M})\varphi^{\mathfrak{M}} \in Fg^{(\mathfrak{Cs}_\alpha^{\mathfrak{M}})}\{\psi^{\mathfrak{M}}\}$,

and now by $(\star\star\star)$ we obtain (iii).

The proof of **(iv)** is similar to the above ones, we leave it to the reader. **QED**

In connection with (iii),(iv) above we note that every Ws_α is weakly subdirectly irreducible (and is also embeddable into a subdirectly irreducible Cs_α). The right hand side of (iv) seems to postulate a strong nonexistence of an approximation of a closed "independent" (from $\Theta\rho\mathfrak{M}$) formula. In this connection we also note that any class K of models of Λ can be "replaced with" a K^+ consisting only of models of the type in (iii) above such that $\Theta\rho(K) = \Theta\rho(K^+)$. This becomes false if we want the elements of K^+ to have complete theories (i.e. type (ii) instead of (iii)).

DEFINITION 17. *(cf. [8])*

(i) $\mathfrak{M}$ *is elementarily equivalent to* $\mathfrak{N}$, *in symbols* $\mathfrak{M} \equiv_e \mathfrak{N}$, *if* $(\forall\varphi \in \Phi\mu^\Lambda)(\mathfrak{M} \models \varphi$ *iff* $\mathfrak{N} \models \varphi)$.

(ii) $\mathfrak{M}$ *is an elementary submodel of* $\mathfrak{N}$, *in symbols* $\mathfrak{M} \prec_e \mathfrak{N}$, *if* $M \subseteq N$ *and* $(\forall\varphi \in \Phi\mu^\Lambda)(\forall k \in {}^\alpha M)(\mathfrak{M} \models \varphi[k]$ *iff* $\mathfrak{N} \models \varphi[k])$.

(iii) $\mathfrak{M}$ *is definitionally equivalent to* $\mathfrak{N}$, *in symbols* $\mathfrak{M} \approx_d \mathfrak{N}$, *if* $M = N$ *and each distinguished relation of* $\mathfrak{M}$ *can be defined in* $\mathfrak{N}$ *and vice versa, i.e.* $\mathfrak{M} \approx_d \mathfrak{N} \iff (\forall i \in I)(\exists\varphi,\psi \in \Phi\mu^\Lambda)[R_i(v_0 ...)^{\mathfrak{M}} = \varphi^{\mathfrak{N}}$ *and* $R_i(v_0 ...)^{\mathfrak{N}} = \psi^{\mathfrak{M}}]$. ■

Below we recall some properties of $\prec_e$, $\equiv_e$ from [3] without proofs. (Their proofs can be found in [3].) We will see that these (and other) notions behave differently in full, variable–rich and in ordinary languages.

(I) Elementary embeddability of infinitary models does not imply elementary equivalence:

$$\mathfrak{N} \prec_e \mathfrak{M} \implies \mathfrak{N} \models \Theta\rho\mathfrak{M} \quad \text{but}$$
$$\mathfrak{N} \prec_e \mathfrak{M} \not\Longrightarrow \mathfrak{N} \equiv_e \mathfrak{M}, \quad \text{even if } \Lambda \text{ is variable–rich.}$$

<u>Remark:</u> The Tarski-Vaught test for elementary submodels works in the case of infinitary models, too. Let $\mathfrak{N} \subseteq \mathfrak{M}$. Then $\mathfrak{N} \prec_e \mathfrak{M}$ iff $(\forall\varphi \in \Phi\mu^\Lambda)(\forall k \in {}^\alpha N)(\forall i \in \alpha)[(\exists m)\mathfrak{M} \models \varphi[k(i/m)] \Rightarrow (\exists n \in N)\mathfrak{M} \models \varphi[k(i/n)]]$. This is Lemma 2.1 in Sain [19].

<u>Notation:</u> $\mathfrak{N} \preceq_e \mathfrak{M}$ iff ($\mathfrak{N} \prec_e \mathfrak{M}$ and $\mathfrak{N} \equiv_e \mathfrak{M}$). Let $Y \subseteq N$. Then $\mathfrak{N}_Y^+ \stackrel{\text{def}}{=} \langle \mathfrak{N}, \{y\}\rangle_{y \in Y}$. See the notational conventions above Prop. 26 later for more detail concerning the language etc. of the new model $\mathfrak{N}_Y^+$.

(II) Adding constants to the models may destroy elementary submodels if Λ is not variable–rich:

$$\mathfrak{N} \precneqq_e \mathfrak{M} \not\Longrightarrow (\mathfrak{N}^+_N \prec_e \mathfrak{M}^+_N \text{ or } \mathfrak{M}^+_N \models \Theta\rho\mathfrak{N}^+_N), \quad \text{but}$$
$$\mathfrak{N} \prec_e \mathfrak{M} \Longrightarrow \mathfrak{N}^+_N \prec_e \mathfrak{M}^+_N \quad \text{if } \Lambda \text{ is variable–rich.}$$
$$\mathfrak{N} \precneqq_e \mathfrak{M} \not\Longrightarrow \mathfrak{N}^+_N \equiv_e \mathfrak{M}^+_N, \quad \text{even if } \Lambda \text{ is variable–rich, and}$$
$$\mathfrak{N}^+_N \equiv_e \mathfrak{M}^+_N \not\Longrightarrow \mathfrak{N} \prec_e \mathfrak{M}, \quad \text{even if } \Lambda \text{ is variable–rich .}$$

($N \subseteq M$ of course)

Proposition 8 generalizes to infinitary models with the exception that in (iv) we have to replace $\prec_e$ by $\precneqq_e$. We state this proposition without proof, because its proof is straightforward.

PROPOSITION 18.

(i) $\mathfrak{Cs}^{\mathfrak{M}}_\alpha = \mathfrak{Cs}^{\mathfrak{N}}_\alpha$ *iff* $\mathfrak{M} \approx_d \mathfrak{N}$.

(ii) $\mathfrak{Cs}^{\mathfrak{M}}_\alpha \cong \mathfrak{Cs}^{\mathfrak{N}}_\alpha$ *iff* $\mathfrak{M} \equiv_e \mathfrak{N}' \approx_d \mathfrak{N}$ *for some* $\mathfrak{N}'$.

(iii) $\mathfrak{Cs}^{\mathfrak{M}}_\alpha$ *is base–isomorphic to* $\mathfrak{Cs}^{\mathfrak{N}}_\alpha$ *iff* $\mathfrak{M} \cong \mathfrak{N}' \approx_d \mathfrak{N}$ *for some* $\mathfrak{N}'$.

(iv) $\mathfrak{Cs}^{\mathfrak{M}}_\alpha$ *is sub–isomorphic to* $\mathfrak{Cs}^{\mathfrak{N}}_\alpha$ *iff* $\mathfrak{M} \precneqq_e \mathfrak{N}' \approx_d \mathfrak{N}$ *for some* $\mathfrak{N}'$.

(v) $rl({}^\alpha M)$: $\mathfrak{Cs}^{\mathfrak{N}}_\alpha \longrightarrow \mathfrak{Cs}^{\mathfrak{M}}_\alpha$ *is an onto homomorphism iff* $\mathfrak{M} \prec_e \mathfrak{N}' \approx_d \mathfrak{N}$ *for some* $\mathfrak{N}'$. ■

Now we are ready to state a model theoretic consequence of our Thm. 7. By Thm. 7 and Prop. 18, there is a variable–rich Λ such that there are two infinite elementarily equivalent models of Λ which have no joint elementary extension. The opposite holds for ordinary languages as well as for full languages. The latter follows from Prop. 18 and [9] I.3.10 which states that $(\forall\mathfrak{A},\mathfrak{B} \in {}_\infty Cs_\alpha)(\exists\mathfrak{C} \in Cs_\alpha)[\mathfrak{A} \cong \mathfrak{B} \Longrightarrow$ both $\mathfrak{A}$ and $\mathfrak{B}$ are sub–base–isomorphic to $\mathfrak{C}]$.

From the proof of Thm. 7 we can easily extract these two models: Assume Λ has three relation symbols, E, R, and S such that E is unary and R, S are ω–ary, and Λ has $\omega + \omega$–many variables. Let

$$M = N = \omega,$$
$$E^{\mathfrak{M}} = E^{\mathfrak{N}} = \{0,1\}, S^{\mathfrak{M}} = S^{\mathfrak{N}} = \{\langle 0 : i \in \omega\rangle\}, \quad \text{and}$$
$$R^{\mathfrak{M}} = \{\langle 0,1,0,1,0,\ldots\rangle\}, \qquad R^{\mathfrak{N}} = \{\langle 1,0,1,0,1,\ldots\rangle\},$$
$$\mathfrak{M} = \langle M, E^{\mathfrak{M}}, R^{\mathfrak{M}}, S^{\mathfrak{M}}\rangle, \quad \text{and} \quad \mathfrak{N} = \langle M, E^{\mathfrak{N}}, R^{\mathfrak{N}}, S^{\mathfrak{N}}\rangle.$$

Now it is proved in Thm. 7 that $\mathfrak{M} \equiv_e \mathfrak{N}$ but for no model $\mathfrak{C}$ of Λ is $\mathfrak{M} \prec_e \mathfrak{C}, \mathfrak{N} \prec_e \mathfrak{C}$ true. We note that by (the proof of) [9] II.3.10, if we consider E to be an ω–ary relation, then the same $\mathfrak{M}$ and $\mathfrak{N}$ will have a joint elementary extension. (we leave it to the reader to find one.) We conjecture that the statement that elementarily equivalent models have isomorphic elementary extensions generalizes from full languages to a considerably broader class of languages. (Perhaps conditions like $(\forall i \in I)\rho(i) \geq \omega$ will suffice.)

Ultraproducts

Next we briefly treat the notion of ultraproducts of infinitary models.

DEFINITION 19 (Compare [8] Def. 3.1.89 and [1] def. in the proof of Thm. 6.5). Let J be a set, F an ultrafilter on J and $\mathfrak{M} = \langle \mathfrak{M}_j : j \in J \rangle$ a system of models of Λ. For all $j \in J$ let $\mathfrak{M}_j = \langle M, R_i^{\mathfrak{M}_j} \rangle_{i \in I}$. We say that c is an $\langle F, \mathfrak{M}, \alpha \rangle$*-choice-function* if

$$c\colon \left(\mathop{\mathrm{P}}_{j \in J} M_j/F \right) \times \alpha \longrightarrow \mathop{\mathrm{P}}_{j \in J} M_j \text{ and } c(q,i) \in q \quad \text{for all} \quad \langle q, i \rangle \in \mathrm{Do}(c).$$

The F, c-product $\mathrm{P}\, \mathfrak{M}_j/F, c$ of $\mathfrak{M}$ is defined as follows:

$$\mathrm{P}\, \mathfrak{M}_j/F, c = \langle N, R_i^{\mathfrak{N}} \rangle_{i \in I}$$

where $N = \mathrm{P}\, M_j/F$, and for all $i \in I$,

$$R_i^{\mathfrak{N}} = \left\{ q \in {}^{\rho(i)}N : \{ j \in J : \langle c(q_k, k)_j : k < \rho(i) \rangle \in R_i^{\mathfrak{M}_j} \} \in F \right\},$$

see Figure 5 below. ∎

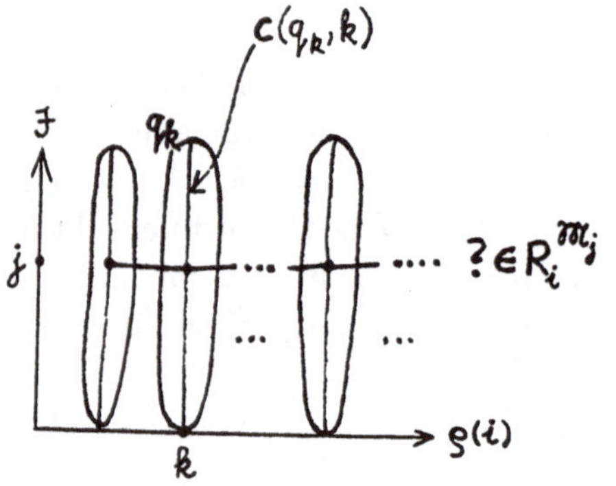

Figure 5

We note that if the $\mathfrak{M}_j$'s are ordinary then this notion coincides with the usual ultraproduct, and the choice-function c does not play any role. But in the case of infinitary models, different choice-functions c lead to different ultraproducts. (This is shown in Example 20 below.) In fact, by choosing c appropriately, we can construct a great variety of ultraproducts of the same models. This versatility of ultraproducts, introduced by the choice-function c, is illustrated in Propositions 21–23 below. By Prop. 22 we have that if $\mathfrak{M} \not\models \varphi$ but $\Theta\rho\mathfrak{M} \cup \{\varphi\}$ is consistent then ${}^J\mathfrak{M}/F, c \models \varphi$ for some choice-function c. By Prop. 21 we have that if $\mathfrak{M} \not\models \varphi$ then ${}^J\mathfrak{M}/F, c \not\models \varphi$ for some c. By Prop. 23, $\mathfrak{M} \precsim_e {}^J\mathfrak{M}/F, c$ for some c.

EXAMPLE 20 Let $\mathfrak{M} = \langle M, R \rangle$ with $M \supseteq \{0, 1\}$ and $R = \{ s \in {}^{\omega}M : |\{ i \in \omega : s_i \neq 0 \}| < \omega \}$. Let F be a nonprincipal ultrafilter on $J \supseteq \omega$ and let c_1, c_2 be $\langle F, {}^J\mathfrak{M}, \alpha\rangle$–choice–functions such that for all $q \in {}^JM/F$ and $i \in \omega$, $c_1(q,i)_n = 0$ if $n \le i$, and $c_2(q,i)_n = 1$, if $n \le i$.

Let $N = {}^JM/F$. Then ${}^J\mathfrak{M}/F, c_1 = \langle N, {}^\omega N\rangle$ while ${}^J\mathfrak{M}/F, c_2 = \langle N, 0\rangle$. I.e. ${}^J\mathfrak{M}/F, c_1 \models R(v_0...)$ while ${}^J\mathfrak{M}/F, c_2 \models \neg R(v_0...)$. ■

We have seen in the above example that formulas not valid in $\mathfrak{M}$ may become valid in the ultraproduct. However, if a formula is valid in $\mathfrak{M}$ then it cannot become non–valid, by Prop. 21 below. I.e., for all F, c and φ, $\mathfrak{M} \models \varphi \Longrightarrow {}^J\mathfrak{M}/F, c \models \varphi$ but ${}^J\mathfrak{M}/F, c \models \varphi \not\Longrightarrow \mathfrak{M} \models \varphi$. Yet, there is a close relationship between the formulas valid in $\mathfrak{M}$ and the ones valid in the ultraproducts, as the following proposition shows. Prop. 21 is a generalization of the Łos–lemma. It is the model–theoretic "equivalent" of [8] 3.1.90–92, but also, the proof of the "ordinary" Łos–lemma extends to a proof of Prop. 21 in a straightforward way.

PROPOSITION 21. *(Łos–lemma for infinitary models) Let J be a set, F an ultrafilter on J, $\mathfrak{M} = \langle \mathfrak{M}_j : j \in J\rangle$ a system of models of Λ and let $\varphi \in \Phi\mu^\Lambda$. Then (i),(ii) below are equivalent.*

(i) $\{j \in J : \mathfrak{M}_j \models \varphi\} \in F$;

(ii) $\mathrm{P}\,\mathfrak{M}_j/F, c \models \varphi$ *for every* $\langle F, \mathfrak{M}, \alpha\rangle$*–choice–function* c.

Proofidea: The usual proof of the Łos lemma extends easily to the proof of the following statement:

($\star^4$) Let J be a set, F an ultrafilter on J, $\mathfrak{M} = \langle \mathfrak{M}_j : j \in J\rangle$ a system of models of Λ, c a $\langle F, \mathfrak{M}, \alpha\rangle$–choice–function, $q \in {}^\alpha(\mathrm{P}\,M_j/F)$, and $\varphi \in \Phi\mu^\Lambda$. Then the following conditions (a) and (b) are equivalent:

(a) $\{j \in J : \mathfrak{M}_j \models \varphi[\langle c(q_k, k)_j : k < \alpha\rangle]\} \in F$.

(b) $\mathrm{P}\,\mathfrak{M}_j/F, c \models \varphi[q]$.

Now (i) $\Longrightarrow$ (ii) follows trivially from ($\star^4$). For (ii) $\Longrightarrow$ (i), assume (ii), but suppose that $\{j \in J : \mathfrak{M}_j \models \varphi\} \notin F$. Thus

$$N \stackrel{\text{def}}{=} \{j \in J : (\exists s \in {}^\alpha M_j)(\mathfrak{M}_j \models \neg\varphi[s])\} \in F.$$

For each $j \in N$ choose ${}_js \in {}^\alpha M_j$ so that $\mathfrak{M}_j \models \neg\varphi[{}_js]$, and for $j \notin N$ choose ${}_js \in {}^\alpha M_j$ arbitrarily. For each $i \in \alpha$ let $q'_i = \langle {}_js_i : j \in \alpha\rangle$, and let $q = \langle q'_i/F : i \in \alpha\rangle$. Then take any $\langle F, \mathfrak{M}, \alpha\rangle$–choice–function c such that $c(q_i, i) = q'_i$ for all $i \in \alpha$. Therefore, $\langle c(q_i, i)_j : i \in \alpha\rangle = \langle {}_js_i : i \in \alpha\rangle = {}_js$; this gives a clear contradiction with ($\star^4$). **QED**

From Prop. 21, compactness of our finitary language of infinitary relations follows. (See Sain [18].) The Löwenheim–Skolem theorems also hold, this is proved in [9], [8] 3.1.45, 94, 95 and improved in Sain [19]. But this language does not enjoy the Beth–definability property, because epis are not surjective in Cs_ω (see Sain [20]). The status of the weak Beth–property (in the sense of Friedman, Makowsky, Shelah) is still open for our languages, see Sain [20]. Finally, we note that in [8] 4.3.23(iii), a complete inference system is given for these languages, for the variable–rich case. It is still a central open problem to find elegant and simple inference systems for arbitrary Λ. The ones available, e.g. in [8] 4.3.65, [1], [2], are not elegant in the sense that inside of the derivations formulas may occur that (depending on the kind of the inference system) either do not belong to the language or are not true. See Problems 4.16,4.1 of [8]. A complete but not very simple inference system can be obtained from 4.1.8–9 of [8] due to J.D. Monk.

Proposition 22 below shows the versatility of the choice–function c in a stronger way. It says that if $\Theta\rho\mathfrak{M}$ does not explicitly "forbid" φ, then φ is valid in some ultrapower of $\mathfrak{M}$. Moreover, it is only the choice of c which depends on that of φ.

PROPOSITION 22. *Let F be any nonprincipal ultrafilter on $\omega = \alpha$, and assume that $(\forall n \in \omega)\mathfrak{M} \not\models \exists v_0 \ldots \exists v_n \neg\varphi$. Then ${}^\omega\mathfrak{M}/F, c \models \varphi$ for some choice–function c.*

Proof: For any $n \in \omega$ let $k_n \in {}^\omega M$ be such that $\mathfrak{M} \not\models \exists v_0 \ldots \exists v_n \neg\varphi[k_n]$. Let the choice–function c be such that $(\forall q \in {}^\omega({}^\omega M/F))(\forall n \in \omega)$

$$c(q, n)_i = k_n(i) \qquad \text{if } i \leq n.$$

Such choice–function exists because F is non–principal. Now it is not difficult to check that ${}^\omega\mathfrak{M}/F, c \models \varphi$. **QED**

The condition $\alpha = \omega$ is not essential in Prop. 22 (we stated this version for simplicity). An easy modification of the proof of Prop. 22 proves the following:

(*) Assume that for any $n \in \omega$ and $i_0, \ldots, i_n \in \alpha, \mathfrak{M} \not\models \exists v_{i_0} \ldots \exists v_{i_n} \neg\varphi$. Then there are an ultrafilter F and a choice–function c such that ${}^J\mathfrak{M}/F, c \models \varphi$.

From (*) we immediately obtain

(**) $\Theta\rho\mathfrak{M} \cup \{\varphi\}$ has a model iff $(\forall n \in \omega)(\forall i_0, \ldots, i_n \in \alpha)\mathfrak{M} \not\models \exists v_{i_0} \ldots \exists v_{i_n} \neg\varphi$.

Now (**) implies that completeness of a set of formulas admits a more traditional formulation. Namely:

(* * *) Σ is complete iff for every φ either $\Sigma \models \varphi$ or $\Sigma \models \neg\forall\overline{v}\varphi$ for some finite sequence $\overline{v}$ of variables.

We note that (**) has a purely algebraic proof, too, using 2.3.10(i) and 3.1.107 of [8].

Recall that $\Sigma \subseteq \Phi\mu^\Lambda$ is *complete* iff for all $\varphi \in \Phi\mu^\Lambda$ either $\Sigma \models \varphi$ or $\Sigma \cup \{\neg\varphi\} \models \mathit{FALSE}$. If $\mathfrak{M}$ is ordinary then $\Theta\rho\mathfrak{M}$ is complete. From (**) we can easily show that $\Theta\rho\mathfrak{M}$ is not necessarily complete in the case of infinitary models. However, $\Theta\rho\mathfrak{M}$ can always be extended to a complete, consistent theory. (We sketch an algebraic proof of this fact here. Every nontrivial algebra has a simple homomorphic image. Let f be a homomorphism from $\mathfrak{Cs}_\alpha^{\mathfrak{M}}$ onto a simple algebra $\mathfrak{C}$. By $\mathbf{H}Cs_\alpha \subseteq \mathbf{SP}Cs_\alpha$ we may assume $\mathfrak{C} \in Cs_\alpha$. By $(\star)$ in the proof of Fact 16 then $\mathfrak{C} = \mathfrak{Cs}_\alpha^{\mathfrak{N}}$ for some $\mathfrak{N}$ with $\mathfrak{N} \models \Theta\rho\mathfrak{N}$. By Fact 16(i) then $\Theta\rho\mathfrak{N}$ is a complete extension of $\Theta\rho\mathfrak{M}$.)

The following proposition is a corollary of [8] 3.1.94.

PROPOSITION 23. *For any ultrafilter F there is a choice–function c such that $\mathfrak{M} \cong \mathfrak{M}' \underset{\equiv}{\prec}_e {}^J\mathfrak{M}/F, c$. Moreover, the usual diagonal embedding sending $m \in M$ to $\langle m, ..., m... \rangle/F$ is the isomorphism $d : \mathfrak{M} \rightarrowtail\!\!\!\!\twoheadrightarrow \mathfrak{M}'$.*

Proof: First we sketch how to prove Prop. 23 by using the Łos–lemma (Prop. 21 and $(\star^4)$ in its proof), then we show how to derive Prop. 23 as a corollary of [8] 3.1.94.

For each $m \in M$ let $d'm = \langle m : j \in J \rangle$, and let $dm = d'm/F$. Let c be any $\langle F, \langle \mathfrak{M} : j \in J \rangle, \alpha \rangle$–choice–function such that $c(dm, i) = d'm$ for all $m \in M$. Then by $(\star^4)$ in the proof of Prop. 21, for any $k \in {}^\alpha M$ and any formula $\varphi \in \mathcal{L}(\mathfrak{M})$ we have $\mathfrak{M} \models \varphi[k]$ iff ${}^J\mathfrak{M}/F, c \models \varphi[d \circ k]$. This immediately yields $\prec_e$. The proof of $\equiv_e$ is similar.

We now want to apply [8] 3.1.94 to give an alternate proof of Prop. 23. (This will also give a hint how to read the "model theoretic content" of theorems in [8].) Let us take $\mathfrak{A}$ of [8] 3.1.94 to be $\mathfrak{Cs}_\alpha^{\mathfrak{M}}$. Then U and ε of [8] 3.1.94 are our M and d respectively. Let c and g be as in [8] 3.1.94, let $\mathfrak{N} = {}^J\mathfrak{M}/F, c$. It is not difficult to check that $g(\varphi^{\mathfrak{M}}) = \varphi^{\mathfrak{N}}$ for all $\varphi \in \mathcal{L}(\mathfrak{M})$. Now 3.1.94(i) implies that d is an isomorphism of $\mathfrak{M}$ onto some $\mathfrak{M}'$, and 3.1.94(ii),(v) then imply $\mathfrak{M}' \underset{\equiv}{\prec}_e \mathfrak{N}$. **QED**

Next we will show that there is a model $\mathfrak{M}$ such that $\mathfrak{M}$ is elementarily equivalent to one of its ultrapowers but cannot be elementarily embedded into this ultraprower (i.e. $\mathfrak{M} \equiv_e \mathfrak{U} = {}^J\mathfrak{M}/F, c$ does not imply that $\mathfrak{M}$ is isomorphic to an elementary submodel of $\mathfrak{U}$). We state and prove this theorem in algebraic form because the algebraic version contains more information.

An algebraic form of the Łos lemma, when restricted to ultrapowers, says that $\mathfrak{Cs}_\alpha^{({}^J\mathfrak{M}/F,c)}$ is a homomorphic image of $\mathfrak{Cs}_\alpha^{\mathfrak{M}}$ along a homomorphism sending $R_i(v_0...)^{\mathfrak{M}}$ to $R_i(v_0...)^{{}^J\mathfrak{M}/F,c}$, for all $i \in I$.

Let $\mathfrak{M}$ be a model, F an ultrafilter on J and c be a choice–function. Let $\mathfrak{U} \overset{\text{def}}{=} {}^J\mathfrak{M}/F, c$. We recall from [9] 3.12, p. 181 that the homomorphism $ud_{F,c} : \mathfrak{Cs}_\alpha^{\mathfrak{M}} \twoheadrightarrow \mathfrak{Cs}_\alpha^{\mathfrak{U}}$ is defined by stipulating:

$$ud_{F,c}(R_i(v_0...)^{\mathfrak{M}}) = R_i(v_0...)^{\mathfrak{U}} \text{ for all } i \in I.$$

Then $\mathfrak{M} \equiv_e \mathfrak{U}$ iff $ud_{F,c} \in \mathrm{Is}(\mathfrak{Cs}_\alpha^{\mathfrak{M}}, \mathfrak{Cs}_\alpha^{\mathfrak{U}})$.

PROPOSITION 24. *Let $\alpha \geq \omega$. Then (i)–(ii) below hold.*

(i) $ud_{F,c} \in \mathrm{Is}(\mathfrak{A}, \mathfrak{B}) \;\not\Rightarrow\; \mathfrak{A}$ *is sub–base–isomorphic to* $\mathfrak{B}$, *for* $\mathfrak{A}, \mathfrak{B} \in Cs_\alpha$. *Moreover,*

(ii) *Let* $K \in \{Cs_\alpha^{reg} \cap Dc_\alpha, Ws_\alpha\}$. *Then* $(\exists \mathfrak{A} \in K)(\exists$ *ultrafilter* F *and a* $\langle F, base(\mathfrak{A}), \alpha\rangle$*–choice–function* $c)$ $ud_c \in \mathrm{Is}(\mathfrak{A}, \mathfrak{B})$ *for some* $\mathfrak{B} \in K$ *but* $\mathfrak{A}$ *is not sub–base–isomorphic to* $\mathfrak{B}$.

Proof: It is enough to prove (ii). First we consider the Ws_α case. For simplicity, let $\alpha \stackrel{\text{def}}{=} \omega \stackrel{\text{def}}{=} U$ but the same proof goes through for arbitrary $\alpha \geq \omega$. Let $\mathfrak{C} \stackrel{\text{def}}{=} \mathfrak{Sb}({}^\alpha\omega^{(\overline{0})})$ and let $\mathfrak{A}$ be the minimal subalgebra of $\mathfrak{C}$. Let F be a nonprincipal ultrafilter on $I \stackrel{\text{def}}{=} \omega$. Let the $\langle F, \omega, \alpha\rangle$–choice–function c be such that

$$(\forall n \in \omega)[c(n, \overline{n}/F) = \langle 0 : i < n\rangle \cup \langle n : i \in \omega \setminus n\rangle \text{ and}$$
$$(\forall k \in {}^IU/F)(k \neq \overline{n}/F \Longrightarrow c(n,k) \restriction n \subseteq \overline{1})].$$

Let $p = \langle \overline{n}/F : n \in \omega\rangle$. Then $ud_c \in \mathrm{Ism}(\mathfrak{A}, \mathfrak{Sb}({}^\alpha Y^{(p)}))$ with $\ker(p) \subseteq Id$. (ud_c is one–one because $\mathfrak{A}$ is simple.) Hence $\mathfrak{A}$ cannot be sub–base–isomorphic to $ud_c{}^*\mathfrak{A}$. This completes the proof of the Ws–case.

Let us turn to the $Cs_\alpha^{reg} \cap Dc_\alpha$ case. Clearly, $\alpha = \omega + \beta$ for some β. For simplicity, we assume $\beta \geq \omega$. The proof for the $\beta < \omega$ case can be obtained from the present one the obvious way. Let $W = \omega + 1$, $R = \{q \in {}^\alpha W : q \restriction \omega \in {}^\omega\omega^{(\overline{0})}\}$ and $\mathfrak{A} = \mathfrak{Sg}^{(\mathfrak{Sb}({}^\alpha W))}\{R\}$. By $\Delta R = \omega$, $\mathfrak{A} \in Dc_\alpha$.

By [8] 3.1.63, $\mathfrak{A}$ is regular. Now choose F and c as in the Ws–case with the obvious modifications, e.g. c is an $\langle F, W, \alpha\rangle$–choice–function now and for $j \in \alpha, k \in {}^IW/F$, $c(j,k)$ is as above if $j < \omega$, $c(j, \overline{w}/F) = \overline{w}$ if $w \in W$ and $j \geq \omega$, else it is arbitrary.

Let $\mathfrak{B} = ud_c^*\mathfrak{A}$, $U = {}^IW/F$, $V = {}^I\omega/F$ and $P = ud_c(R)$. Let p be as in the Ws–case. Then it can be checked that $P = \{q \in {}^\alpha U : q \restriction \omega \in {}^\omega V^{(p)}\}$. Thus $\mathfrak{B} \in Cs_\alpha$ with base U by [8] 3.1.91(v) and $\mathfrak{B}$ is regular by [8] 3.1.63.

Now we show that $\mathfrak{A}$ is not sub–base–isomorphic to $\mathfrak{B}$. Any sub–base–isomorphic image, say, $h(R)$ of R contains some Y of the form $Y = \{q \in {}^\alpha U_1 : q \restriction \omega \in {}^\alpha U_0^{(r)}\}$ for some $U_0 \subsetneq U_1 \subseteq U$ and $r \in {}^\alpha U_0$ such that $\ker(r \restriction \omega) = \omega \times \omega$. Moreover, $h(R) \cap {}^\alpha U_1 = Y$. Let $Q = \{q \in {}^\alpha U : q \restriction \omega = p\}$. Now $\Delta(Q) = 0$ and $p_0 = rl(Q)$, $p_1 = rl(-Q)$ provide a subdirect decomposition of $\mathfrak{B}$ such that $p_1^*(\mathfrak{B})$ is a minimal algebra by $p_1(P) = 0$. Clearly, $Q \cap Y = 0$. Thus $-Q \supseteq Y$, hence $p_1(h(R)) = -Q \cap h(R) \supseteq Y = h(R) \cap {}^\alpha U_1 = -Q \cap h(R) \cap {}^\alpha U_1$. Let $Z = {}^\alpha U_1 \setminus Q$. Then $Z \cap h(R) = Y$. Clearly $\Delta^{[Z]}Y \supseteq \omega$ where $\Delta^{[Z]}$ is understood in the algebra $\mathfrak{Sb}(Z)$, since $C_n^{[Z]}(Y) \neq Y$ for all $n < \omega$, by ${}^\alpha U_1^{(r)} \subseteq Z$. But then $\Delta^{[-Q]}(Y) \supseteq \omega$, hence $\Delta(p_1(hR)) \supseteq \omega$ by $Z \cap p_1(hR) = Y$. This contradicts our previous observation that $p_1^*\mathfrak{B}$ is minimal, proving that $\mathfrak{A}$ is not sub–base–isomorphic to $\mathfrak{B}$.

It remains to show that $ud_c \in \mathrm{Is}(\mathfrak{A})$. Let $x \in A, x \neq 0$. We want to show that $ud_c(x) \neq 0$. Let $Q' = \{q \in {}^\alpha W : q \restriction \omega \in {}^\omega W^{(\overline{0})}\}$. Then exactly as above, one can show that $rl(Q') \in \mathrm{Is}(\mathfrak{A})$, i.e. $x \cap Q' \neq 0$. Let $s \in x \cap Q'$.

Let $\Gamma \subseteq_\omega \alpha$ be the set of indices used when generating x from R (i.e. let $x \in Sg^{(\mathfrak{R}\mathfrak{d}_\Gamma \mathfrak{A})}\{R\}$). Then $\Delta(x) \subseteq \omega \cup \Gamma$ and by using [9] 4.7.1.2, one can prove the following statement for all $\Delta \subseteq \alpha \smallsetminus \Gamma$:

$$(*) \qquad (\forall r \in {}^\Delta W)[(\forall i \in \Delta)(r_i \in \omega \Leftrightarrow s_i \in \omega) \Longrightarrow s[\Delta/r] \in x],$$

where $s[\Delta/r]$ is as defined in [9]. Let now $\Sigma \subseteq_\omega \alpha$ be such that $\Sigma \supseteq \Gamma \cup \{i \in \omega : s_i \neq 0\}$. Let $q \in {}^\alpha U$ be such that $q_i = \overline{s_i}/F$ if $i \in \Sigma$ and $q_i = \bar{i}/F$ if $i \in \omega \smallsetminus \Sigma$. Then using $(*)$, regularity of x together with $\Delta(x) \subseteq \omega \cup \Gamma$, and the properties of c, it can be checked that $q \in ud_c(x)$. Thus $ud_c(x) \neq 0$ showing that $ud_c \in \mathrm{Is}\,\mathfrak{A}$. **QED**

To conclude the part dealing with ultraproducts we note that a more general algebraization of the Łos lemma can be obtained by combining the above homomorphism $ud_{F,c}$ and 4.3.67 of [8]. This generalization of $ud_{F,c}$ will map $\mathrm{P}_{j\in J}\, \mathfrak{Cs}_\alpha^{\mathfrak{M}_j}/F$ homomorphically onto a Cs_α containing $\mathfrak{Cs}_\alpha^{(P\mathfrak{M}_j/F,c)}$. By adjusting the definitions, this can be improved to saying that $\mathfrak{Cs}$ is a functor (from models to CA's) which commutes with taking ultraproducts.

Saturated and universal models

Now we turn to saturated models.
Notation: Let $\Sigma \subseteq \Phi\mu^\Lambda$. Then

(i) $\mathrm{freevar}(\Sigma) \stackrel{\mathrm{def}}{=} \{i : i$ is an ordinal and v_i occurs freely in $\Sigma\}$
(ii) $(\bigwedge \Sigma)$ is the usual infinitary formula denoting the conjunctions of the elements of Σ. ■

Note that $(\bigwedge \Sigma)$ is *not* in $\Phi\mu^\Lambda$, but its meaning is known from standard infinitary logic $L_{\infty\omega}$.

DEFINITION 25. $\mathfrak{M}$ is κ–saturated $\stackrel{def}{\Longleftrightarrow}$

$$(\forall \beta < \kappa)(\forall \Sigma \subseteq \Phi\mu^{\langle \alpha\cup\kappa, I, \rho\rangle})\Big[\mathrm{freevar}(\Sigma) \subseteq \beta \Longrightarrow (\forall k \in {}^\beta M)(\forall i < \beta)$$
$$\Big((\forall \Sigma_0 \subseteq_\omega \Sigma)\mathfrak{M} \models (\exists v_i \bigwedge \Sigma_0)[k] \Longrightarrow \mathfrak{M} \models (\exists v_i \bigwedge \Sigma)[k]\Big)\Big]. \quad ■$$

Theorems 1, 3(i) in §3.1 are the "ordinary" counterparts of the following:

THEOREM 8. *Let $\gamma \leq \alpha$ and $\mathfrak{M}$ a model of $\langle \gamma, I, \rho\rangle$. Then (i)–(iii) below hold.*

(i) *Assume $|\alpha \smallsetminus \gamma| \geq |\gamma|$. Then $\mathfrak{M}$ is $|\alpha|^+$–saturated $\Longleftrightarrow$ $\mathfrak{Cs}_\alpha^{\mathfrak{M}}$ is saturating.*
(ii) *Assume $\gamma < |\alpha|$. Then $\mathfrak{M}$ is $|\alpha|$–saturated $\Longleftrightarrow$ $\mathfrak{Cs}_\alpha^{\mathfrak{M}}$ is $|\alpha|$–saturating.*
(iii) *Assume $\gamma < \kappa \leq |\alpha|^+$ and $|\alpha \smallsetminus \gamma| \geq |\gamma|$. Then $\mathfrak{M}$ is κ–saturated $\Longleftrightarrow$ $\mathfrak{Cs}_\alpha^{\mathfrak{M}}$ is κ–saturating.*

Proof: It is enough to prove (iii), because any Crs_α is saturating iff it is $|\alpha|^+$–saturating.

(iii): Let $\mathfrak{A} \stackrel{\text{def}}{=} \mathfrak{Cs}_\alpha^{\mathfrak{M}}$. Assume $\gamma < \kappa \leq |\alpha|^+$ and $|\alpha \smallsetminus \gamma| \geq |\gamma|$. Then $|\gamma| \leq |\alpha| = |\alpha \smallsetminus \gamma|$.

(1) Assume $\mathfrak{M}$ is κ–saturated. We want to prove that $\mathfrak{A}$ is κ–saturating. Let $F \in FlNr_\beta \mathfrak{A}$ for some β with $\beta < \kappa$. We may assume $\gamma < \beta$ by $\gamma < \kappa \cap (\alpha + 1)$ and by $\delta < \eta \Longrightarrow Nr_\delta \mathfrak{A} \subseteq Nr_\eta \mathfrak{A}$. For every $x \in F$ there is $\varphi_x \in \Phi\mu^{\langle\alpha\ldots\rangle}$ with $x = \varphi_x{}^{\mathfrak{M}}$. Since for every atomic formula ψ we have freevar$(\psi) \subseteq \gamma$, we know that φ_x has at most finitely many free variables of index $\not\leq \beta$. We can replace φ_x with its universal closure under these variables since $x \in Nr_\beta \mathfrak{A}$. So, from now on freevar$(\varphi_x) \subseteq \beta$. Therefore

$$\Sigma \stackrel{\text{def}}{=} \{\varphi_x : x \in F\}$$

satisfies the condition freevar$(\Sigma) \subseteq \beta$ in Def. 25 above. Thus by $\mathfrak{M}$ being κ–saturated and by $\beta < \kappa$, we conclude that

(*) $\quad\Sigma$ satisfies the "compactness property" in that definition.

Let $f \in \bigcap\{c_i x : i \in F\}$ (for some fixed $i < \beta$). Let $k \stackrel{\text{def}}{=} f \upharpoonright \beta$. Then by $f \in c_i x$ we conclude $\mathfrak{M} \models \exists v_i \varphi_x[f]$ hence $\mathfrak{M} \models \exists v_i \varphi_x[k]$ by freevar$(\varphi_x) \subseteq$ Do(k). Since F is downward directed, $(\forall \Sigma_0 \subseteq_\omega \Sigma)\mathfrak{M} \models (\exists v_i \bigwedge \Sigma_0)[k]$ holds, too. Now, by (*), $\mathfrak{M} \models (\exists v_i \bigwedge \Sigma)[k]$. Thus $\mathfrak{M} \models (\exists v_i \bigwedge \Sigma)[f]$. This means $f \in c_i \bigcap F$. We proved $(c_i \bigcap F) \supseteq (\bigcap c_i{}^* F)$. The other direction always holds. Thus $\mathfrak{A}$ is κ–saturating.

(2) Conversely, assume that $\mathfrak{A}$ is κ–saturating. Let $\beta < \kappa, i < \beta$ and $\Sigma \subseteq \Phi\mu^{\langle\alpha\cup\kappa\ldots\rangle}$ with freevar$(\Sigma) \subseteq \beta$ and

(**) $\quad(\forall \Sigma_0 \subseteq_\omega \Sigma)\mathfrak{M} \models (\exists v_i \bigwedge \Sigma_0)[k]$, for some fixed $k \in {}^\beta M$.

<u>Case (i)</u> Assume $\beta \leq \alpha$.

If $\varphi \in \Sigma$, φ may contain only finitely many variables not in α, and all these will be bound (not free) by $\beta \leq \alpha$. By $|\alpha \smallsetminus \gamma| \geq \omega$, we may rename these variables such that they will be in $\alpha \smallsetminus \gamma$. So the new, equivalent version of φ is in $\Phi\mu^{\langle\alpha\ldots\rangle}$ hence $\varphi^{\mathfrak{M}} \in \mathfrak{A}$. Thus we may assume that $\Sigma \subseteq \Phi\mu^{\langle\alpha\ldots\rangle}$. Let F be the filter generated in $Nr_\beta \mathfrak{A}$, by $Y \stackrel{\text{def}}{=} \{\varphi^{\mathfrak{M}} : \varphi \in \Sigma\} (\subseteq Nr_\beta A)$. Let $f \in {}^\alpha M$ with $k \subseteq f$. By (**) (and by $\mathfrak{M} \models (\exists v_i \bigwedge \Sigma_0)[k] \Longrightarrow \mathfrak{M} \models (\exists v_i \bigwedge \Sigma_0)[f]$, $(\forall x \in F) f \in c_i x$.

Since $\mathfrak{A}$ is κ–saturating and $\beta < \kappa \cap (\alpha + 1)$, then $f \in c_i \bigcap F$. Thus $f \in c_i \bigcap Y$. By the definition of Y then $\mathfrak{M} \models (\exists v_i \bigwedge \Sigma)[f]$. Thus $\mathfrak{M} \models (\exists v_i \bigwedge \Sigma)[k]$.

<u>Case (ii).</u> Assume $\alpha < \beta$.

Then $\alpha < \beta < |\alpha|^+$. We will show how to reduce this case to Case (i). Clearly, $|\beta| = |\alpha| = |\alpha \smallsetminus \gamma|$. Let $h \colon (\beta \smallsetminus \alpha) \rightarrowtail (\alpha \smallsetminus \gamma)$ with $|(\alpha \smallsetminus \gamma) \smallsetminus Rg(h)| = |\alpha|$. Then there is $t \colon (\alpha \smallsetminus \gamma) \rightarrowtail (\alpha \smallsetminus \gamma)$ with $Rg(t) \cap Rg(h) = 0$. Let $p \stackrel{\text{def}}{=} (Id \upharpoonright \gamma) \cup (h \cup t)$. Then $p \colon \beta \rightarrowtail \alpha$. We may assume $Rg(p) = \alpha$. Let

$k^+ = k \circ p^{-1}$. Then $k^+ \in {}^{\alpha}M$. Let Σ^+ be obtained from Σ by systematically renaming all variables (both free and bound) in Σ by their p–images. I.e. $\Sigma^+ = \Sigma[v_i/v_{p(i)} : i \in \beta]$. Then $\Sigma^+ \subseteq \Phi\mu^{\langle \alpha \cup \kappa \ldots \rangle}$ because p is identity on the variables occurring in the atomic formulas. But now, freevar$(\Sigma^+) \subseteq \alpha$.

Now, we can apply Case (i) with the β of Case (i) to be chosen α. Then we get $\mathfrak{M} \models (\exists v_{p(i)} \bigwedge \Sigma^+)[k^+]$. By the definition of satisfaction, this yields $\mathfrak{M} \models (\exists v_i \bigwedge \Sigma)[k]$. **QED(Theorem 8)**

Now, using Thm. 8 above, we can formulate model–theoretic corollaries of our algebraic theorems Thms. 4–7. E.g., Thm. 4(iv) (more precisely, its proof) has the following corollary: There is a variable–rich language $\Lambda = \langle \alpha, I, \rho \rangle$ which has two elementarily equivalent, $|\alpha|^+$–saturated but not isomorphic models both with universe $\{0, 1, 2\}$. We do not go into more detail here, but the reader is invited to use Thm. 8 for deriving further (infinitary) model theoretic corollaries from the results in §3.2. By providing the model theoretic equivalent (κ–valuation–compactness) of κ–compact Cs's, Def. 28 below might be useful in deriving such model theoretic corollaries. One of these corollaries is the existence of κ–saturated but not κ–valuation–compact models or κ–saturated but not κ–universal ones (contrasting with Thm. 5.1.12 of Chang–Keisler [5]). Another one is, that every $\mathfrak{M}$ has a κ–saturated elementary extension (actually, an ultrapower). Moreover, if Λ is full, then this elementary extension is at the same time κ–valuation–compact, too.

On the Usual Definition of κ-Saturatedness

For ordinary languages and infinite κ, our definition of κ–saturatedness above is equivalent with the usual one given in §5.1 in Chang–Keisler [5]. (For $\kappa < \omega$, a model is κ–saturated in [5]'s sense iff it is $\kappa + 1$–saturated in our sense.) For infinitary models, we can rewrite our above definition to make the analogy with [5]'s definition more explicit by considering addition of extra constants to our models as usual. We deviated here from this style only because it is slightly awkward to treat constants in our present formalisms $\Phi\mu^{\Lambda}$. Now we will briefly look into the equivalence of our definition of κ–saturatedness with the "constants–oriented" one, e.g. in §5.1 in [5]. For a (possibly infinitary) model $\mathfrak{M}$ and $Y \subseteq M$ recall that

$$\mathfrak{M}_Y^+ \stackrel{\text{def}}{=} \langle \mathfrak{M}, \{a\} \rangle_{a \in Y}.$$

I.e., $\mathfrak{M}_Y^+$ is the expansion of $\mathfrak{M}$ with the elements of Y as constants (cf. [5] p. 68). Here we simulate the constant $a \in Y$ with the unary relation $\{a\}$, because in our formalism $\Phi\mu^{\Lambda}$ we have relation symbols only and we have to simulate everything by them (this is quite usual in logic).

The following proposition indicates a way of translating our definition of κ–saturatedness to the usual constant oriented wording. (In passing we note that we cannot restrict ourselves to types $\Sigma(v_0)$ having a single variable only, simply because, if an atomic formula has infinitely many free variables, our formalism does not permit getting rid of them. This does not seem to be an essential change, cf. Lemma 5.1.1(ii) of Chang–Keisler [5].) In a sense, the proposition below seems to be an infinitary counterpart of Lemma 5.1.1(iii) of [5].

PROPOSITION 26. *Assume $\kappa \geq \omega$. Then $\mathfrak{M}$ is κ–saturated $\iff$ $(\forall Y \subseteq_\kappa M)\mathfrak{M}_Y^+$ is κ–saturated.*

Proof: The direction $\Longleftarrow$ is obvious. If $\kappa = \omega$, we have to deal with ordinary formulas only, hence the proof is easy. Assume therefore $\kappa > \omega$, $\mathfrak{M}$ is κ-saturated and $Y \subseteq_\kappa M$. Let $\beta < \kappa$ and $\Sigma \subseteq \mathcal{L}(\mathfrak{M}_Y^+)$ using the first κ variables and with freevar$(\Sigma) \subseteq \beta$. Since freevar$(\Sigma) \subseteq \beta$ and each formula contains only finitely many quantifiers, there is $\beta_1 < \kappa$ such that the infinitary atomic formulas occurring in Σ have variables in $\beta_1 \geq \beta$. So the variables not in β_1 but occurring in Σ are auxiliary ones only and each formula of Σ contains finitely many of them. We can rename these bound variables such that all of them come from some β_2 with $\beta \leq \beta_1 \leq \beta_2 < \kappa$. Therefore we may assume that Σ *contains variables only from* $\beta_2 < \kappa$.

Let $k \in {}^{\beta_2}M$ and $i \in \beta$ with $(\forall \Sigma_0 \subseteq_\omega \Sigma)\mathfrak{M} \models (\exists v_i \bigwedge \Sigma_0)[k]$. There is an ordinal $\delta < \kappa$ with $\beta_2 < \delta$ and a bijection $y : Y \rightarrowtail\!\!\!\!\rightarrow (\delta \smallsetminus \beta)$. Let $k^+ \overset{\text{def}}{=} k \cup (y^{-1})$. Then $k^+ \in {}^{\delta}M$. Let Σ^+ be obtained from Σ by replacing, for every $a \in Y$, every occurrence of $\{a\}(v_0)$ in Σ with $v_0 = v_{y(a)}$. Then $\Sigma^+ \subseteq \Phi\mu^{\langle\kappa,I,\rho\rangle}$ and freevar$(\Sigma^+) \subseteq \delta$. Define ${\Sigma_0}^+$ the same way for any $\Sigma_0 \subseteq \Sigma$. Clearly for any $\Sigma_0 \subseteq \Sigma$ we have

$$(\star) \qquad \mathfrak{M} \models (\exists v_i \bigwedge {\Sigma_0}^+)[k^+] \iff \mathfrak{M}_Y^+ \models (\exists v_i \bigwedge \Sigma_0)[k],$$

since $i < \beta_2$. Therefore, $(\forall \Gamma \subseteq_\omega \Sigma^+)\mathfrak{M} \models (\exists v_i \Gamma)[k^+]$. Since $\mathfrak{M}$ is κ–saturated and $\delta < \kappa$, this implies $\mathfrak{M} \models \exists v_i \bigwedge \Sigma^+[k^+]$. By $(\star)$ then $\mathfrak{M}_Y^+ \models \bigwedge \Sigma[k]$ as desired. **QED**

The corollary below is an algebraic counterpart (in generalized form) of Lemma 5.1.1 of Chang–Keisler [5].

For a Cs_α $\mathfrak{A}$ with base U and a set $X \subseteq U$ the new Cs_α $\mathfrak{A}_X$ was defined above the statement of Corollary 9. To bring back familiarity with this concept, we note that $\mathfrak{A} \subseteq \mathfrak{A}_X$ and that $\mathfrak{Cs}_\alpha^{(\mathfrak{M}_Y^+)} = (\mathfrak{Cs}_\alpha^{\mathfrak{M}})_Y$ for any model $\mathfrak{M}$ and $Y \subseteq M$.

COROLLARY 27. *Assume $\omega \leq \kappa \leq |\alpha^+|$, $\gamma < \kappa$, $|\alpha \smallsetminus \gamma| \geq |\gamma|$ and that $\mathfrak{A}$ is a Cs_α generated by regular elements all of which are in $Nr_\gamma\mathfrak{A}$. Then for any $Y \subseteq_\kappa base(\mathfrak{A})$ we have*

$$\mathfrak{A} \textit{ is } \kappa\textit{-saturating} \iff \mathfrak{A}_Y \textit{ is } \kappa\textit{-saturating.}$$

Proof: Assume $\mathfrak{A}$ is κ–saturating. Under our assumptions, $\mathfrak{A} = \mathfrak{Cs}_\alpha^{\mathfrak{M}}$ for some $\mathfrak{M}$ of language $\langle \gamma, \ldots \rangle$. By Thm. 8(iii), then $\mathfrak{M}$ is κ–saturated. Then by Prop. 26, $\mathfrak{M}_Y^+$ is κ–saturated. By Thm. 8(iii) then $\mathfrak{A}_Y = \mathfrak{Cs}_\alpha^{(\mathfrak{M}_Y^+)}$ is κ–saturating. The other direction is trivial. **QED(Corollary 27)**

DEFINITION 28 Let $\mathfrak{M}$ be a model of $\langle\gamma, I, \rho\rangle$.

(i) $\mathfrak{M}$ is said to be κ–valuation–compact iff

$$(\forall\beta < \kappa)(\forall\Sigma \subseteq \Phi\mu^{\langle\gamma\cup\kappa, I,\rho\rangle})[\text{freevar}(\Sigma) \subseteq \beta \implies$$
$$((\forall\Sigma_0 \subseteq_\omega \Sigma)(\exists k \in {}^\beta M)\mathfrak{M} \models \bigwedge\Sigma_0[k] \implies (\exists k \in {}^\beta M)\mathfrak{M} \models \bigwedge\Sigma[k])].$$

(ii) $\mathfrak{M}$ is κ–hereditarily compact iff $(\forall Y \subseteq_\kappa M)\mathfrak{M}^+_Y$ is κ–valuation–compact. ∎

FACT 29. *If $\mathfrak{M}$ is ordinary and $\kappa \geq \omega$ then*

$$\mathfrak{M} \textit{ is } \kappa\textit{-saturated} \iff \mathfrak{M} \textit{ is } \kappa\textit{-hereditarily compact.}$$

Proof: is well known from usual model theory, cf. e.g. [5]. ∎

This κ–hereditary compactness is another possible generalization of κ–saturatedness from ordinary models to our infinitary ones. So, one would like to see how it relates to the other model theoretic properties investigated so far in the present §3.3, and then one would like to see its algebraic counterpart, cf. Def. 31 below, (to see how this counterpart relates to κ–saturating and κ–compact Cs_α's etc).

For simplicity, instead of "$\mathfrak{Cs}^{\mathfrak{M}}_\alpha$ is κ–saturating", we will sometimes write "$\mathfrak{M}$ is κ–saturating". We will do this only if α is recoverable form the context.

PROPOSITION 30.

(i) *If $\kappa \geq \omega$ then $\mathfrak{M}$ is κ–hereditarily compact $\implies$ $\mathfrak{M}$ is κ–saturated and κ–valuation compact.*

(ii) *In (i) we have $\not\Leftarrow$. Moreover, for any $\kappa > \omega$ and any non–ordinary $\Lambda = \langle\alpha...\rangle$ with $\alpha \geq \omega$, there is an $\mathfrak{M}$ which is both κ–saturating and κ^+–valuation–compact but not κ–hereditarily compact.*

ProofIdea: We will prove **(ii)** only for the case $I = \{R\}$ and $\rho(R) = \omega$. The general case goes with practically the same argument.

Let $U = V \cup W$ with $V \cap W = 0$ and $|V| = |W| > \kappa^+$. Let $R \overset{\text{def}}{=} \{q \in {}^\omega U : |\{i : q_i \notin V\}| < \omega\}$. Let $\mathfrak{M} = \langle U, R\rangle$.

Claim 1 $\mathfrak{Cs}^{\mathfrak{M}}$ is κ–saturating and κ^+–compact.

This claim follows form Lemma 13.

Claim 2 $\mathfrak{M}$ is not ω^+–hereditarily compact.

To see this, let $u \in V$. Let $\Sigma \overset{\text{def}}{=} \{\{u\}(v_0), \neg R(v_i)_{i<\omega}\} \cup \{v_i = v_0 : i < \omega\}$. Then Σ is in $\mathcal{L}(\langle\mathfrak{M}, \{u\}\rangle)$. Clearly, Σ is finitely satisfiable, but not satisfiable in $\mathfrak{M}$. This proves (ii).

To prove **(i)**, assume that $\mathfrak{M}$ is κ–hereditarily compact. Let $\beta < \kappa$, freevar$(\Sigma) \subseteq \beta$ and $\Sigma \subseteq \Phi\mu^{\langle\kappa...\rangle}$. Let $k \in {}^\beta M$, $i \in \beta$ and assume $(\forall\Sigma_0 \subseteq_\omega \Sigma)\mathfrak{M} \models \exists v_i \bigwedge\Sigma_0[k]$. Let $\Sigma^+ \overset{\text{def}}{=} \{\{k_j\}(v_j) : j \in \beta \text{and} j \neq i\} \cup \Sigma$. Then $(\forall\Gamma \subseteq_\omega \Sigma^+)\Gamma$ is satisfiable in $\mathfrak{M}^+ = \langle\mathfrak{M}, \{k_j\}\rangle_{j\in\beta}$. Indeed, letting $\Sigma_0 = \Gamma \cap \Sigma$, there is $a \in M$ with $\mathfrak{M} \models \bigwedge\Sigma_0[k(i/a)]$ and then $\mathfrak{M} \models \Gamma[k(i/a)]$, too. Then by κ–valuation–compactness of $\mathfrak{M}^+$ we have $\mathfrak{M}^+ \models \bigwedge \Sigma^+[g]$ for some $g \in {}^\beta M$. By the definition of Σ^+ then $(\forall j \neq i) g_j = k_i$. Hence, $\mathfrak{M} \models \bigwedge \Sigma[k(i/g_i)]$ proving $\mathfrak{M} \models \exists v_i \bigwedge \Sigma[k]$. Thus $\mathfrak{M}$ is κ–saturated. **QED(Proposition 30)**

DEFINITION 31. A Cs_α $\mathfrak{A}$ is said to be κ–hereditarily compact iff $(\forall Y \subseteq_\kappa base(\mathfrak{A}))\mathfrak{A}_Y$ is κ–compact. ∎

COROLLARY 32. *Let $\kappa \geq \omega$ and $\mathfrak{A}$ a Cs_α. Assume, $\mathfrak{A}$ is generated by regular elements in $Nr_\gamma\mathfrak{A}$ for some $\gamma < \kappa$. Then (i), (ii) below hold.*

(i) *$\mathfrak{A}$ is κ-hereditarily compact* $\not\Longleftrightarrow$
[$\mathfrak{A}$ is κ-saturating and κ-compact].

(ii) *$\mathfrak{A}$ is ω^+-hereditarily compact* $\not\Longleftarrow$
[$\mathfrak{A}$ is κ-saturating and κ-compact].

Proof: uses the above proposition, and otherwise is similar to that of our Corollary 27. ∎

FACT 33. *Let $\mathfrak{M}$ be ordinary and $\omega \leq \kappa \leq |\alpha|^+$. Then*

$$\mathfrak{M} \text{ is } \kappa\text{-saturated} \iff \mathfrak{Cs}_\alpha^{\mathfrak{M}} \text{ is } \kappa\text{-hereditarily compact.} \quad ∎$$

FACT 34. *Let $\mathfrak{A} \in Cs_\alpha^{reg} \cap Lf_\alpha$ and $\omega \leq \kappa \leq |\alpha|^+$. Then*

$$\mathfrak{A} \text{ is } \kappa\text{-saturating} \iff \mathfrak{A} \text{ is } \kappa\text{-hereditarily compact.} \quad ∎$$

We note that

$$\kappa\text{-hereditarily compact} \implies \kappa^+\text{-compact}$$

does *not* generalize from $Cs_\alpha^{reg} \cap Lf_\alpha$ to $Cs_\alpha \cap Lf_\alpha$.

By Corollary 32 above, $|\alpha|^+$–hereditary compactness is a strictly stronger property of Cs_α's than $|\alpha|^+$–saturatedness and compactness together (although they agree on $Cs_\alpha^{reg} \cap Lf_\alpha$). It would be nice to know more about these algebraic counterparts (more precisely, generalized counterparts) of κ–saturatedness of models. For example, it would be nice to know if $\mathfrak{Cs}_\alpha^{\mathfrak{M}}$'s (or $\mathfrak{M}$'s) being κ–hereditarily compact implies κ–universality of $\mathfrak{M}$ (in the usual model theoretic sense) under some cardinality conditions (e.g. $\omega < \kappa$, $\mathfrak{M}$ of language $\langle \gamma, I, \rho \rangle$ with $\gamma < \kappa$, $|\alpha \smallsetminus \gamma| \geq |\gamma|$ and then we look at $\mathfrak{Cs}_\alpha^{\mathfrak{M}}$). Or perhaps the result would not be κ–universality but only universality w.r.t. a smaller cardinal, e.g. δ–universality for any δ with $\delta^{|\gamma|} < \kappa$, and only under some additional conditions (compare the conditions of the algebraically generalized downward Löwenheim–Skolem theorem in [8] and in Sain [19]). After having completed this paper, we heared that Serény has important new results saying that $|\alpha|^+$–hereditarily compact Cs's have nice properties pointing in the direction that κ–hereditary compactness might be a promising candidate for being the algebraic counterpart of κ–saturatedness of models. This direction seems to be worthy of further investigation.

In the other direction we quote from Serény [23], [21] p. 35 Statement ($\star$), that if $|M| \geq \omega$ and Λ is full then $\mathfrak{M}$ does have a κ–universal elementary extension. (The proof uses the proof method of 3.1.106 [8].) We conjecture that the condition requiring Λ to be full can be substantially weakened.

Investigations similar to the ones concerning "saturatedness" in this section can be pursued in connection with κ–compact Cs's and κ–universal, κ–valuation–compact infinitary models. Many interesting things have been done in this direction by G. Serény, his results are available in Serény[21–23]. In passing we note that very interesting connections between model theoretic definability theory and algebraic logic are found in Sain [20].

Throughout this paper we ignored a very important and active (see, for example [25]) branch of cylindric algebraic logic, namely the case of $\alpha < \omega$. This amounts to investigating first–order logic with finitely many variables. The above quoted papers of Sain and Serény investigate the $\alpha < \omega$ case too. In passing, we note that *quasi–polyadic* algebraic model theory (cf. p. 266 in §5.6 of [8] and §V.7 (pp. 120–) of [6]) is almost the same as the one outlined herein. Especially, the results concerning variable rich languages are practically the same, cf. [1]. An obvious source of difference is that equality can be dropped in the quasi–polyadic case but not here.

REMARK 35. (On terminology) In several publications ([1], Sain [18], [19], p. 311 of [9] etc.) our presently investigated logics were denoted by ${}_cL^t_F$. The translation goes as follows. Let $\Lambda = \langle \alpha, I, \rho \rangle$. Then the logic $\langle \Phi\mu^\Lambda_r, \{\, \mathfrak{M} : \mathfrak{M} \text{ is a model of } \Lambda \,\}, \models \rangle$ is called in those publications "${}_cL^\rho_F$ with α many variables".

Summing up, in ${}_cL^t_F$, t is basically the same as our Λ except that it does not contain α, hence α has to be indicated separately. ■

PROBLEM 1. *What are the Löwenheim–Skolem–Tarski numbers for our "infinitary" logics? That is, let $\alpha \geq \omega$ be fixed. Then*

(i) *For which cardinals κ, λ is it true that for every $\Lambda = \langle \alpha \ldots \rangle$, every model of Λ with cardinality κ is elementarily equivalent with a model of cardinality λ? (Cf. Problem 3.1 of [8].)*

(ii) *Let $\Lambda = \langle \alpha, I, \rho \rangle$, $\mathfrak{M}$ a model of Λ. Assume $|I| \leq \kappa$ and $\kappa \geq 2^{|M \cup \alpha|}$. Does then $\mathfrak{M}$ have an elementary extension of cardinality κ?*

(iii) *What is the answer to Problem 11 of Sain [19]? As mentioned already, the logic denoted there by ${}_cL^t_F$ is the same as our present one with its set of formulas $\Phi\mu^\Lambda$ where Λ is the same similarity type as t.* ■

PROBLEM 2. *Do our "infinitary" logics, e.g. ${}_cL^t_F$ have the weak Beth definability property? For terminology see Sain [20]. By [20], the usual Beth property fails for these logics.*

Acknowledgments: I am grateful to the referee for substantial improvement of the paper.

This work has been supported by the Hungarian National Foundation for Scientific Research Grant No 1810.

Thanks are due to I. Sain and G. Serény for their very extensive help in preparing this paper. Many of the ideas and results herein are due to them. I am also grateful for their making available their unpublished manuscripts in this area for me.

References

[1]. H. Andréka–T. Gergely–I. Németi, *On universal algebraic construction of logics*, Studia Logica Vol **36** (1977), 9–47.

[2]. H. Andréka–I. Németi, *Dimension complemented and locally finite dimensional cylindric algebras are elementarily equivalent*, Algebra Universalis **13** (1981), 157–163.

[3]. H. Andréka–I. Németi, *Connections between cylindric algebras and logic*, Manuscript (1984).

[4]. H. Andréka–R.J. Thompson, *A Stone-type representation theorem for algebras of relations of higher rank*, Trans. Amer. Math. Soc., **309** 2 (1988), 671–682.

[5]. C.C. Chang–H.J. Keisler, "Model Theory," North–Holland, 1973.

[6]. P. R. Halmos, "Algebraic Logic," Chelsea Publ. Co., New York, 1962.

[7]. L. Henkin, *The representation theorem for cylindric algebras*, in "Mathematical interpretation of formal systems," North–Holland, Amsterdam, 1955, pp. 85–97.

[8]. L. Henkin–J.D. Monk–A. Tarski, "Cylindric Algebras Part I and Part II," North–Holland, 1971, 1985.

[9]. L. Henkin–J.D. Monk–A. Tarski–H. Andréka–I. Németi, "Cylindric Set Algebras," Lecture Notes in Mathematics Vol 883, Springer–Verlag, 1981.

[10]. L. Henkin–A. Tarski, *Cylindric Algebras*, in "Lattice Theory, Proceedings of symposia in pure mathematics Vol 883," Springer–Verlag, 1981.

[11]. S. Jaskowski, *Sur les variables propositionelles dependantes*, Studia Soc. Sci. Torunensis Sec. A. **1** (1948), 17–21.

[12]. B. Jónsson–A. Tarski, *Boolean algebras with operators Part I*, Amer. J. Math. **73** (1951), 891–939.

[13]. H.J. Keisler, *A complete first-order logic with infinitary predicates*, Fundamenta Mathematicae Vol **52** (1963), 177–203.

[14]. I. Németi, *Cylindric-relativized set algebras have strong amalgamation*, The Journal of Symbolic Logic **50** 3 (Sept. 1985), 689–700.

[15]. I. Németi, *The equational theory of cylindric-relativized set algebras is decidable*, Preprint No 63/1985, Math. Inst. Budapest (Sept. 1985). see also [16]

[16]. I. Németi, ***Free algebras and decidability in algebraic logic***, Dissertation for "Habilitation" or DSc with the Hungarian Academy of Sciences, Budapest (1985).

[17]. D. Resek–R.J. Thompson, ***An equational characterization of SCr_α***. Submitted, 45pp

[18]. I. Sain, ***Finitary logics of infinitary structures are compact***, Abstracts of Amer. Math. Soc **3** 3 (April 1982), p. 252.

[19]. I. Sain, ***Cylindric algebra versions of the downward Löwenheim–Skolem theorem***, Notre Dame J. of Formal Logic **29** 3 (1988), 332–344.

[20]. I. Sain, ***Beth's and Craig's properties via epimorphisms and amalgamation in algebraic logic***, This volume (1988).

[21]. G. Serény, ***Compact cylindric set algebras***, (In Hungarian) Dissertation with Eötvös Loránd University, Budapest (1986).

[22]. G. Serény, ***Compact cylindric set algebras***, Bulletin of the Section of Logic (Warsaw–Lodz) **14** 2 (June 1985), 57–64.

[23]. G. Serény, ***Compact cylindric algebras and their isomorphisms.*** Manuscript, 1989

[24]. S. Shelah, ***Isomorphic but not base–isomorphic base–minimal cylindric algebras***, Algebraic Logic, Colloq. Math. Soc. J. Bolyai, North–Holland (1988) (to appear).

[25]. A.Tarski–S.Givant, "A formalization of set theory without variables," Amer. Math. Soc. Colloquium Publications, 1987.

Dynamic Algebras as a Well-Behaved Fragment of Relation Algebras

Vaughan Pratt[1]

ABSTRACT The varieties **RA** of relation algebras and **DA** of dynamic algebras are similar with regard to definitional capacity, admitting essentially the same equational definitions of converse and star. They differ with regard to completeness and decidability. The **RA** definitions that are incomplete with respect to representable relation algebras, when expressed in their **DA** form are complete with respect to representable dynamic algebras. Moreover, whereas the theory of **RA** is undecidable, that of **DA** is decidable in exponential time. These results follow from representability of the free intensional dynamic algebras.

1 Introduction

1.1 Overview

Binary relations have proved a fruitful framework in both logic and computer science. In logic they have served as the eliminator of variables [TG87], and in computer science as the illuminator of software [dBdR72, Pra76]. One finds the algebraic versions of these topics today under the respective rubrics of relation algebra [TG87] and dynamic algebra [Koz79c, Pra79a]. The nonalgebraic origins of the former lie in the subject of foundations of mathematics, and of the latter in that of logics of programs [KT89].

When the organizers of this conference very kindly asked me to talk on a subject of my choice it seemed a foregone conclusion that a conference organized by relation algebraists would expect a talk on dynamic algebras. Not having worked in this area since 1981 however, I would have preferred to talk about my more recent work on concurrent behavior. With dynamic algebras already well covered at the conference by Dexter Kozen, I hoped at the start of the meeting that this might be possible. The day before my talk I took an informal poll, whose outcome determined the topic of my talk and ultimately the unexpected results of this paper.

At the time I knew little of either the results or the history of relation algebra. I was vaguely aware that Tarski had shown the equational theory of representable relation algebras to be undecidable in the 1940's, and that the equational theory of that class was not finitely based, but I was unaware

[1]Dept. of Computer Science, Stanford, CA 94305

of the larger finitely axiomatized variety **RA**.

On learning more of the background of relation algebras from various helpful sources, especially George McNulty and Roger Maddux, it occurred to me that it would be a nice idea to organize this paper as a comparison of the merits of relation and dynamic algebras. It also seemed a good idea to build up these notions from Boolean monoids and Boolean modules [Bri81] respectively, with the former mingling logical and relative notions in a single sort and the latter keeping them segregated.

My initial impression was that modules improved on monoids in the areas of definitional force, completeness, representability, and decidability, while monoids had the advantages of homogeneity of sort and expressive power. Both form finitely based varieties.

In the course of making the case for these claims I learned to my surprise that the dynamic algebra definitions of converse and star were equally effective, suitably modified, in a Boolean monoid, that is, an ordered monoid whose partial order is a Boolean algebra. What had misled me about the suitability of Boolean monoids as a medium for defining converse was that every published equational axiomatization of **RA** had four equations mentioning converse, and that the equational theory of the representable relation algebras was not finitely axiomatizable, making the axiomatization appear an ad hoc attempt to deal with an impossible situation. Moreover since none of these axiomatizations mentioned star I assumed that a satisfactory axiomatization of star must be similarly out of the question. I only recently learned of the Ng-Tarski equations for star [NT77,Ng84].

As it turns out, in a Boolean monoid converse and star can each be defined with a single equation. Each equation abstracts the essence of the dynamic algebra definition of that operation. The equation for star does not mention converse (unlike the three-equation Ng-Tarski axiomatization), and vice versa.

In this account of **RA**, converse and star become siblings, as they are in **DA**. In fact, although converse appeared in the first dynamic logic paper four pages ahead of star [Pra76], the relative importance of star to programmers had caused converse to temporarily disappear from dynamic logic by the time of Segerberg's axiomatization of propositional dynamic logic [Seg77]. It was restored and straightforwardly axiomatized the following year by Parikh [Par78], whose axioms have here become our single-equation definition of **RA** converse.

The surviving advantages of **DA** remain those of decidability of the equational theory, and representability, as per Kozen [Koz80] and Brink [Bri81] but extended to star at least for the free algebras [Pra79a,Pra80a], from which follows equational completeness of **DA** with respect to the Kripke model.

I attribute these advantages to the "maintenance of a suitable distance" between the Boolean and monoidal sorts. Too far apart (complete independence of the two sorts) and one is left with a Boolean algebra and a monoid, in neither of which can either converse or star be defined equationally. **RA** represents the other extreme, in which the two sorts are identified. The Boolean module organization of **DA** keeps the sorts distinct but lets them communicate via operations diamond $\diamond : K \times B \to B$ and test $? : B \to K$. This is close enough to permit the equational definitions of star and converse, yet not so close as to compromise representability, completeness, or decidability of the associated equational theory.

For chronological completeness let me mention here the work of my group during the past several years on concurrent processes. The direction this work has taken has been heavily influenced by the insights of both dynamic logic and dynamic algebra. The passage from relation algebras to process algebras may be described as inverse abstraction (back to relations as sets of pairs), two generalizations (from pairs as labeled linearly ordered doubletons to labeled partial orders [Pra86], and from partial orders to generalized metrics [CCMP89]), and abstraction back to algebra, with the resulting logic having models far removed from binary relations, yet remaining remarkably like **RA**. The Boolean algebra is relaxed to a lattice and the monoid becomes commutative (it now represents collision instead of composition). Converse and star remain meaningful but lack some of the vitality they show in **RA** and **DA**.

1.2 Conventions

A *variety* is the class of models of an equational theory; equivalently, a class closed under homomorphisms, subalgebras, and direct products (HSP closure). A *universal Horn* formula has the form $s_1{=}t_1$ & ... & $s_n{=}t_n \to s{=}t$. A *quasivariety* is the class of models of a universal Horn theory; equivalently the closure ISP(K) of a pseudoelementary class K under isomorphisms, subalgebras, and direct products. We supply more details about "pseudoelementary" in conjunction with the examples of Boolean monoids.

Given a class C of similar algebras we denote by *type* C the similarity type to which it belongs, by Θ_C its equational theory, and by Φ_C its universal Horn theory. All theories are presumed to have infinitely many variables, for definiteness countably many.

We denote reducts to type A by superscript A. For example a dynamic algebra $\mathcal{D} = (\mathcal{B}, \mathcal{K}, \diamond, ?)$, of type **DA**, has a reduct $\mathcal{B} = (B, \vee, 0, ^-)$ of type $\mathbf{DA}^B$, and another $\mathcal{K} = (K, 0, +, ;, ^\smile, ^*)$, of type $\mathbf{DA}^K$. The equational theory Θ_{RA} of relation algebras has a reduct Θ_{RA}^{BM} to Boolean monoids, defined as those theorems of **RA** that are in the language of **BM**, i.e. that omit converse. Thus Θ_A must by definition be a conservative extension of Θ_A^B, whence to say that Θ_{RA}, as an extension of Θ_{BM}, is not conservative over Θ_{BM} is to say that Θ_{RA}^{BM} strictly includes Θ_{BM}.

The equational theory Θ_{DA} of the two-sorted class **DA** of dynamic algebras partitions as $\Theta_{DA} = \Theta_{DA}^B + \Theta_{DA}^K$ consisting respectively of equations between Boolean terms and between Kleenean terms. On the other hand, although Φ_{DA} has reducts Φ^B_{DA} and Φ^K_{DA}, these do not exhaust Φ_{DA}, which may have equations of both sorts in the one formula. The sets V^B and V^K of variables of each sort are disjoint. We reserve the sort names themselves for the underlying sets of each sort, assumed disjoint; thus the set of individuals of a dynamic algebra $\mathcal{D}$ partitions as $D = B + K$.

1.3 Logical and Relative

The two theories of this paper, those of relation algebras and dynamic algebras, arise out of the following dilemma.

An n-ary relation on a set X is a subset of X^n. If p is a unary relation we write $x \in p$ as $p(x)$, while for a binary relation a we write $(x, y) \in a$ as xay.

That a relation is a set imbues it with a *logical* character: for any n the set 2^{X^n} of all n-ary relations on X forms a Boolean algebra in the usual way. That pairs can be linked, xay and ybz forming $xa;bz$, and reversed, xay becoming $ya^{\smile}x$, confers on binary relations their *relative* character: the binary relations on X form a monoid under composition or relative product $a;b$ with identity 1', with the additional operations of converse, $a^{\smile}$, and star, a^*.

A calculus with both a logical and a relative character would seem very useful for both the foundations of mathematics [TG87] and logics of programs [KT89]. Since relations exhibit both characteristics they should form an excellent basis for such a calculus.

It is clear that unary relations cannot supply the relative part. However either unary or binary relations can supply the logical part. Depending on which way we resolve this dilemma we obtain one of two finitely axiomatized varieties, Boolean monoids or Boolean modules [Bri81]. Each may be equipped independently with converse and star, each definable with a single equation in each case. A Boolean monoid equipped with converse is called a relation algebra, and one equipped with converse and star is called an **RAT** [NT77,Ng84], a relation algebra with transitive closure. Brink [Bri81] defines Boolean modules to be equipped with converse. A Boolean module with converse and star is a dynamic algebra [Koz79a,Pra79a].

2 Boolean Monoids

As pointed out by Brink [Bri81], Boolean monoids predate Boolean modules by a decade, 1860 (De Morgan) vs. 1870 (Peirce), so it is appropriate that we consider them first. They achieve economy of concept through homogeneity of data: a single sort as opposed to the two sorts of Boolean modules. Our purpose in treating Boolean monoids in this paper is twofold: to prime the reader already acquainted with **RA** for the perspective from which we shall view **DA**; and to embellish the standard **RA** story with some details, some imported from **DA**, some resulting from extensive discussions with Roger Maddux and George McNulty, and some just filling gaps.

An *ordered monoid* $\mathcal{A} = (A, \leq, ;)$ is a set A which is both a partial order $(A, \leq)$ and a monoid $(A, ;)$ whose *composition* or *relative product* $a;b$ is monotone with respect to $\leq$ in each argument. We denote the unit of the monoid by 1'. A *Boolean monoid* is an ordered monoid whose partial order is a Boolean algebra. We denote join and complement by $a+b$ and a^- respectively. To make the class of Boolean monoids a variety we treat a Boolean monoid as an algebra $(A, +, ^-, ;, 1')$. The class **BM** of Boolean monoids is then definable by finitely many equations.

We take as logical abbreviations $a \leq b$ for $a + b = b$, $a \leq b \leq c$ for $a+b = bc$, ab for $(a^- + b^-)^-$, 1 for $1' + 1'^-$, 0 for 1^-, and $a \rightarrow b$ for $a^- + b$. For relative abbreviations we take $a \dagger b$ for $(a^-; b^-)^-$ (relative sum), 0' for $1'^-$, $a \hookrightarrow b$ for $a^- \dagger b$ (relative implication), and $a \hookleftarrow b$ for $a \dagger b^-$ (relative coimplication).[2]

Example 1. The motivating example of a Boolean monoid is the set of all binary relations on a *base set* X, under the usual composition of binary relations. We refer to this algebra as **BM** X. **BM** X is a *simple* algebra, one with only two congruences, the identity and the clique. We refer to those Boolean monoids isomorphic to a subalgebra of **BM** X for some X as the *simple representable Boolean monoids*, forming the class **SRBM**.

A *pseudoelementary class* is a reduct of an elementary class [Mak64, Ekl77, 4.3]. That is, it is obtainable by omitting some of the sorts and operations of some class definable with a first order theory.

Proposition 1 SRBM *is pseudoelementary.*

Proof: Let X and R be the sorts of the following two-sorted first-order theory of the set R of all binary relations over a set X. The language of this theory has all the Boolean monoid operations, as operations on R, together with a ternary relation (x, a, y) expressing that elements x and y of X are related by the binary relation a, an element of R. We require that if $a \neq b$ then there exist x, y such that exactly one of (x, a, y) and (x, b, y) hold; this ensures that each element of R acts like a distinct binary

[2] De Morgan [DM64] writes ab, ab', and $a{,}b$ for $a;b$, $a \hookleftarrow b$, and $a \hookrightarrow b$ respectively, construing them as "an a of a b of", "an a of every b of", and "an a of none but b's of," and asserting their sufficiency. Peirce [Pei33, 3.242,1880] observed their interdefinability, notating them ab, a^b, ${}^a b$ and adjoining a fourth connective for relative sum, which he subsequently notated $a\dagger b$; the notation $a \dagger b$ we use here is due to Schröder [Sch95], subsequently (1897) adopted in modified form (the "scorpion tail") by Peirce. For $0, 1, 0', 1'$ Peirce writes $0, \infty, \mathbf{n}, 1$ respectively, and calls relations $a \leq 1'$, $a \geq 0'$, $a \geq 1'$, and $a \leq 0'$ respectively *concurrent*, *opponent*, *self-relative*, and *alio-relative*.

relation, i.e. a distinct subset of X^2. It is easy to write down axioms that say that R includes the identity relation (there exists a such that for all x, y, (x, a, y) if and only if $x = y$) and the empty relation (there exists a such that for all x, y, $\neg(x, a, y)$), and is closed under the Boolean operations (with complement being relative to X^2) as well as composition. Now by "forgetting" the set X and the ternary relation (x, a, y) we are left with a one-sorted relation algebra isomorphic to a subalgebra of **BM** X, that is, a simple representable Boolean monoid. ∎

Corollary 2 *ISP(***SRBM***) is a quasivariety.*

Proof: Every pseudoelementary class closed under isomorphisms, subalgebras, and direct products is a quasivariety, and these operations preserve the property of being pseudoelementary. ∎

I am indebted to George McNulty for the idea of using pseudoelementary classes in this argument.

We call the members of ISP(**SRBM**) *representable Boolean monoids*, forming the class **RBM**. Up to isomorphism the representable Boolean monoids are formed as subalgebras of direct products of simple representable ones. The top element 1 of the direct product of simple representable Boolean monoids amounts to an equivalence relation on the disjoint union of the base sets of those Boolean monoids, namely the relation $x \equiv y$ which holds just for those x, y coming from the same base set.

Example 2. In any Boolean algebra, take composition to be meet, and hence relative sum to be join, called a *cartesian* Boolean monoid since it is also a cartesian closed category [Mac71, IV.6-1(b)]. The same concept is termed "Boolean" in the relation algebra literature, e.g. Jónsson [Jon82], which presents the obvious conflict with the present terminology.

A Boolean monoid is called *normal* when it satisfies $a; 0 = 0 = 0; a$, and *additive* when it satisfies $a; (b+c) = a; b + a; c$ and $(a+b); c = a; c + b; c$. We shall call it *Peircean* when it satisfies $(a \dotplus b); c \leq a \dotplus (b; c)$ and $a; (b \dotplus c) \leq (a; b) \dotplus c$ [Pei33, 3.334]. Examples 1 and 2 enjoy all three of these properties.

Example 3. The *dual* of a Boolean monoid is obtained by exchanging relative product and relative sum, itself a Boolean monoid.

The duals of examples 1 and 2 are not in general normal, additive, or Peircean.

Example 4. The set of all subsets of any equivalence relation E on a set X is closed under the usual composition of binary relations and hence forms a Boolean monoid. We have already encountered these above as the representable Boolean monoids up to isomorphism.

When E is the identity relation on X the resulting representable Boolean monoid can be seen to be cartesian. Conversely every cartesian Boolean monoid is so representable via the Stone embedding of a Boolean algebra into a power set and the embedding of X in X^2 as the latter's diagonal.

We now treat the operations of converse and star in turn.

2.1 RESIDUATION

An ordered monoid is called a *residuated order* when it has operations $a\backslash c$ and c/b defined as

$$a;b \leq c \quad \leftrightarrow \quad b \leq a\backslash c \tag{RKR}$$

$$a;b \leq c \quad \leftrightarrow \quad a \leq c/b \tag{LKL}$$

These operations are called respectively the right and left *residuals* of c *over* a, b respectively [WD39,Dil39,Fuc63,Bir67,Jon82]. They may also be viewed as yet another pair of implications, bringing the number of implications we have now encountered to five, namely $a \rightarrow b$, $a \hookrightarrow b$, $a \hookleftarrow b$, $a\backslash b$, and a/b. In the case of commutative monoids, where $a;b = b;a$, this reduces via $a \hookrightarrow b = b \hookleftarrow a$ and $a\backslash b = b/a$ to just three implications.

We will find it convenient later on to break down (RKR) and (LKL) into four universal Horn formulas,

$$a;b \leq c \quad \rightarrow \quad b \leq a\backslash c \tag{KR}$$

$$b \leq a\backslash c \quad \rightarrow \quad a;b \leq c \tag{RK}$$

$$a;b \leq c \quad \rightarrow \quad a \leq c/b \tag{KL}$$

$$a \leq c/b \quad \rightarrow \quad a;b \leq c. \tag{LK}$$

The letter K in the names of these formulas connotes the theorem De Morgan refers to as Theorem K [DM64]. This theorem, which was brought to my attention by Roger Maddux, amounts to the assertion of KR and KL; RK and LK can be derived from these given $a^{\smile\smile} = a$. The L and R refer to the left and right residuals respectively, and their placement relative to K indicates the direction of the implication.

Residuated orders enjoy a number of useful properties [Bir67, Theorem XIV-4], all of which are easily proved. In particular $a;b$ preserves arbitrary sups in each argument, e.g. if the empty sup or least element 0 exists then $a;0 = 0 = 0;a$, and if the sup $a+b$ of a and b exists then $(a+b);c = a;c+b;c$ and $c;(a+b) = c;a+c;b$. Residuation of a over b (on either side) is monotone in a and antimonotone in b. Furthermore $a\backslash b$ (and likewise b/a) preserves arbitrary infs in the b argument, e.g. $a\backslash 1 = 1$ and $a\backslash(bc) = (a\backslash b)(a\backslash c)$. It also "antipreserves" arbitrary sups in the a argument in that it maps them to the corresponding infs, e.g. $0\backslash a = 1$ and $(a+b)\backslash c = (a\backslash c)(b\backslash c)$. And residuation is axiomatizable inequationally, namely via

$$a;(a\backslash b) \quad \leq \quad b \tag{rK}$$

$$b \quad \leq \quad a\backslash(a;b) \tag{Kr}$$

$$(b/a);a \quad \leq \quad b \tag{lK}$$

$$b \le (b;a)/a. \tag{Kl}$$

In **BM** each of these inequalities is equivalent to the universal Horn formula with the corresponding upper case identifier: (rK) to (RK), etc.

We say that a is *reflexive* when $1' \le a$, and *transitive* when $a;a \le a$. These properties are of interest in their own right, but are of particular interest in the following section on star.

Proposition 3 *In a residuated order, $a\backslash a$ and a/a are each both reflexive and transitive.*

Proof: We show this just for $a\backslash a$. Evidently $a;1' \le a$, whence $1' \le a\backslash a$, showing reflexivity. Now

$$\begin{array}{rcl} a\backslash a & \le & a\backslash a \\ \text{Hence } a;(a\backslash a) & \le & a \\ \text{so } a;(a\backslash a);(a\backslash a) & \le & a;(a\backslash a) \\ & \le & a \\ \text{Thus } (a\backslash a);(a\backslash a) & \le & a\backslash a \end{array}$$

∎

When an ordered monoid has the structure of a lattice or a Boolean algebra, the corresponding residuated order is called a residuated lattice or residuated Boolean algebra respectively. A residuated cartesian lattice ($a;b = ab$), often assumed also to have a least element 0, is called a *Heyting algebra*.

3 Converse

A *relation algebra* (RA) is a residuated Boolean algebra with a unary operation called *converse*, notated $a^{\smile}$, satisfying $a\backslash b = (a^{\smile};b^{-})^{-}$ and $a/b = (a^{-};b^{\smile})^{-}$. This definition of the class **RA** is the content of Theorem 2.2 of Chin and Tarski [CT51], which together with the axioms for a Boolean monoid suffice to axiomatize **RA**.[3]

It is customary in giving equational axiomatizations of **RA** to mention converse in at least four equations, which invariably include

$$a^{\smile\smile} = a$$

[3]Monotonicity of $a;b$ in a and b is implied by (LKL) and (RKR) respectively. Thus a complete axiomatization of **RA** need consist just of these two plus the axioms for a Boolean algebra and for a monoid, omitting monotonicity. The inequational versions do not imply monotonicity and hence must accompany the full theory of **BM**.

$$(a;b)^\smile = b^\smile;a^\smile$$
$$(a+b)^\smile = a^\smile + b^\smile$$

along with (any) one of (Kr), (Kl), (rK), or (lK). For a change of pace here is a one-equation definition of converse.

Proposition 4 *The equations for a Boolean monoid, together with the equation*

$$((b^-;a^\smile)^-;a) + b = b(a^\smile;(a;b)^-)^-,$$

constitute a complete equational axiomatization of **RA**.

This equation can thus be regarded as an equational definition of converse.

Proof: Examination reveals this equation just to be (lK) and (Kr) combined via the equivalence $x \le y \le z$ iff $x + y = yz$. The essential novelty here is that just two of the four inequalities (Kr), (rK), (Kl), and (lK) suffice to completely axiomatize converse and hence to prove the other two inequalities.

From (lK) and (Kr) we can obtain $a^{\smile\smile} = a$, which we decompose as follows.

$$a^{\smile\smile} \le a \qquad \text{(I)}$$
$$a \le a^{\smile\smile} \qquad \text{(C)}$$

We obtain (I) from (Kr) thus.

$$\begin{array}{rcll} b & \le & (a^\smile;(a;b)^-)^- & \text{(Kr)} \\ \text{Hence } 1' & \le & (a^\smile;a^-)^- & \text{(setting } b = 1') \\ \text{so } a^{\smile\smile} & \le & a^{\smile\smile};(a^\smile;a^-)^- & \\ & \le & a & \text{(Kr)} \end{array}$$

We obtain (C) from (lK) as follows.

$$\begin{array}{rcll} (b^-;a^\smile)^-;a & \le & b & \text{(lK)} \\ \text{Hence } a^{\smile-};a & \le & 0' & (b \to 0') \\ \text{so } a^{\smile\smile-};a^\smile & \le & 0' & (a \to a^\smile) \\ \text{whence } 1' & \le & (a^{\smile\smile-};a^\smile)^- & \text{(contrapos.)} \\ \text{Therefore } a & \le & (a^{\smile\smile-};a^\smile)^-;a & \\ & \le & a^{\smile\smile} & \text{(lK)} \end{array}$$

Substituting $a^\smile$ for a in Kr, taking the contrapositive, and using $a^{\smile\smile} = a$ then yields rK. Kl is obtained similarly from lK.

Finally we obtain KR from Kr. Assume $a;b \le c$. Then $b \le a\backslash(a;b) \le a\backslash c$. Similarly KL follows from Kl, while RK follows from rK, and LK from lK. But by Theorem 2.2 of Chin and Tarski [CT51] this and the equations for Boolean monoids constitute a complete axiomatization of **RA**. ∎

The relationships we have obtained are summarized by the following lattice of inclusions between varieties of Boolean monoids. Each variety is labeled with the list of equations defining it, with BM implicit (so I denotes BM together with $a^{\smile\smile} \leq a$). Kr+rK is abbreviated to R (the theory of "right-handed" relation algebras) and Kl+lK to L (ditto for left handed), and Kl+Kr+lK+rK to RA. (Hence Theorem 2.2 [CT51] amounts to the statement that a relation algebra is a right-handed relation algebra that is also left handed.) Meet in this lattice corresponds to intersection of the corresponding varieties. I have not verified whether all 13 varieties are distinct, but I conjecture that they are.

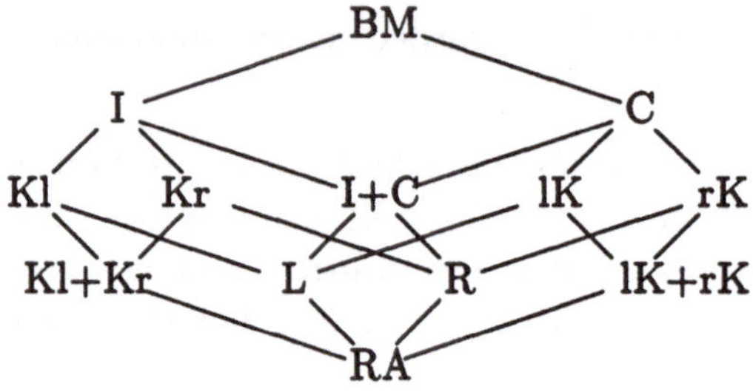

The equivalences (RKR) and (LKL) have from the very beginning been a staple of writings on the algebra of binary relations. Roger Maddux brought to my attention that these are given as a property of converse by all three of the major 19th century writers on the algebra of binary relations, De Morgan [DM64, Theorem K][4], Peirce [Pei33, 3.147(170)], and Schröder [Sch95, §17,2)]. In the form

$$(c; b^{\smile})a = 0 \quad \equiv \quad (a; b)c = 0 \quad \equiv \quad (a^{\smile}; c)b = 0$$

it is taken as the defining characteristic of converse and hence of **RA** in Theorem 2.2 of Chin and Tarski and Definition 4.1 of Part II of Jónsson and Tarski [JT52], foreshadowed by Tarski [Tar41, XVI]. (An advantage of the latter form is that it is defined in any lower semilattice with a least element, rendering it applicable to considerably more general structures than Boolean algebras, e.g. Heyting algebras and relevant logics.)

We write **Rel** X for the result of equipping **BM** X with the standard operation of converse for binary relations. We write **Eqv** E for the relation algebra of all subsets of an equivalence relation E.

A *representable relation algebra* (RRA) is a representable Boolean monoid equipped with converse having its usual meaning for binary relations.

[4]De Morgan gives only KR and KL, but also was well aware of the involutary nature of converse, which as we have seen entails KR→RK and KL→LK. In nominating Peirce rather than De Morgan as the "creator of the modern theory of relations" Tarski [Tar41] appears not to have taken Theorem K into account. Peirce gave an equivalent equational characterization of converse in 1870, [Pei33], but was no better equipped than De Morgan to appreciate its completeness.

Equivalently it is a subalgebra of an algebra **Eqv** E. The class **RRA** of such forms a variety [Tar55].

3.1 Star

The operation star, or ancestral, or reflexive transitive closure, which we shall notate a^*, resembles converse in some respects. It is a unary operation definable in any ordered monoid. Whereas $a^\smile$ is a reversed, a^* is a iterated indefinitely. Under progressively stronger assumptions about the ordered monoid progressively more may be said about star. For residuated Boolean algebras there are a number of equivalent definitions of star, including more than one equational definition.

An *ordered monoid with star* is an ordered monoid $(A, \leq, ;, 1')$ such that for each $a \in A$ there exists an element of A, denoted a^*, such that

$$1' \leq a^* \tag{S1}$$

$$a; a^* \leq a^* \tag{S2}$$

$$a; b \leq b \;\rightarrow\; a^*; b \leq b \tag{S3}$$

We say that a *converges at* b when $a; b \leq b$. Thus (S2) asserts that a converges at a^*, and (S3) asserts that a^* converges wherever a does.

Proposition 5 *The following hold in any ordered monoid with star.*

$$a^*; a^* \leq a^*; \tag{Tr}$$

$$a \leq a^*; \tag{Gr}$$

$$1' \leq b \;\&\; b; b \leq b \;\&\; a \leq b \quad \rightarrow \quad a^* \leq b \tag{Cl}$$

Proof: Setting b to a^* in (S3) and using (S2) gives (Tr). (That is, a^* converges at itself.) By "multiplying" (S1) on the left by a we obtain $a \leq a; a^*$, which with (S2) yields (Gr). For (Cl), if $a \leq b$ and $b; b \leq b$ then $a; b \leq b$, whence by (S3) $a^*; b \leq b$. If moreover $1' \leq b$ then $a^* = a^*; 1' \leq a^*; b \leq b$, giving (Cl). ∎

We define the *reflexive transitive closure* of a to be the least reflexive transitive element greater or equal to a, when it exists.

Proposition 6 *In an ordered monoid with star, a^* is implicitly defined as the reflexive transitive closure of a.*

Proof: (S1) asserts that a^* is reflexive, (Tr) that it is transitive, and (Gr) that it is greater or equal to a. (Cl) asserts that it is the least such. There can be at most one least such, whence a^* is uniquely defined. ∎

Star, ancestral, or reflexive transitive closure, is variously notated a^* [Kle56], $^*a$, and a_0 [Ded01,Sch95], while transitive closure has been written a^+, a^ω [NT77], and a_{00} [Sch95]. The first-mentioned in each of these lists is the notation universally used in computer science and is adopted here. In any ordered monoid a^+ is definable in terms of a^* via the equation $a^+ = a; a^*$. In an ordered monoid in which the operation $a + 1'$ of reflexive closure is defined, and hence in a semilattice monoid, a^* is definable via either of the equations $a^* = a^+ + 1'$ or $a^* = (a + 1')^+$.

Star was first studied in detail by Schröder [Sch95], who notated it a_0 following Dedekind's 1888 notation [Ded01] for the "chain" of a "transformation" a. There Dedekind gave three axioms for chains, in paragraphs numbered respectively 45-47. Schröder translated Dedekind's axioms into the language of binary relations, numbering them (D45)-(D47), as follows.

$$b \le a^*; b \tag{D45}$$

$$a; a^*; b \le a^*; b \tag{D46}$$

$$a; c + b \le c \quad \rightarrow \quad a^*; b \le c \tag{D47}$$

Proposition 7 (S1) *is equipollent with* (D45), (S2) *with* (D46), *and* (S3) *with* (D47).

Proof: We may recover (S1)-(S3) from (D45)-(D47) respectively by setting b to 1' in (D45) and (D46), and setting c to b in (D47) and simplifying. In the other direction, (D45) and (D46) can be obtained by multiplying (S1) and (S2) respectively on the right by b. Substituting c for b in (S3) yields $a; c \le c \rightarrow a^*; c \le c$, which in turn implies $a; c + b \le c \rightarrow a^*; c \le c$. But $b \le c$ implies $a^*; b \le a^*; c$, yielding (D47). ■

Our definition of star is not equational because (S3) is not an equation, only a universal Horn formula. Ng and Tarski [NT77,Ng84] define the class **RAT** of relation algebras with star[5] and give a finite equational axiomatization of **RAT**. Their proof appeals to the transitivity of a/a and thus relies on converse. Here we show that the larger class of Boolean monoids with star also constitutes a variety. In the absence of converse we shall depend on complement, reformulating Segerberg's induction axiom for star in dynamic logic [Seg77] as an equational property of Boolean monoids.

Recall that normal means $a; 0 = 0 = 0; a$ and additive that $a; (b + c) = a; b + a; c$ and $(a + b); c = a; c + b; c$.

[5] Ng and Tarski treat transitive closure rather than reflexive transitive closure, whence the T in **RAT**, but as we have already noted these are interdefinable equationally, and thus the same variety is obtained whether the operation is taken to be star or transitive closure.

Proposition 8 *The class of normal additive Boolean monoids with star is a finitely axiomatized variety, with star defined by the equations* (S1), (S2), *and*

$$a^*;b \;\le\; b + a^*;((a;b)b^-). \tag{Ind}$$

Proof: (Ind→S3) This is where we use normality. Given $a;b \le b$ we wish to show $a^*;b \le b$. From the hypothesis we infer $(a;b)b^- = 0$, whence

$$\begin{aligned} a^*;b &\le b + a^*;((a;b)b^-) \\ &= b + a^*;0 \\ &= b. \end{aligned}$$

(S3→Ind) Here we depend on additivity. It suffices to show that substituting the right hand side of (Ind) for b in (S3) satisfies the hypothesis $a;b \le b$ of (S3), assured by the following calculation.

$$\begin{aligned} a;(b + a^*;((a;b)b^-)) &\le b + (((a;b)b^-) + a;a^*;((a;b)b^-)) \\ &\le b + a^*;((a;b)b^-) \qquad ((\text{S1}),(\text{S2})) \end{aligned}$$

∎

Equation (Ind) viewed suitably is relational induction. To see this take its contrapositive

$$b \wedge a^* \hookrightarrow (b \rightarrow a \hookrightarrow b) \le a^* \hookrightarrow b, \tag{Ind'}$$

writing $a \rightarrow b$ for $\neg a + b$ ("static" or material implication as opposed to "dynamic" implication $a \hookrightarrow b$). Read b as an induction hypothesis, $a^* \hookrightarrow$ as "after any number of a's", and $a \hookrightarrow$ as "after one a". Then (Ind') in English is "if b holds and after any number of a's the truth of b implies that b still holds after one more a, then b holds after any number of a's."

We may relate this to mathematical induction as follows. Take X to be the set N of natural numbers and take $a \subseteq N^2$ to be the successor relation, whence a^* is $\le_N$. Restrict attention to those b satisfying $b \le 1'$, allowing us to view b as an arbitrary predicate $b(x)$ on X defined as holding at x just when $(x,x) \in b$. (Ind') then asserts that if $b(x)$, and if for all $y \ge x$ $b(y)$ implies $b(y+1)$, then $b(y)$ for all $y \ge x$. Hence (Ind') holds just for "upwardly closed" b, namely those predicates b such that $b(x)$ and $x \le y$ implies $b(y)$. In particular $b(0)$ implies that b holds everywhere. This is precisely the content of mathematical induction.

Here we wrote $a \hookrightarrow$ and $a^* \hookrightarrow$ everywhere in preference to $a\backslash$ and $a^*\backslash$ so as to define star independently of converse. We could however have used $a\backslash$ in place of $a \hookrightarrow$: all occurrences of $a \hookrightarrow$ and $a^* \hookrightarrow$ in (Ind') can be made $a\backslash$ and $a^*\backslash$. This has the effect of taking the converse of a, computing its star, then taking the converse of the result, but the answer still comes out in the end to a^*. This observation may prove useful for nonclassical logics like **RA** but in which $a\backslash b$ is given as a primitive implication without a separate notion of converse, e.g. linear entailment in Girard's linear logic [Gir87].

As with converse a one-equation definition of star is possible: the reader may verify that

$$(1' + a^* + a^+); b \;=\; (a^*; b)(b + a^*; ((a; b)b^-))$$

is equipollent with S1-S3, where a^+ abbreviates $a; a^*$.

We have already remarked on the Ng-Tarski variety **RAT** of relation algebras with star (equivalently, with transitive closure). Relation algebras offer more opportunities to axiomatize star equationally than do Boolean monoids, as the five equivalences of the following proposition indicate.

Proposition 9 *For relation algebras with an operation a^* satisfying* (S1) *and* (S2), *the following five formulas are equivalent.*

$$a; b \leq b \;\rightarrow\; a^*; b \leq b \tag{S3}$$

$$a^*; b \;\leq\; b + a^*; ((a; b)b^-) \tag{Ind}$$

$$1' \leq b \;\&\; b; b \leq b \;\&\; a \leq b \quad\rightarrow\quad a^* \leq b \tag{Cl}$$

$$a^* \;\leq\; (a + b)^* \;\&\; (a\backslash a)^* \;=\; a\backslash a \tag{Ta}$$

$$a^* \;\leq\; (a + b)^* \;\&\; (a(a\backslash a))^+ \;=\; a(a\backslash a) \tag{Ng}$$

where a^+ abbreviates $a; a^*$.

Proof: (Ta) appears in [NT77] (in the equivalent form $a; (a^+ = a)$, and is attributed in [Ng84] to Tarski. (Ng) appears in [Ng84] (stated there for a/a in place of $a\backslash a$), where the equivalence of (Cl), (Ta), and (Ng) are treated. We showed the equivalence of (S3) and (Ind) above, and that (S3) implied (Cl), assuming (S1) and (S2).

It suffices therefore to show that (Ta) implies (S3). Assume $a; b \leq b$. Then $a \leq b\backslash b$, whence $a^* \leq (b\backslash b)^* = b\backslash b$ by respectively the first and second equations of (Ta). Hence $a^*; b \leq b$, verifying (S3). ∎

Hence the class of relation algebras with star can be shown to be a finitely axiomatized variety using (S1), (S2), and any one of (Ind), (Ta), or (Ng).

This proposition can be extended to residuated Boolean algebras in a straightforward manner. A key observation is that $a\backslash a$ is transitive in any residuated order, as noted in the section on residuation.

I do not know whether (S1)-(S2) and (Cl) implies (S3) in an arbitrary Boolean monoid, though I conjecture it does not. I further conjecture that the obvious "left-handed" version of (S3) is not equipollent with (S3), though with (S1)-(S2) it implies (Cl) for the same reasons as (S3) does, and has a corresponding left-handed version of (Ind).

One noteworthy asymmetry between converse and star is that, whereas converse *produces* the equations $a; 0 = 0$ and $a; (b + c) = a; b + a; c$, star *consumes* them, requiring $a; 0$ (normality) to define reflexive transitive closure via (Ind) and $a; (b + c) = a; b + a; c$ (additivity) in order that (Ind) define no more than reflexive transitive closure. Thus in this equational microeconomy, conservatively regulated by the direction of motion (back one for converse and forward many for star), supply meets demand. In the absence of converse however we must meet star's demand artificially by adding normality and additivity to the basis.

4 Dynamic Algebras

4.1 Boolean Modules

Whereas a Boolean monoid lumps the logical and relative aspects of the calculus of relations into a single sort, a Boolean module [Bri81] maintains the distinction. The resulting system is a little more complex, but has several advantages: free models that are representable, and an equational theory that is decidable and finitely axiomatizable yet complete for the representable modules.

The passage from Boolean monoid to relation algebra involves taking on the operations converse and star. We will set up Boolean modules and dynamic algebras along roughly the same lines, obtaining dynamic algebras from Boolean modules by adding some operations defined *relative to the Boolean module.* This approach differs from that of Kozen [Koz79b,Koz79a] and Brink [Bri81], who assume those operations are given as an algebra of scalars prior to the installation of that algebra in the module.

A Boolean module $(\mathcal{B}, K, \diamond)$ consists of a Boolean algebra $\mathcal{B} = (B, \vee, \neg)$ (following the customary notation used in the dynamic logic literature), a set K of names of operators, and an operation $\diamond : K \times B \to B$ such that $\diamond(a, p)$, also written $\langle a \rangle p$, is the result of applying to p the Boolean operator named by a. Following Jonsson and Tarski [JT48,JT51], a Boolean operator is a function $f : B \to B$ on a Boolean algebra such that $f(0) = 0$ and $f(p \vee q) = f(p) \vee f(q)$. We write $\langle a \rangle$ for the operator named by a, and write $\Diamond : K \to (B \to B)$ for the function defined by $\Diamond(a)(p) = \diamond(a, p)$.

4.2 Kleenean Algebras

The *type* of a Kleenean algebra $\mathcal{K} = (K, +, 0, ;, \breve{}, ^*)$ consists of five operations with respective arities 2,0,2,1,1 respectively. We abbreviate 0^* to λ, the Kleenean algebra notation for the unit of the monoid notated 1' in **RA**.

The motivating example is a Kleenean algebra of relations on a set X. This is an algebra $\mathcal{K} = \langle K, +, 0, ;, ^*, \breve{} \rangle$ where $K = 2^{X^2}$ and the operations have their standard interpretations for binary relations. Thus a Kleenean algebra of relations on X is obtained as a reduct of **Rel** X by dropping complement, then expanding by adding $*$ defined as reflexive transitive closure of binary relations. A *representable Kleenean algebra* is an algebra isomorphic to a subalgebra of a Kleenean algebra of relations. The delicate distinction that arose for Boolean monoids between the algebra **Rel** X of subsets of the complete relation X^2 and the algebra **Eqv** E of subsets of an arbitrary equivalence relation E does not come up here since we no longer have either intersection or a top element. Any representable relation algebra with star becomes a representable Kleenean algebra when complement is dropped.

Redko [Red64] has shown that the equational theory of the representable Kleenean algebras without converse is not finitely axiomatizable. Moreover Conway [Con71] has enumerated several *finite* models of this theory which do not satisfy axiom (S3) (given in the section on star for Boolean monoids) expressing that a^* is the least reflexive transitive element dominating a. By *Conway's Leap* I shall mean the 4-element Kleenean algebra in which $K = \{0,1,2,3\}$ with both $+$ and ; interpreted as numeric max except for $a;0 = 0 = 0;a$. Take $a^* = a + even(a)$, i.e. $0^* = 1^* = 1$, $2^* = 3^* = 3$. This satisfies the equational Kleenean theory for representable Kleenean algebras, yet $2^0 \vee 2^1 \vee 2^2 \vee \ldots 2^i = 2$ for $i \leq 3$ (and hence beyond) while $2^* = 3$, defeating S3. Hence if Kleenean algebras satisfy both the equational theory of representable Kleenean algebras and (S3) they do not form a variety.

The question then arises as to the appropriate definition of a Kleenean algebra. The previous paragraph notwithstanding, we shall define Kleenean algebras equationally, albeit in an indirect way that circumvents the above difficulties.

4.3 Dynamic Algebras

A dynamic algebra is a Boolean module $(\mathcal{B}, \mathcal{K}, \diamond)$ in which the set K has been expanded to a Kleenean algebra, satisfying the following five equations, one for each Kleenean operation.[6] In this approach the Kleenean algebra is defined as an integral part of a dynamic algebra, as opposed to being defined externally to a dynamic algebra, as is customary with algebras of scalars.

These equations are essentially the Segerberg axioms [Seg77] for Fischer and Ladner's propositional dynamic logic [FL79], translated into equational form. The differences from Segerberg's system are the equations for converse, which are due to Parikh [Par78], and the weakening to monotonicity of the equations that would have expressed normality and finite additivity, namely $\langle a \rangle 0 = 0$ and $\langle a \rangle (p \vee q) = \langle a \rangle p \vee \langle a \rangle q$, which can be recovered from the equation for converse.

We use the abbreviations $p \leq q$ for $p \vee q = q$, $p \wedge q$ for $\neg(\neg p \vee \neg q)$, and

[6] Strictly speaking converse is not a Kleenean operation. However it augments Kleene's four operations for regular expressions [Kle56,HU79] without any fuss.

$[a]p$ for $\neg(\langle a\rangle\neg p)$.

$$\langle 0\rangle p = 0 \tag{D1}$$

$$\langle a+b\rangle p = \langle a\rangle p \vee \langle b\rangle p \tag{D2}$$

$$\langle a;b\rangle p = \langle a\rangle\langle b\rangle p \tag{D3}$$

$$\langle a\rangle[a^{\smile}]p \leq p \leq [a^{\smile}]\langle a\rangle p \tag{D4}$$

$$p \vee \langle a\rangle\langle a^*\rangle p \leq \langle a^*\rangle p \leq p \vee \langle a^*\rangle(\langle a\rangle p \wedge \neg p) \tag{D5}$$

A dynamic algebra *with test* is one with an operation $? : B \rightarrow K$, notation $?(p) = p?$, satisfying

$$\langle p?\rangle q = p \wedge q. \tag{D?}$$

The first three equations define $\langle 0\rangle$ to be the constantly zero operator, $\langle a+b\rangle$ to be the pointwise disjunction of $\langle a\rangle$ and $\langle b\rangle$, and $\langle a;b\rangle$ to be the composition of $\langle a\rangle$ and $\langle b\rangle$. If we translate $\langle a\rangle p$ as $a;b$ then equation (D4) can be seen to be exactly the equation rK,Kr of relation algebra, which as can be seen from the lattice diagram for those axioms entails the theory R of right-handed relation algebras. Similarly equation (D5) so translated can be seen to be the relation algebra equation for star.

The arguments showing that the relation algebra equations for converse and star implicitly define those operations carry over to this situation without difficulty; see [Pra79a,Pra80a] for the case of star. Since $[a^{\smile}]$ is the right adjoint of $\langle a\rangle$ [Mac71, Thm IV.5-1], $\langle a^{\smile}\rangle$ is the *dual* right adjoint of $\langle a\rangle$. $\langle a^*\rangle$ is the reflexive transitive closure of $\langle a\rangle$, satisfying $\langle a\rangle \leq \langle a^*\rangle$, $\langle a^{**}\rangle = \langle a^*\rangle$, $\langle a\rangle \leq \langle b\rangle$ implies $\langle a^*\rangle \leq \langle b^*\rangle$, and $\langle a\rangle$ is reflexive and transitive if and only if $\langle a^*\rangle = \langle a\rangle$. (An operator f is reflexive when $p \leq f(p)$ and transitive when $f(f(p)) \leq f(p)$.)

A ***preKleenean*** algebra is the Kleenean algebra of a dynamic algebra. A ***separable dynamic algebra*** [Koz79a] is one for which $\Diamond$ is injective, satisfying the Π^0_2 sentence $\forall p[\langle a\rangle p = \langle b\rangle p] \rightarrow a = b$. An ***intensional*** dynamic algebra is an algebra isomorphic to a subalgebra of a separable dynamic algebra. A ***Kleenean algebra*** is the Kleenean algebra of an intensional dynamic algebra. We denote the corresponding classes **SDA**, **IDA**, and **KA** respectively.

Proposition 10 **IDA** *is a quasivariety.*

Proof: Since **SDA** is defined by equations and a Π^0_2 sentence it is an elementary class closed under direct products. Hence its closure **IDA** under isomorphisms and subalgebras is a quasivariety. ∎

4.4 EXAMPLES

In the following example the Boolean and Kleenean elements are respectively unary and binary relations on a set X. This is the motivating example of a dynamic algebra, and corresponds to Example 1 of a Boolean monoid, as well as to our notion of a Kleenean algebra of relations.

Example 1. Let **Kri** $X = (\mathcal{B}, \mathcal{K}, \diamond)$ consist of the Boolean algebra $\mathcal{B} = (2^X, \vee, 0, \wedge, 1, \neg)$ of unary relations on X, and the Kleenean algebra $\mathcal{K} = (2^{X^2}, +, 0, ;, ^*, ^\smile)$ of binary relations on X, with symbols interpreted as for **Rel** X, and such that star is reflexive transitive closure, $\langle a \rangle p = \{x \mid \exists y[xay \wedge p(y)]\}$. This example may be extended to a dynamic algebra with test by adjoining the operation $p? = \{(x,x) | p(x)\}$.

A *Kripke structure on* X is any subalgebra of **Kri** X. The Boolean elements of a Kripke structure are unary relations and the Kleenean binary. It may be verified that Kripke structures satisfy all the equations, those for star and converse being the most challenging. A *representable dynamic algebra* (RDA) is a dynamic algebra isomorphic to a Kripke structure. These form the class **RDA**.

Proposition 11 RDA *is a quasivariety* ([Nem82]).

Proof: It suffices to show that the direct product of a family (**Kri** X_i) of Kripke structures is an RDA. Its Boolean component is the power set of $\sum_i X_i$, the disjoint union of the X_i's. Its Kleenean component is isomorphic to the power set of the equivalence relation on $\sum_i X_i$ definable as $\sum_i X_i^2$, relating just those pairs of elements from the same X_i. ∎

Proposition 12 RDA $\subseteq$ IDA.

Proof: Every **Kri** X is separable. **RDA** is the ISP closure of the **Kri** X while **IDA** is the ISP closure of **SDA**. ∎

It follows from a result of Kozen [Koz81] that the converse does not hold.

In the next example the Boolean and Kleenean elements are languages as sets of strings over a common alphabet X. For languages we must omit converse and test, but see the section on "model robustness" for an approximation to **Lan** X containing converse.

Example 2. Let **Lan** $X = (\mathcal{B}, \mathcal{K}, \diamond)$ consist of the Boolean algebra $\mathcal{B} = (2^{X^\omega}, +, \emptyset, \sim)$ of infinitary languages (sets of infinite-to-the-right strings on X), and the Kleenean algebra $\mathcal{K} = (2^{X^*}, 0, +, ;, ^*)$ of all sets of finite strings, with operations (omitting converse) having their usual meaning for languages [Kle56,HU79]. Take $\langle a \rangle p$ to be the concatenation of languages a and p. We now have a dynamic algebra *without converse*.

All axioms save (D5b) (right hand inequality of (D5)) are easily verified. For (D5b), given any string $s \in \langle a^* \rangle p$ find the least n such that $s = a_1 \ldots a_n t$ for strings $a_i \in a$ and $t \in p$. If $n = 0$ then $s = t$ and $s \in p$.

Otherwise $a_i \ldots a_n t \notin p$ for $1 \leq i \leq n$ or we could find a smaller n. In particular $a_n t \notin p$ whence $a_n t \in \langle a \rangle p - p$, so $s \in \langle a^* \rangle(\langle a \rangle p - p)$.

From any nonempty infinitary language L not containing the symbol 0 we may construct the language $p = 0L$ as a universal separator. If s is in a but not b then $s0L$ is a nonempty subset of $\langle a \rangle p$ but is disjoint from $\langle b \rangle p$. This makes **Lan** X separable.

4.5 Representability and Completeness

Thus far dynamic algebras seem very much like relation algebras, using essentially the same equations to essentially the same effect. However, whereas the relation algebra axioms incompletely axiomatize the representable relation algebras, the dynamic algebra axioms completely axiomatize the representable dynamic algebras. We may put this more graphically as follows.

A representable Boolean algebra is an algebra isomorphic to a field of sets. That is, the class **RBA** of such algebras is the quasivariety ISP(2) generated by the two-element Boolean algebra. A Boolean algebra is a complemented distributive lattice, conditions expressible with finitely many equations and thus making the variety **BA** of Boolean algebras finitely axiomatized. Evidently **RBA** $\subseteq$ **BA**. It is one of nature's little pranks that **RBA** is a variety, but it is a bigger prank that **RBA** = **BA** [Sto36].

The class **RRA** of representable relation algebras is the quasivariety generated by algebras **Rel** X. The variety **RA** of relation algebras is finitely based, with **RRA** $\subseteq$ **RA**. We again have the little prank, that **RRA** is a variety [Tar55]. The difference is that we no longer have the big prank: **RA** $\neq$ **RRA** [Lyn50], and although there are infinitely many finitely axiomatizable varieties between these two, **RRA** itself is not finitely axiomatizable [Mon64].

With dynamic algebras the situation is in between these two. The class **RDA** of representable dynamic algebras is the quasivariety generated by algebras **Kri** X. The variety **DA** of dynamic algebras is finitely based, with **RDA** $\subseteq$ **IDA** $\subseteq$ **DA**. Now the free IDA's are residually finite [Pra79a] and moreover are representable [Nem82], whence the equational theory of **IDA** completely axiomatizes **RDA**, i.e. **RDA** and **IDA** have the same equational theory. Also **RDA**, **IDA**, and **DA** have the same Boolean equational theory. This is a weaker connection than that of **RBA**=**BA**, but a stronger one than **RRA** $\subseteq$ **RA**, where the equational theory of **RA** is strictly less than that of **RRA**.

To show that every free IDA is representable, let us consider the effect of the passage from dynamic algebras to intensional dynamic algebras on the equational theory. This effect is quite striking: the Boolean theory does not change while the Kleenean theory is transformed at one stroke from the vacuous theory to the theory of Kleenean algebras.

Since the only axioms of **DA** are Boolean, Θ^K_{DA} is trivial, consisting just of all equations $a = a$. Hence the Kleenean variety generated by the preKleenean algebras is just the anarchic variety consisting of all word algebras of type $\mathbf{DA}^K$ and their quotients. As we shall see momentarily however the preKleenean algebras are considerably more organized than their vacuous theory might suggest.

Proposition 13 *A Boolean module has at most one expansion to a separable dynamic algebra.*

Proof: In an SDA, the elements of K serve as distinct names for functions on B. Axioms (D1)-(D3) are easily seen to define the corresponding three operations uniquely, as respectively the constantly zero operation, pointwise disjunction, and composition. Axioms (D4) and (D5) also uniquely define converse and star respectively, for the same reasons as do the corresponding axioms for these operations in **RA**. For completeness we give the proof in full here.

The left half of (D4) can be written equivalently as $p \leq [a]\langle a^{\smile}\rangle p$ (p is universally quantified here). The right half says that for any q such that $p \leq [a]q$ (and we have just seen that $\langle a^{\smile}\rangle p$ is such a q), we must have $\langle a^{\smile}\rangle p \leq q$, i.e. $\langle a^{\smile}\rangle p$ is the least q for which $p \leq [a]q$. To see this, let q satisfy $p \leq [a]q$. Then by monotonicity $\langle a^{\smile}\rangle p \leq \langle a^{\smile}\rangle[a]q$. But the latter is bounded by q, i.e. $\langle a^{\smile}\rangle p \leq q$. There can only be one least such q, whence $\langle a^{\smile}\rangle$ is uniquely determined.

A similar argument, given in [Pra79a,Pra80a], obtains for star. The left half of (D5) says that $\langle a^*\rangle p$ is among those q's satisfying $p \vee \langle a\rangle q \leq q$. But any such q satisfies $p \leq q$, so by monotonicity $\langle a^*\rangle p \leq \langle a^*\rangle q$. But the latter is bounded by $q \vee \langle a^*\rangle(\langle a\rangle q \wedge \neg q)$, and $\langle a\rangle q \wedge \neg q$ vanishes (since $\langle a\rangle q \leq q$, whence so does $< a^* > (\langle a\rangle q \wedge \neg q)$, yielding $\langle a^*\rangle p \leq q$. Thus $\langle a^*\rangle p$ is the least such q. But there can only be one least such q, whence $\langle a^*\rangle$ is uniquely determined. ∎

We now provide a sense in which the preceding result extends to all **DA**'s. In the following, by a ***homomorphism*** we mean an operation-preserving function between algebras of the same type, or in the case of an algebra with n sorts, then n "parallel" such functions. For the two sorts of dynamic algebras a homomorphism $f : \mathcal{D} \to \mathcal{D}'$ must be a pair (f_B, f_K) consisting of Boolean and Kleenean homomorphisms $f_B : \mathcal{B} \to \mathcal{B}'$ and $f_K : \mathcal{K} \to \mathcal{K}'$ satisfying $f_K(a + b) = f_K(a) +' f_K(b)$, $f_B(p \vee q) = f_B(p) \vee' f_B(q)$, $f_B(\langle a\rangle p) = \langle f_K(a)\rangle' f_B(p)$, $f_K(p?) = f_B(p)?'$, and similarly for the other operations. By a ***quotient*** of an algebra $\mathcal{D}$ we mean as usual the equivalence class of all homomorphic images of an algebra isomorphic to a particular such image, or equivalently, the representative $\mathcal{D}/\cong$ of that class whose elements are the congruence classes of some congruence $\cong$ on $\mathcal{D}$ (namely the kernel of the above "particular image").

Proposition 14 $\Diamond$ *is a preKleenean homomorphism.*

Proof: If we denote pointwise disjunction in $B \to B$ by $+'$, then the equation defining $+$ merely asserts $\Diamond + = +' \Diamond$, and similarly for the other four Kleenean operations, bearing in mind the preceding result that these operations are uniquely determined for $B \to B$. But this is then just the assertion that $\Diamond : K \to (B \to B)$ is a homomorphism of preKleenean algebras K and $B \to B$. ∎

It follows that the kernel of $\Diamond$ is a preKleenean congruence. This was first observed for $*$-continuous dynamic algebras by Kozen [Koz79c,Koz80] (see the history section). It was generalized to the weaker Segerberg notion of star by the author [Pra80a].

Proposition 15 *Every dynamic algebra $\mathcal{D}$ has a unique quotient $\mathcal{D}'$ in* **IDA** *such that $\mathcal{B} = \mathcal{B}'$.*

Proof: By the preceding proposition, the unique factorization of the function $\Diamond$ as the composition of an injection $\Diamond'$ with a surjection q is such that q is a quotient. This yields a dynamic algebra $\mathcal{D}'$ as a quotient of $\mathcal{D}$; no finer quotient will land in **IDA**. For uniqueness, the action of $\Diamond$ must be preserved by any homomorphism (via the natural isomorphism relating $\Diamond : K \to (B \to B)$ to $\diamond : K \times B \to B$, "homomorphism" being defined so as to preserve the latter). Hence if $\mathcal{B}$ does not change, distinct operators must remain distinct, whence no coarser quotient will preserve $\mathcal{B}$. ∎

Proposition 16 $\Theta^B_{IDA} = \Theta^B_{DA}$.

Proof: Since **IDA** $\subseteq$ **DA** it suffices to show $\Theta^B_{IDA} \subseteq \Theta^B_{DA}$. From the preceding theorem we infer that every dynamic algebra $\mathcal{D}$ is isomorphic to a subalgebra of the product of an intensional dynamic algebra $\mathcal{D}'$ with a "Boolean-trivial" dynamic algebra $\mathcal{D}''$, namely the quotient of $\mathcal{D}$ fixing K and collapsing B to a point. Hence all equations in Θ_{IDA} hold of $\mathcal{D}'$, and trivially of $\mathcal{D}''$, and hence of $\mathcal{D}$. ∎

But whereas postulating injectivity of diamond does not increase the Boolean theory, it takes the Kleenean theory from the vacuous theory to the theory of Kleenean algebras!

Proposition 17 *An intensional dynamic algebra $(\mathcal{B}, \mathcal{K}, \Diamond)$ for which $\mathcal{B}$ is complete and atomic is representable. In particular every finite IDA is representable.*

Proof: Taking X as the set of atoms of $\mathcal{B}$, interpret $\mathcal{B}$ as 2^X, each $a \in K$ as the relation $paq \equiv (q \le \langle a \rangle p)$, and $\Diamond$ as in a Kripke structure. Since a complete and atomic Boolean algebra is isomorphic to the power set of its atoms, and since all joins exist and are preserved by $\langle a \rangle$, this Kripke structure is isomorphic to the given dynamic algebra. ∎

By *free dynamic algebra* we shall understand a free algebra of the variety generated by dynamic algebras. Conway's Leap, as defined in the section on Kleenean algebras, shows that this variety contains algebras that do not match our intuition about dynamic algebras. The free algebras of this variety turn out not to so violate intuition, and indeed are not only dynamic algebras, justifying the name, but even more importantly are representable. Showing this will be our main goal, allowing for the occasional digression.

Proposition 18 *Every free dynamic algebra is a subdirect product of finite dynamic algebras* [Pra80a].

Proof: With the assumption of freedom we are able to translate into dynamic algebra terminology Fischer and Ladner's filtration construction [FL79], whereby from any Kripke structure satisfying a particular formula they construct a finite Kripke structure satisfying it. Nothing in their proof makes essential use of attributes of representable dynamic algebras not already possessed by arbitrary dynamic algebras. The reader is referred to the proof of Theorem 5 [Pra80a] for the short (half a page) details of this translation. The case of a free dynamic algebra on the empty set of Boolean generators is treated by Németi [Nem82]. ∎

Corollary 19 *HSP(***RDA***)=HSP(***IDA***). That is,* $\Theta_{RDA} = \Theta_{IDA}$.

Proof: Every intensional dynamic algebra is a quotient of a free intensional dynamic algebra. Each of these in turn is a subdirect product of finite intensional dynamic algebras, which in turn are representable. HSP preserves equations, whence the theory of Kripke structures is a subset of that of intensional dynamic algebras. ∎

Corollary 20 *The Segerberg axioms are sound and complete relative to Kripke structures* [Pra80a]. *That is,* $\Theta^B_{RDA} \subseteq \Theta^B_{IDA}$.

Corollary 21 *Every free intensional dynamic algebra is representable.* [Nem82]

(With the previous corollary as my goal I overlooked this nice strengthening in [Pra80a].)

Proof: For any quasivariety K the free algebras of HSP(K) belong to K. **RDA** is a quasivariety and HSP(**RDA**)=HSP(**IDA**). ∎

4.6 Computational Complexity

Theorem 22 *There exist* $1 < c < d$ *such that* Θ_{DA} *and its complement* $\bar{\Theta}_{DA}$ *are not in* $DTIME(c^n)$ [FL79] *but are in* $DTIME(d^n)$ [Pra79b].

That is, the time required to deterministically test either satisfiability or validity of dynamic algebra equations is one exponential in the number n of occurrences of variables in the formula, a bound that cannot be improved by more than by a polynomial of degree $\log_c d$. For comparison, the best deterministic procedure known for pure Boolean equations, i.e. propositional calculus, requires time $2^{n/4}$ or 1.1892^n [VG88], down to 1.093^n for equations $t = 0$ when t is in conjunctive normal form. Fischer and Ladner do not supply a specific value for c, but their proof is constructive and if pushed hard might conceivably yield a c as high as 1.01. Thus we are still some distance from knowing whether dynamic logic is any harder to decide in practice, i.e. deterministically, than Boolean logic, though close enough that these two bounds may well pass each other within the coming decade.

4.7 The Language Model

It can be shown [Pra79b] that the equational theory of Example 2 is that of **IDA**, in the absence of converse.

We could alternatively have taken $\mathcal{B} = 2^{X^*}$. This is still an intensional dynamic algebra, with any single string serving as a universal separator. The difference is that the equational theory of the resulting class is strictly larger than that of **IDA**. In particular $\langle a^*\rangle([a]p \vee [a]\neg p)$ is now a theorem for any a and p. This asserts that for any predicate it is possible to run any program sufficiently often that at its next execution it is deterministic with respect to that predicate.

We may extend Example 2 to include converse by embedding $\mathcal{K}$ (an algebra of languages) in the larger algebra $\mathcal{K}^+$ of all (normal additive) operators on $\mathcal{B}$. We then define $\langle a^\smile\rangle p = \{s \mid \langle a\rangle s \wedge p \neq 0\}$. This defines $\langle a^\smile\rangle$ as an operator but it does not define $a^\smile$ as a language.

To verify (D4a) (left hand inequality of (D4)), consider any string $t \in \langle a\rangle[a^\smile]p$. Then there exists $s \in [a^\smile]p$ such that $t \in \langle a\rangle s$. So $s \notin \langle a^\smile\rangle\neg p$, whence $\langle a\rangle s \wedge \neg p = 0$. Hence $t \notin \neg p$, i.e. $t \in p$. For (D4b), suppose $s \in p$. Then $\langle a\rangle s \leq \langle a\rangle p$, that is, $\langle a\rangle s \wedge \neg\langle a\rangle p = 0$, whence $s \notin \langle a^\smile\rangle\neg\langle a\rangle p$, i.e. $s \in [a^\smile]\langle a\rangle p$.

In order to add the test operation $p?$ to this example we evidently require that the structure satisfy the sentence $\forall p\exists a\forall q[\langle a\rangle q = p \wedge q]$. But Example 2 falsifies this at $p = 0X^\omega$. For if $a \neq 0$ then take $q = 1X^\omega$ making $p \wedge q = 0 \neq \langle a\rangle q$, while if $a = 0$ then take $q = p$.

I do not have a completely satisfactory solution. Here is as much as I have been able to do. Modify $\mathcal{B}$ to satisfy the sentence as follows. The infinite string $(AB\ldots YZ)^\omega$ has 26 distinct suffixes, take $\mathcal{B}$ to be the Boolean algebra consisting of the 2^{26} sets of such suffixes. Take $\mathcal{K}$ as before but with X chosen to include the 26 letters. Define $\langle a\rangle p = ap \cap 1$ (1 being the top of $\mathcal{B}$, i.e. the set of 26 suffixes) and $p? = \pi_{26}(p)$ where $\pi_{26}(p)$ is the set of all length-26 prefixes of strings of p. Axiom $(D?)$ is now easily verified, and the arguments for the other axioms are easily modified to accommodate this change to diamond.

But now we have lost separability. By modifying $\mathcal{K}$ along the same lines, changing $a; b$ from concatenation to something more discriminating, we could restore it. But this completely loses the spirit of more **Lan** X. This raises the somewhat vague question, is there a way to define test for a **Lan**-like dynamic algebra?

The beginning of Example 2 (no converse and test) appears in [Pra80a]. The modification used to define test is essentially the result of "the LAN construction" [Pra79b] used to show that the theory of **Lan** X coincides with the theory of dynamic algebras, the difference being that we did not cater for test there, allowing complement to be taken relative to X^ω rather than to L^ω as here.

5 Reflections

For a change of pace let us reflect on some of the philosophical issues bearing on dynamic algebras and their relationship to relation algebras.

5.1 ANALYSIS

Why do Boolean modules and dynamic algebras have so many properties that relation algebras lack? I like to think of it in terms of holding the two essential ingredients of dynamic and relation algebras at the proper distance. Too far apart and all you have is a Boolean algebra and a monoid. Too close and they interfere destructively.

Brink [Bri81] argues that Boolean modules are relatively well-behaved compared to relation algebras. I make a similar point in the context of regular algebras versus dynamic algebras [Pra79a,Pra79b,Pra80a]. Redko [Red64] has shown that the equational theory of regular algebras has no finite basis. Conway [Con71] has observed that this theory has a three-element model in which $x^0 + x^1 + \ldots + x^n$ is constant with increasing $n > 0$ yet $x*$ is not that constant, a discontinuity we refer to as *Conway's Leap*. Replacing one-sorted regular algebras by two-sorted dynamic algebras disposes of both these aberrations, as we will see later in the section on properties of dynamic algebras.

The common idea here seems to be that intersection and composition in too close proximity only "fight" each other. If instead each is moved to an appropriate sort, a logical sort accommodating the Boolean operations and a relative sort for the Kleenean operations, the separation seems to encourage cooperation instead of competition.

5.2 Star, Converse, and Test

Star has turned out to be converse's long-lost fraternal twin. Star should have been included in relation algebras from the outset. Converse without star is a piston without a crankshaft, or $\cos(x)$ without e^{ix}.

Test restores strong connectivity of information flow around the algebra. This flow obtains vacuously in one-sorted relation algebras.

Star and test are standard features of any imperative programming language. Star provides iteration, while test enables the rational performance of choice and iteration, expressed deterministically with **if** p **then** a **else** b and **while** p **do** a respectively, and more generally with guarded commands [Dij76].

The logic-of-programs significance of dynamic algebra is as follows. The set X is viewed as the states of a computer. Binary relations are viewed as programs: the meaning of (x, y) as an element of a program is that when that program starts in state x it *may* stop in y. A *deterministic* program is one that is a partial function. The program $a; b$ performs a then b. The program $a + b$ nondeterministically chooses to perform one of a or b. The program a^* nondeterministically chooses an $i \geq 0$ and performs a^i, that is, $a; a; \ldots; a$ i times. The test program $p?$ changes nothing but stops only if p holds, otherwise it is said to *block*. The programming construct "**if** p **then** a **else** b" can then be expressed as $(p?; a) + (\neg p?; b)$, while "**while** p **do** a" can be written $(p?; a)^*; \neg p$. The program $a^{\smile}$ "runs a backwards;" a may be deterministic without $a^{\smile}$ being deterministic, e.g. the deterministic program that replaces x by its square when run backwards nondeterministically replaces x by one of its square roots.

5.3 Merits of Decidability

I would like to pass judgment on the value of decidability in mathematical theories. One may with considerable precedent take the position that undecidability, if not lack of a finite basis, is necessary in any theory rich enough to serve as a foundation of mathematics [TG87]. After all is not all of mathematics founded on Zermelo-Fränkel set theory, evidently an undecidable theory?

I would like here to question the inevitability of undecidability.

First, if we really did need a universal theory to serve as a foundation for mathematics I would grant that such a theory should be undecidable. I question however the premise that a universal theory is needed in the first place. The link between mathematics and foundations seems more potential than actual. That is, it is a tenet of faith that "conventional" mathematical proofs can be expanded out to a purely set theoretic argument, yet this is almost never done. Moreover category theory has in recent years posed a challenge to set theory as an alternative and strikingly different foundation, indicating the nonuniqueness of such expansions. The possibility then arises that no such foundation is needed. Instead we may consider any given argument as being conducted in one or more relatively small and localized theories.

Second, the purpose of theory is to organize thought, not to drown it, to be constructive without being oracular. There is something of a movement in programming to make programs more like proof systems, and computations more like proofs. Coming in the other direction there is similar enthusiasm in logic for making proof systems more constructive, and proofs more like computations. There is however the distinct possibility that the two movements will rush right past each other and find that they have merely switched places!

Instead of founding mathematics on a single theory such as ZF or RA, why not view mathematics as a large collection of domain-specific theories? The whole of mathematics founded on a single small theory may have the real estate advantages of an inverted pyramid but it also has its structural disadvantages.

I propose that the proper notions of constructivity in a logic are its computational complexity and its human surveyability. These elements should be present in proportions suited to the application, mainly the former for a mechanical theorem prover, mainly the latter for computer aided instruction, and in more even proportions for a mathematician's mechanical apprentice.

This then speaks for computational tractability as an important criterion for judging the merits of any theory. If there is any distinction at all to be made between computation and logic it may well be the respective thresholds of polynomial time and exponential time as criteria for tractability!

5.4 Induction as Termination at Convergence

Kozen's original definition of dynamic algebra included $*$-continuity as a condition. The terminology subsequently standardized on by Kozen and myself is "dynamic algebra" for the equational class with the term "$*$-continuous" added to denote the condition that $\langle a^*\rangle p = p \vee \langle a\rangle p \vee \langle a\rangle\langle a\rangle p \vee \ldots = \bigvee_{i<\omega} a^i; p$.

Kozen [Koz81], p.175. argues for the practical value of the additional $*$-continuity condition as follows. "Looping is inherently infinitary and nonequational; ... Thus the equational approach must eventually be given up if we are ever to bridge the gap between algebraic and operational semantics." To my recent query as to whether he still held this view he replied, "Strictly speaking, no. Practically speaking, yes." I would like to take this opportunity to offer my position on the relative appropriateness of the two definitions of star.

The conditions differ only on the question of when iteration terminates. Under the equational definition, iteration terminates at convergence, that is at q satisfying $\langle a\rangle q \leq q$, whilst under the stronger $*$-continuity condition it terminates at ω.

I prefer convergence rather than ω as the place to stop because ω is not first-order definable (in the same sense that one may say that finiteness is not first-order definable), it is uneconomic for short iterations, it is needlessly restrictive for long iterations, and it is a potential Achilles heel for nonclassical dynamic logics.

While the class of dynamic algebras based on termination at convergence forms a quasivariety, that based on termination at ω is not even first-order-definable. Just as one cuts hair at different lengths for better appearance, and opens electrical circuits slowly for less electrical noise, so should one terminate iteration at convergence to achieve the tameness of a quasivariety rather than always exactly at ω, which goes beyond first-order logic.

In the classical formulation of dynamic logic adhered to in this paper, the economic argument reduces to an esthetic quibble, there being no charge for "gedanken-iterations." In nonclassical frameworks however $*$-continuity may prove unsound. I have no feeling for whether $*$-continuity will prove compatible with intuitionistic dynamic logic, but its nonconservatism seems quite opposed to, and hence likely to be unsound for, the explicit conservatism of (intuitionistic) linear [Gir87] dynamic logic.

I see no point in banning iteration beyond ω, in theory or in practice. The place in mathematics of iteration beyond ω has long since been secured. Computer science has clung more recently than mathematics to the superstition that all its practically accessible objects are finite. However any questions as to the worldly meaning of iteration beyond ω in computation have surely been dispelled by now by such applications as Manna and Dershowitz's multiset orderings, involving iteration up to ϵ_0, and Schwichtenberg's use of infinite ordinals to give an elegant short description of certain rapidly growing functions and hence very large numbers. Today's high level programming languages cannot afford to maintain the fiction that there is no iteration after ω. If the compiler's target language proscribes iteration beyond ω, it should be the compiler's duty to shield the programmer from this low-level restriction.

5.5 Intensionality

Extensionality occasionally gets in the way, and **SDA** would appear to constitute a simple but well-motivated example of this phenomenon. The Kleenean elements of an SDA are extensional in that separability connotes extensionality. Yet in the passage from **SDA** to **IDA** extensionality is lost.

If Kleenean elements were Gödel numbers then we would obtain intensionality as a basic property of acceptable Gödel numberings, that there is no bijective Gödel numbering of partial recursive functions. But any such connection with Gödel numbering or effectiveness can only be made via the computer science origins of dynamic algebra, not via its inherent structure, nothing in which hints of such a connection. We have only an abstract algebra of programs combined with imperative control structures, with no reason to suppose that the programs are not distinct partial recursive functions, or at least relations.

Instead intensionality arises here despite our efforts to achieve extensionality, when we pass from **SDA** to the quasivariety **IDA** by taking subalgebras.

Almost the same transition is made by Kozen [Koz81] when he passes from separable to inherently separable dynamic algebras, where the essential advantages of extensionality are preserved without preserving the extensionality itself. Kozen defines an inherently separable dynamic algebra as one sharing its Kleenean component with a separable dynamic algebra. The notion of intensional dynamic algebra introduced in this paper makes it in effect an inherently separable dynamic algebra for which the sharing is mediated via an inclusion between the two Boolean algebras, this being a roundabout way of describing a subalgebra of a separable dynamic algebra. Clearly intensional implies inherently separable for dynamic algebras, but I do not know whether the converse holds; if it does the "almost" at the start of this paragraph may be removed.

5.6 Origins of Dynamic Algebra

My adoption of universal algebra in 1979 represented for me the transfer to logic of a principle I previously understood only as a programmer. Unsound logic, meaning a discrepancy between a theory T and a class of (real or fictitious) worlds W, is in programming terms a bug. The connection between programming and logic can be made by substituting abstract program for theory, concrete program for proof system, computation for proof, and instruction step for proof step, leaving the notion of model unchanged.

Whereas a beginning programmer fixes a bug by fixing the program, experience teaches the principle that fixing the world is also an option. Unix for example was perceived by some in the early 1980's as a bug in the world of operating systems. This bug has been fixed, or at least attenuated, by adapting both Unix and the world to each other.

So it goes with logic. Given a proof system denoting a theory T and a class W of typical worlds, to prove T complete for W, an amateur logician such as myself would think only to compare T to other T's while holding W fixed. The first completeness proof [Par78] of Segerberg's axiomatization [Seg77] of PDL however proceeded by changing both, reflecting Parikh's extensive logical background.

Failing to understand Parikh's proof, and also wanting to understand the relationship of the proof theory to the problem of deciding PDL theoremhood, I undertook to find a structure that would work for me both as an understandable completeness proof and a decision method. This resulted in [Pra78], extensively revised as [Pra80b]. The former axiomatized dynamic logic in the language of Gentzen sequents, the latter extended this approach to a theory whose atomic formulas were $u \models p$ meaning "state u satisfies proposition p" and $u\langle a\rangle v$ meaning "from state u program a can halt in state v." In these proofs I held Kripke structures themselves to be the only models and all the rest as various proof systems of varying distances from Kripke structures, with completeness proved for the remoter ones via those closer to Kripke structures where completeness was more obvious.

It was at about this time in 1978 that Dexter Kozen conceived the notion of dynamic algebra. He mentioned the concept to me, not by name that I recall, towards the end of 1978 when I visited IBM Yorktown Heights, and said there might be a representation theorem there. At the time I saw no connection with algorithms and completeness proofs for dynamic logic. I had no idea then of the role played by representation theorems in completeness proofs that work with many W's and one T.

At STOC-79 in April I recounted an intriguing equational derivation to a group of about eight dynamic logic enthusiasts, without however being able to formulate an associated theorem. (It turned out to be the proof that the inductive definition of star entailed the definition as local reflexive transitive closure.) Thinking that the derivation might play a role in an algebraic completeness proof for an equational theory, but never having seen an equational completeness proof before, I asked Janos Makowsky's advice. He recommended Henkin's paper on the logic of equality, a pedagogically drawn-out universal algebra proof that the equational theory of the monoid of natural numbers is completely axiomatized by the theory of commutative monoids. I found it the perfect Rosetta stone for learning how to translate syntactic insights about proofs into semantic ones.

This led to my formulation and submission to FOCS-79 in early May of a semantic proof that half a dozen models of programming logic were all completely axiomatized by the equational theory of PDL [Pra79b]. The proof proceeded by showing that any algebra of those classes could be constructed from the algebras of a neighboring class by homomorphisms, subalgebras, and direct products, thereby establishing a strongly connected graph of inclusions between the equational theories of the classes. One of the classes consisted of the models of Segerberg's axioms, which I then called Hoare algebras. That this semantic proof method is complete is an immediate corollary of Birkhoff's theorem [Bir35] that every class closed under homomorphisms, subalgebras, and direct products forms a variety.

The one step in this proof that I did not supply was the inclusion $\Theta_{FKRI} \subseteq \Theta_{DA}$, FKRI denoting the class of finite Kripke structures. This was not a lacuna in the proof since it is the statement of completeness of the Segerberg axioms, which by then had no shortage of published proofs. Nevertheless I thought it would be nice to make the whole proof purely algebraic by finding the proper HSP formulation of the completeness proof. This however I was unable to do in time for the FOCS-79 deadline.

Shortly after submitting that paper I received a manuscript from Dexter Kozen [Koz79c]. It gave the full details of Kozen's dynamic algebra (a term

I subsequently adopted for the proceedings version of my paper). Kozen's notion was the same as mine in most respects. The biggest difference from mine was that Kozen, like Brink, modeled his definition on that of an R-module over a ring R, where the notion of a ring is presumed to be given *a priori*. Thus Kozen took K to be what amounted to a semiring with star satisfying $a^* = \bigvee_{i<\omega} a^i$, with $a; b^* = \bigvee a; b^i$ and $a^*; b = \bigvee a^i; b$. In contrast my definition satisfied *no* Kleenean equations. Since that paper was about completeness of the Boolean theory the missing Kleenean theory presented no problem.

This led two months later to my proof [Pra79a] that every free separable dynamic algebra was a subdirect product of finite separable, hence representable, dynamic algebras. I later rewrote this to reduce the length of the proof proper to only half a proceedings page [Pra80a], at which point I felt I had a good grip on why Segerberg's axioms were complete.

Further reflection on the meaning of Segerberg's induction axiom led me to propose a formulation of the least-fixpoint or μ-calculus for Boolean modules [Pra81]. This would appear to be the first time that the notions of least fixpoint and Boolean module were brought together. This juxtaposition has since enjoyed considerable attention from the computer science community, most notably in its expression as Kozen's $L\mu$ calculus [KP83].

Acknowledgments. I am very grateful to Dexter Kozen for his insights in 1979 which were most helpful to me in clarifying my thinking on this subject. Much email traffic and long phone calls to George McNulty and Roger Maddux turned up many interesting ideas and facts bearing on this material. I am grateful to Don Pigozzi for encouraging me to write this paper for this volume. Roger and Don made many valuable suggestions for improvements to the paper.

6 References

[Bir35] G. Birkhoff. On the structure of abstract algebras. *Proc. Cambridge Phil. Soc*, 31, 1935.

[Bir67] G. Birkhoff. *Lattice Theory*. Volume 25, A.M.S. Colloq. Publications, 1967.

[Bri81] C. Brink. Boolean modules. *Journal of Algebra*, 71:291–313, 1981.

[CCMP89] R.T Casley, R.F. Crew, J. Meseguer, and V.R. Pratt. Temporal structures. In *Proc. Conf. on Category Theory and Computer Science, LNCS*, Springer-Verlag, Manchester, September 1989.

[Con71] J.H. Conway. *Regular Algebra and Finite Machines*. Chapman and Hall, London, 1971.

[CT51] L.H. Chin and A. Tarski. Distributive and modular laws in the arithmetic of relation algebras. *Univ. Calif. Publ. Math.*, 1:341–384, 1951.

[dBdR72] J.W. de Bakker and W.P. de Roever. A calculus for recursive program schemes. In M. Nivat, editor, *Automata, Languages and Programming*, pages 167–196, North Holland, 1972.

[Ded01] R. Dedekind. *Essays on the Theory of Numbers.* Open Court Publishing Company, 1901. Translation by W.W. Beman of *Stetigkeit und irrationale Zahlen* (1872) and *Was sind und was sollen die Zahlen?* (1888), reprinted 1963 by Dover Press.

[Dij76] E.W. Dijkstra. *A Discipline of Programming.* Prentice-Hall, Englewood Cliffs, N.J., 1976.

[Dil39] R.P. Dilworth. Noncommutative residuated lattices. *Trans. AMS*, 46:426–444, 1939.

[DM64] A. De Morgan. On the syllogism, no. IV, and on the logic of relations. *Trans. Cambridge Phil. Soc.*, 10:331–358, 1864.

[Ekl77] P.C. Eklof. Ultraproducts for algebraists. In *Handbook of Mathematical Logic*, pages 105–137, North Holland, 1977.

[FL79] M.J Fischer and R.E. Ladner. Propositional dynamic logic of regular programs. *JCSS*, 18(2), 1979.

[Fuc63] L. Fuchs. *Partially Ordered Algebraic Systems.* Pergamon Press, 1963.

[Gir87] Jean-Yves Girard. Linear logic. *Theoretical Computer Science*, 50:1–102, 1987.

[HU79] J.E. Hopcroft and J.D. Ullman. *Introduction to Automata Theory, Languages, and Computation.* Addison-Wesley, 1979.

[Jon82] B. Jónsson. Varieties of relation algebras. *Algebra Universalis*, 15:273–298, 1982.

[JT48] B. Jónsson and A. Tarski. Representation problems for relation algebras. *Bull. Amer. Math. Soc.*, 54:80,1192, 1948.

[JT51] B. Jónsson and A. Tarski. Boolean algebras with operators. Part I. *Amer. J. Math.*, 73:891–939, 1951.

[JT52] B. Jónsson and A. Tarski. Boolean algebras with operators. Part II. *Amer. J. Math.*, 74:127–162, 1952.

[Kle56] S.C. Kleene. Representation of events in nerve nets and finite automata. In *Automata Studies*, pages 3–42, Princeton University Press, Princeton, NJ, 1956.

[Koz79a] D. Kozen. On the duality of dynamic algebras and Kripke models. In E. Engeler, editor, *Proc. Workshop on Logic of Programs 1979, LNCS 125*, pages 1–11, Springer-Verlag, 1979.

[Koz79b] D. Kozen. *A representation theorem for models of $*$-free PDL.* Technical Report RC7864, IBM, September 1979.

[Koz79c] D. Kozen. A representation theorem for models of $*$-free PDL. May 1979. Manuscript.

[Koz80] D. Kozen. A representation theorem for models of *-free PDL. In *Proc. 7th Colloq. on Automata, Languages, and Programming*, pages 351–362, July 1980.

[Koz81] D. Kozen. On induction vs. *-continuity. In D. Kozen, editor, *Proc. Workshop on Logics of Programs 1981, LNCS 131*, pages 167–176, Spring-Verlag, 1981.

[KP83] D. Kozen and R. Parikh. A decision procedure for the propositional μ-calculus. In E. Clarke and Kozen D., editors, *Proc. Workshop on Logics of Programs 1983, LNCS 164*, pages 313–325, Springer-Verlag, 1983.

[KT89] D. Kozen and J. Tiuryn. *Logics of Programs.* Technical Report 89-962, Dept. of Computer Science, Cornell University, 1989. To appear in: van Leeuwen (ed.), *Handbook of Theoretical Computer Science,* North Holland, Amsterdam, 1989.

[Lyn50] R.C. Lyndon. The representation of relational algebras. *Ann. of Math., Ser 2*, 51:707–729, 1950.

[Mac71] S. Mac Lane. *Categories for the Working Mathematician.* Springer-Verlag, 1971.

[Mak64] M. Makkai. On PC_Δ classes in the theory of models. *Publications of Math. Inst. Hung. Acad. Sci.*, 9:159–194, 1964.

[Mon64] J.D. Monk. On representable relation algebras. *Michigan Math. J.*, 11:207–210, 1964.

[Nem82] I. Németi. Every free algebra in the variety generated by the representable dynamic algebras is separable and representable. *Theoretical Computer Science*, 17:343–347, 1982.

[Ng84] K.C. Ng. *Relation Algebras with Transitive Closure.* PhD thesis, University of California, Berkeley, 1984. 157+iv pp.

[NT77] K.C. Ng and A. Tarski. Relation algebras with transitive closure, Abstract 742-02-09. *Notices Amer. Math. Soc.*, 24:A29–A30, 1977.

[Par78] R. Parikh. A completeness result for a propositional dynamic logic. In *LNCS 64*, pages 403–415, Springer-Verlag, 1978.

[Pei33] C.S. Peirce. Description of a notation for the logic of relatives, resulting from an amplification of the conceptions of Boole's calculus of logic. In *Collected Papers of Charles Sanders Peirce. III. Exact Logic*, Harvard University Press, 1933.

[Pra76] V.R. Pratt. Semantical considerations on Floyd-Hoare logic. In *Proc. 17th Ann. IEEE Symp. on Foundations of Comp. Sci.*, pages 109–121, October 1976.

[Pra78] V.R. Pratt. A practical decision method for propositional dynamic logic. In *Proc. 10th Ann. ACM Symp. on Theory of Computing*, pages 326–337, San Diego, May 1978.

[Pra79a] V.R. Pratt. *Dynamic Algebras: Examples, Constructions, Applications.* Technical Report MIT/LCS/TM-138, M.I.T. Laboratory for Computer Science, July 1979.

[Pra79b] V.R. Pratt. Models of program logics. In *20th Symposium on foundations of Computer Science*, San Juan, October 1979.

[Pra80a] V.R. Pratt. Dynamic algebras and the nature of induction. In *12th ACM Symposium on Theory of Computation*, Los Angeles, April 1980.

[Pra80b] V.R. Pratt. A near optimal method for reasoning about action. *Journal of Computer and System Sciences*, 2:231–254, April 1980. Also MIT/LCS/TM-113, M.I.T., Sept. 1978.

[Pra81] V.R. Pratt. A decidable mu-calculus. In *Proc. 22nd IEEE Conference on Foundations of Computer Science*, pages 421–427, October 1981.

[Pra86] V.R. Pratt. Modeling concurrency with partial orders. *International Journal of Parallel Programming*, 15(1):33–71, February 1986.

[Red64] V.N. Redko. On defining relations for the algebra of regular events (Russian). *Ukrain. Mat. Z.*, 16:120–126, 1964.

[Sch95] E. Schröder. *Vorlesungen über die Algebra der Logik (Exakte Logik). Dritter Band: Algebra und Logik der Relative.* B.G. Teubner, Leipzig, 1895.

[Seg77] K. Segerberg. A completeness theorem in the modal logic of programs. *Notices of the AMS*, 24(6):A–552, October 1977.

[Sto36] M. Stone. The theory of representations for Boolean algebras. *Trans. Amer. Math. Soc.*, 40:37–111, 1936.

[Tar41] A. Tarski. On the calculus of relations. *J. Symbolic Logic*, 6:73–89, 1941.

[Tar55] A. Tarski. Contributions to the theory of models. III. *Indag. Math*, 17:56–64, 1955.

[TG87] A. Tarski and S. Givant. *A Formalization of Set Theory Without Variables*. American Math. Soc., 1987.

[VG88] A. Van Gelder. A satisfiability tester for non-clausal propositional calculus. *Information and Computation*, 79(1), October 1988.

[WD39] M. Ward and R.P. Dilworth. Residuated lattices. *Trans. AMS*, 45:335–354, 1939.

All Recursive Types Defined Using Products and Sums Can Be Implemented Using Pointers

Eric G. Wagner[1]

ABSTRACT This paper presents an algebraic formulation, and proof, of the folk theorem to the effect that all the recursive types defined using "products and sums" (e.g., NATURAL-NUMBERs, STACKs, TREEs, etc.) can be implemented using pointers.

1 Introduction

This paper presents an algebraic formulation, and proof, of the folk theorem to the effect that all the recursive types defined using "products and sums" (e.g., NATURAL-NUMBERs, STACKs, TREEs, etc.) can be implemented using pointers. We give an algebraic formulation of recursive types and their operations. A collection of recursive types, together with their inherent operations, form a special kind of algebra. We also give an algebraic formulation of imperative programming languages with pointers and variables. Roughly speaking, a memory state corresponds to an algebra, and any operation which "creates pointers" corresponds to a transformation on algebras. Thus the two mathematical frameworks are rather different. What we show though is that "the usual" implementation of recursive types by pointers works, from a mathematical point of view, because, in a sense to be explained below, it induces an algebra which is a homomorphic preimage of the desired algebra of recursive types.

Defining types recursively using sums and products is the same as defining them using recursive RECORDs and VARIANTs. While RECORDs are a common programming construct, VARIANTs are not all that common. For the sake of completeness, and comparison, we will give an an informal description of both RECORDs and VARIANTs.

A typical declaration of a RECORD type `ITEM` would be of the form

```
ITEM = RECORD
```

---

[1]Mathematical Sciences Department, IBM Research Division, T. J. Watson Research Center, Yorktown Heights, NY 10598

This paper is in final form and no version of it will be submitted for publication elsewhere.

```
        field_1 : type_1
        field_2 : type_2
                :
        field_n : type_n
    END
```

The string `field_i` is a *field name*, while the string `type_i` is a *type name*. The field names must be distinct, the type names need not be distinct. The string `ITEM` is also a type name, and the RECORD is *recursive* if there exists i such that `type_i = ITEM`. Intuitively, a variable of type `ITEM` will take values which are n-tuples of components, the ith component being of type `type_i`, for each $i = 1, \ldots, n$. For each field name there will be operation for accessing the corresponding component (*field*) of the record. There will also be an operation for inserting a value into a record as-a-whole.

The relationship between RECORDs and products is quite straight forward. If T_i is the set of values of `type_i` then the set of all RECORDs of type `ITEM` is isomorphic to the product

$$T_1 \times T_2 \times \ldots \times T_n,$$

the accessing operations are just the corresponding product projections, while the insert operation can be viewed as a mediating morphism for the product (see Section 2 for terminology).

As noted above, VARIANTs are not as common a programming construct as are RECORDs. The union objects of ALGOL68 [Van69] variant-records of PASCAL [WJ78] are related to the construct of VARIANT that we have in mind, but differ from it in important ways. A typical declaration of a VARIANT type `ITEM` would be of the form

```
    ITEM = VARIANT
        tag_1   : type_1
        tag_2   : type_2
                :
        tag_n   : type_n
    END.
```

The string `tag_i` is a *tag*, while the string `type_i` is a *type name*. The tags must be distinct but the type names need not be distinct. The string `ITEM` is also a type name, and the variant is *recursive* if there exists i such that `type_i = ITEM`. Intuitively, a variable of type `ITEM` will have a value of one of the types corresponding to the type names `type_1, ..., type_n`. Furthermore, this value will be "tagged" by the appropriate tag `tag_i`. For each tag `tag_i` there will be a corresponding operation for inserting a value of `type_i` into the VARIANT. Accessing of the VARIANT is done by means of *case-statements* which specify appropriate operations for each tag and corresponding type.

VARIANTs are closely related to sums, where we take sum to mean coproduct (see Section 2 for terminology). The relationship between VARIANTs and coproducts is quite straight forward. If T_i is the set of values of `type_i` then the set of all VARIANTs of type `ITEM;` is isomorphic to the coproduct

$$T_1 + T_2 + \ldots + T_n,$$

the inserting operations are just the corresponding coproduct injections, while the accessing operations can be viewed as a mediating morphisms for the coproduct.

The application of recursive types defined using sums and products, or VARIANTs and RECORDs, is illustrated by the example of the data type "Trees-of-D", that is, ordered binary trees with leaves labeled by elements of the set D. Writing **Tr** for "Tree-of-D", we can specify this data type by the recursion equation

$$\mathbf{Tr} \cong D + (\mathbf{Tr} \times \mathbf{Tr})$$

or, essentially equivalently, by the pair of equations

$$\mathbf{Tr} \cong D + \mathbf{Pr},\ \mathbf{Pr} \cong \mathbf{Tr} \times \mathbf{Tr}$$

which, by the introduction of an additional type variable (**Pr** for Pairs-of-trees), separate the sums and products. We shall generally use separated equations in what follows because they simplify the analysis. What is very important here is that the corresponding product projections and coproduct injections are very natural operations related to Trees-of-D. This becomes clearer if we provide them with natural names.

`leaf`:$D \rightarrow D+\mathbf{Tr}$

`join`:$\mathbf{Pr} \rightarrow D+\mathbf{Tr}$

`left`:$\mathbf{Tr}\times\mathbf{Tr} \rightarrow \mathbf{Tr}$

`right`:$\mathbf{Tr}\times\mathbf{Tr} \rightarrow \mathbf{Tr}$

Intuitively, `leaf` is the operation which takes a element d of D to the one-vertex tree labeled by D; `join` is the operation which takes two trees, t_1 and t_2 and makes them into a new tree by adjoining a common root; `left` is the operation which given a pair of trees gives you the left element of the pair, and `right` is similarly defined.

It is a straight forward matter to take the above pair of equations and transform them into a definition of the same data type in terms of recursive RECORDs and VARIANTs

```
TREE = VARIANT
     leaf    : D
     join    : PAIR
```

```
END;

PAIR = RECORD
      left    : TREE
      right   : TREE
END.
```

There exist programming languages in which recursive types are defined by means of type declarations that look like the above. However this is somewhat misleading. While the equations $\mathbf{Tr} \cong D + \mathbf{Pr}$ and $\mathbf{Pr} \cong \mathbf{Tr} \times \mathbf{Tr}$ define infinite sets, the corresponding declarations do not necessarily create any trees. Rather the declarations set up some apparatus for producing trees as needed. In general this will be done using "pointers". To a first approximation, and from a programmer's point of view, a pointer is the address of a storage cell or memory location. Take the phrase, "contains a pointer to X", to mean, "contains the address of the the storage cell containing X". Pointers could be utilized, as follows, to implement Trees-of-D:

> A leaf, that is a tree consisting of a single vertex labeled by an element d of D, is represented by a location containing the value d.
>
> A tree consisting of the join of two subtrees, t_1 and t_2, is represented by a pointer to a binary record whose first field contains a pointer to the root of t_1, and whose second field contains a pointer to the root of t_2.

In some programming languages, such as PASCAL [WJ78] and ADA, [ADA82,HP83] implementations of recursive types via pointers are always done explicitly, that is, rather than the above declarations, we would find declarations that are more of the form:

```
TREE = VARIANT
      leaf    : D
      join    : ↑PAIR
END;

PAIR = RECORD
      left    : ↑TREE
      right   : ↑TREE
END.
```

Here `↑PAIR` is the type consisting of pointers to objects of type `PAIR`, and similarly, `↑TREE` is the type consisting of pointers to objects of type `TREE`. The implementation of the operations on `TREE`s will then be done by programs which, among other things, explicitly "create" pointers as needed.

Our goals are, one, to abstract the above algebraically so as to get away from informal concepts such as storage cells and addresses, and, two, to show that the pointers implement the recursive types. The following overview of the paper sketches our route to these goals.

Section 2 explains some notation and reviews the formal definitions for products and coproducts. In Section 3 we give a precise formulation of the recursive types defined using sums and products. We define the notion of a recursive type specification and show how each such specification has a natural interpretation as an algebra consisting of the desired sets of types and the natural operations upon them. In Section 4 we show how we can exploit the mediating morphisms associated with the sums and products to define a considerably broader notion of derived operation than the usual notion from universal algebra. In particular we show that there are natural ways to define closure under both the conditional and the while-do constructs of programming languages. Of course our eventual goal is to show that we can realize the same derived operations using pointers. Section 5 introduces a general "memory model" that provides the needed abstraction of storage cells, pointers etc. The model, as presented, is more general than needed for this paper. In Section 6 we specialize the model to a model of the pointer implementation of recursive types. To each recursive type specification $\mathcal{S}$ there corresponds a class of "pointer algebras", each algebra corresponding to a finite approximation the type specified by $\mathcal{S}$. The creation of pointers is captured by an operation NEW on pointer algebras. As a first step in reconciling pointer algebras and recursive type definitions we define an "algebra **P** of pointer algebras" which we show is a homomorphic preimage of the desired algebra of recursive types. This homomorphism shows that the pointer algebras correspond to finite sets of recursive types but does not show us how to manipulate them. This is done in Section 7, where we show how the derived operations on recursive types are mirrored by operations on pointer algebras.

2 Preliminaries

In this paper we will write composition of functions and morphisms in diagramatic order, that is, given $f : A \to B$ and $g : B \to C$ we write their composite as $f \bullet g : A \to C$. Given a category **C** we will write $|\,\mathbf{C}\,|$ for the class of its objects and **C** for its class of morphisms. For more background on categories see [Mac71] or [Wag].

The sums and products used in this paper are, for the most part, the categorical coproducts and products from the category **Set** of sets and total functions. But the way we use them exploits their categorical definitions rather than their set-theoretical definitions. As an aid to the reader we give both definitions.

Definition 1 Let A and B be sets, then a *categorical product* for A and B is a set, denoted $A \times B$, together with functions $p_{A,B} : A \times B \to A$, and $q_{A,B} : A \times B \to B$, with the property that, for any set C and any functions $f : C \to A$, and $g : C \to B$ there exists a unique morphism, $\langle f, g \rangle$, from C to $A \times B$, such that $\langle f, g \rangle \bullet p_{A,B} = f$ and $\langle f, g \rangle \bullet q_{A,B} = g$. We call $A \times B$ the *product object* for A and B, we call $p_{A,B}$ and $q_{A,B}$ the *projections* for A and B, and we call $\langle f, g \rangle$ the *product mediator* for f and g.

The *Cartesian product* of two sets A and B is the set

$$A \otimes B = \{\langle a, b \rangle \mid a \in A, b \in B\}$$

It is easy to see that when we equip $A \otimes B$ with the mappings

$$\begin{aligned} \pi_1 : A \otimes B &\to A \\ \langle a, b \rangle &\mapsto a \end{aligned}$$

$$\begin{aligned} \pi_2 : A \otimes B &\to B \\ \langle a, b \rangle &\mapsto b \end{aligned}$$

that it is a product in **Set**. □

Definition 2 Let A and B be sets, then a *categorical coproduct* for A and B is a set, denoted $A + B$, together with functions $\kappa_{A,B} : A \to A + B$, and $\lambda_{A,B} : B \to A + B$, with the property that, for any set C and any functions $f : A \to C$, and $g : B \to C$ there exists a unique morphism, $[f, g]$, from $A + B$ to C such that $\kappa_{A,B} \bullet [f, g] = f$ and $\lambda_{A,B} \bullet [f, g] = g$. We call $A + B$ the *coproduct object* for A and B, we call $\kappa_{A,B}$ and $\lambda_{A,B}$ the *injections* for A and B, and we call $[f, g]$ the *coproduct mediator* for f and g.

The *disjoint sum* of two sets A and B is the set

$$A \oplus B = (A \otimes \{1\}) \cup (B \otimes \{2\})$$

It is easy to see that when we equip $A \oplus B$ with the mappings

$$\begin{aligned} \iota_1 : A &\to A \oplus B \\ a &\mapsto \langle a, 1 \rangle \end{aligned}$$

$$\begin{aligned} \iota_2 : B &\to A \oplus B \\ b &\mapsto \langle b, 2 \rangle \end{aligned}$$

that it is a coproduct in **Set** □

Let **Pfn** be the category of sets and partial functions.

Fact 1 *The category* **Pfn** *also has products and coproducts. The coproducts are the same as in* **Set**, *but the products are different from those in* **Set**. □

3 Recursive Types

We begin by defining the concept of a recursive type specification $\mathcal{R}$ and its semantics, the algebra $\mathcal{R}(\alpha)$.

Definition 3 A *recursive type specification*, $\mathcal{R}$, consists of disjoint finite sets G, R, and V, a designated object $\mathbf{1} \in G$, and, where $S = G \cup R \cup V$, a mapping $\mu : R \cup V \to S \times S$. Intuitively the elements of G are *given* or *ground*, types; and the elements of R are recursively defined *record* types, and the elements of V are recursively defined *variant* types. If $r \in R$ and $\mu(r) = \langle s_1, s_2 \rangle$ then the type r will be the product of the types s_1 and s_2. Similarly, if $v \in V$ and $\mu(v) = \langle s_1, s_2 \rangle$ then the type v will be the coproduct of the types s_1 and s_2.

Define $\Sigma(\mathcal{R})$, the *signature for* $\mathcal{R}$, to be the $(S * \times S)$-indexed family of sets $\Sigma(\mathcal{R}) = \langle \Sigma(\mathcal{R})_{w,s} \mid w \in S *, s \in S \rangle$ where, for each $r \in R$, $\Sigma(\mathcal{R})_{r,\mu(r)_1} = \{p_r : r \to \mu(r)_1\}$, $\Sigma(\mathcal{R})_{r,\mu(r)_2} = \{q_r : r \to \mu(r)_2\}$; for each $v \in V$, $\Sigma(\mathcal{R})_{\mu(v)_1,v} = \{\kappa_r : \mu(v)_1 \to v\}$, $\Sigma(\mathcal{R})_{\mu(v)_2,v} = \{\lambda_r : \mu(v)_2 \to v\}$; and $\Sigma(\mathcal{R})_{w,s} = \emptyset$, for all other w and s. □

The above definition restricts us to binary RECORDs and VARIANTs, that is, the RECORDs have exactly two fields and the VARIANTs have exactly two tags. This restriction simplifies the proofs in this paper by reducing the number of cases that need to be considered. We will give a more general treatment of RECORDs and VARIANTs in a forthcoming paper.

Proposition 1 *Given an* assignment $\alpha : G \to \mid \mathbf{Set} \mid$, *there exists a least* $\Sigma(\mathcal{R})$*-algebra* $\mathcal{R}(\alpha)$, *such that,* $A_{\mathbf{1}}$ *is "the" singleton set; for each* $r \in R$, $A_r = A_{\mu(r)_1} \times A_{\mu(r)_2}$ *with projections* $(p_r)_A : A_r \to A_{\mu(r)_1}$ *and* $(q_r)_A : A_r \to A_{\mu(r)_2}$*; and, for each* $v \in V$, $A_v = A_{\mu(v)_1} + A_{\mu(v)_2}$ *with injections* $(\kappa_v)_A : A_{\mu(v)_1} \to A_v$ *and* $(\lambda_v)_A : A_{\mu(v)_2} \to A_v$. □

Example 1 Take $G = \{\mathbf{1}\}$, $R = \emptyset$, $V = \{\mathbf{N}\}$, and $\mu(\mathbf{N}) = \{\langle \mathbf{1}, \mathbf{N} \rangle\}$, then there is but one choice for α, and $\mathcal{R}(\alpha)$ is the algebra of the natural numbers in that $\mathcal{R}(\alpha) \cong \omega = \{0, 1, 2, \ldots\}$, $\lambda_{\mathcal{R}(\alpha)} = suc$, the successor function, and $\kappa_{\mathcal{R}(\alpha)} = 0$ in the sense that it is the function from $\mathcal{R}(\alpha)_{\mathbf{1}} \to \mathcal{R}(\alpha)_{\mathbf{N}}$ taking the unique element of $\mathcal{R}(\alpha)_{\mathbf{1}}$ to 0. □

Example 2 Take $G = \{\mathbf{1}\}$, $R = \emptyset$, $V = \{\mathbf{B}\}$, and $\mu(\mathbf{B}) = \langle \mathbf{1}, \mathbf{1} \rangle$. Then $\mathbf{B} = \mathbf{1}+\mathbf{1}$, and writing `true` for $\kappa_B : \mathbf{1} \to \mathbf{B}$, and `false` for $\lambda_B : \mathbf{1} \to \mathbf{B}$, we see that $\mathcal{R}(\alpha)$ can be interpreted as the Boolean type. □

We could, of course, combine the above two examples by taking $G = \{\mathbf{1}\}$, $R = \emptyset$, $V = \{\mathbf{N}, \mathbf{B}\}$, and μ such that $\mu(\mathbf{N}) = \langle \mathbf{1}, \mathbf{N} \rangle$, $\mu(\mathbf{B}) = \langle \mathbf{1}, \mathbf{1} \rangle$.

We end this section by giving two examples where R and V are both non-empty.

Here is the tree example from the introduction.

Example 3 Take $G = \{\mathbf{1}, D\}$, $R = \{\mathbf{P}\}$, $V = \{\mathbf{T}\}$, and μ such that $\mu(\mathbf{P}) = \langle \mathbf{T}, \mathbf{T} \rangle$, and $\mu(\mathbf{T}) = \langle D, \mathbf{P} \rangle$. □

Example 4 Of course we must give the example of a Stack-of-D, (or, equivalently, of a List-of-D). Take $G = \{\mathbf{1}, D\}$, $R = \{\mathbf{P}\}$, $V = \{\mathbf{S}\}$, and μ such that $\mu(\mathbf{P}) = \langle D, \mathbf{S} \rangle$, and $\mu(\mathbf{S}) = \langle \mathbf{1}, \mathbf{P} \rangle$. I claim that the natural names for the projections and injections are as follows:

$$\kappa_S = \mathtt{empty} : \mathbf{1} \to \mathbf{S}$$

$$\lambda_S = \mathtt{push} : \mathbf{P} \to \mathbf{S}$$

$$p_P = \mathtt{pop} : \mathbf{P} \to \mathbf{S}$$

$$q_P = \mathtt{top} : \mathbf{P} \to D$$

□

4 Derived Operations

We can exploit the special structure of $\mathcal{R}(\alpha)$ to define *derived operations* in addition those defined in terms of composition and product tupling. Before spelling out the precise notion of derived operation that we have in mind, we give some motivating examples. The first example illustrates the exploitation of the coproduct mediator to give a kind of definition-by-cases.

Example 5 We can define the predecessor function on $\mathbf{N}$, as the coproduct mediator $\mathtt{pred} = [0, 1_N] : (\mathbf{1} + \mathbf{N}) \to \mathbf{N}$.

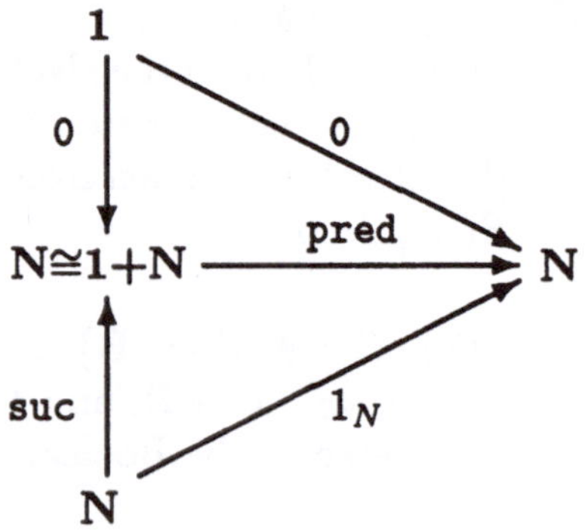

□

The following Proposition provides the means for extending definition-by-cases to functions of more that one variable.

Proposition 2 *For any sets* $A, B, C \in |\,\mathbf{Set}\,|$,

$$A \times C \xrightarrow{\kappa_{A,B} \times 1_C} (A+B) \times C \xleftarrow{\lambda_{A,B} \times 1_C} B \times C$$

is a coproduct in **Set** *and so*

$$((A+B) \times C) \cong ((A \times C) + (B \times C))$$

□

How this can be exploited to define a function of two arguments is illustrated by the definition of the operation "$\wedge$" in the following example.

Example 6 Take $G = \{\mathbf{1}\}$, $R = \emptyset$, $V = \{\mathbf{B}\}$, and $\mu(\mathbf{B}) = \langle \mathbf{1}, \mathbf{1} \rangle$ Then $\mathbf{B} = \mathbf{1}+\mathbf{1}$, and writing `true` for $\kappa_B : \mathbf{1} \to \mathbf{B}$, and `false` for $\lambda_B : \mathbf{1} \to \mathbf{B}$, we see that $\mathcal{R}(\alpha)$ can be interpreted as the Boolean type. Furthermore the negation operation, $\neg$, and the conjunctive operation, $\wedge$, are easily defined as follows:

$$\neg = [\mathtt{false},\ \mathtt{true}] : \mathbf{B} \to \mathbf{B}$$

and

$$\wedge = [q_{1,B},\ p_{1,B} \bullet \mathtt{false}] : \mathbf{B} \times \mathbf{B} \to \mathbf{B}$$

where the latter makes use of the fact that

$$\mathbf{B} \times \mathbf{B} = (\mathbf{1}+\mathbf{1}) \times \mathbf{B} \cong (\mathbf{1} \times \mathbf{B}) + (\mathbf{1} \times \mathbf{B})$$

The following diagram shows the construction pictorially

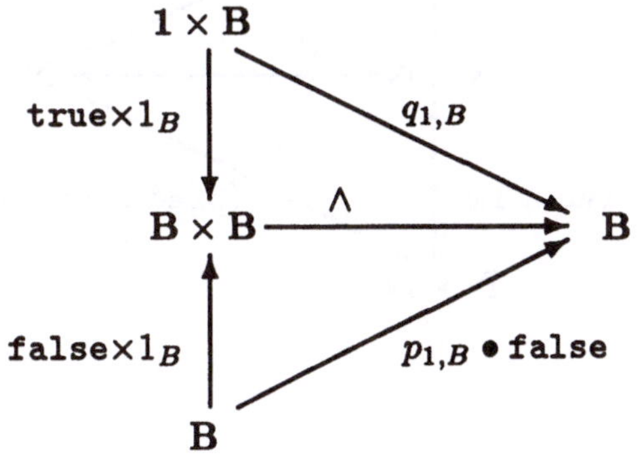

□

The derived operations that we have defined so far are rather simple, to get more interesting operations we must introduce some form of iteration. From the point of programming theory, the natural candidate is some form of WHILE-DO. WHILE-DO is easy to define in the current context, but, since it allows us to define derived operations that are partial functions, it requires some caution. In particular, in order to have Proposition 2 still hold, we must continue to use products from **Set** rather than using the products from **Pfn**.

The following application of WHILE-DO to define the addition of natural numbers illustrates both the informal and the formal constructs.

Example 7 Given the successor and predecessor functions, we can write the following WHILE-DO program for addition:

```
Proc add(A,B:NAT):NAT /* add:NAT×NAT→NAT */
   WHILE A≠0 DO
      A, B := pred(A), suc(B) /* simultaneous assignment */
   OD;
   Return B;
   END.
```

We claim that the desired function add : $\mathbf{N} \times \mathbf{N} \to \mathbf{N}$ equals $W \bullet p_{N,N}$ where $W : \mathbf{N} \times \mathbf{N} \to \mathbf{N} \times \mathbf{N}$ is the unique function such that

$$W = [0 \times 1_N, (\mathtt{pred} \times \mathtt{suc}) \bullet W].$$

Study of the following diagram may help clarify why this is so.

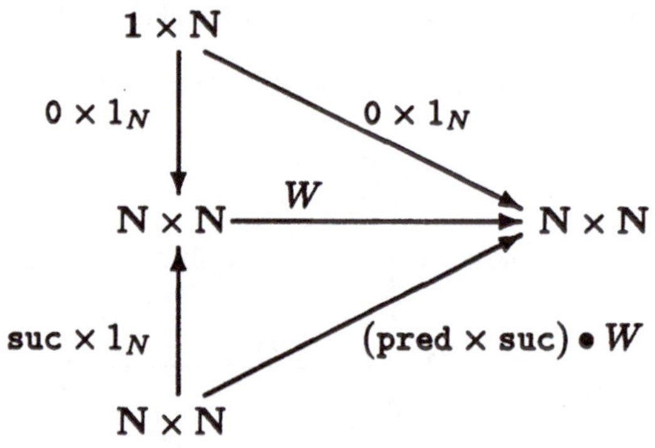

□

The basic results for the WHILE-DO construct are summed up in the next proposition.

Proposition 3 *Let A, B and C be sets, and let $g : B \times C \rightharpoonup (A+B) \times C$ be a partial function, then there exists a least partial function $f : (A+B) \times C \rightharpoonup (A+B) \times C$ such that*

$$f = [\kappa_{A,B} \times 1_C,\ g \bullet f].$$

Furthermore, for every $x \in (A+B) \times C$, either $f(x)$ is undefined or there exists $n_x \geq 0$ such that

$$f(x) = [\kappa_{A,B},\ g]\ n_x(x).$$

□

The ideas behind the above examples can be drawn together to yield the following generalized concept of derived operation.

Definition 4 Given $\mathcal{R} = \langle G, R, V, 1, \mu\rangle$ and $\alpha : G \rightarrow\mid \mathbf{Set} \mid$, let $A = \mathcal{R}(\alpha)$ be the $\Sigma(\mathcal{R})$-algebra constructed as above. Let $A = \langle A_s \mid s \in S = G \cup R \cup V\rangle$ be the family of carriers of A. For any $u = s_1 \cdots s_n \in S*$ let $A\ u = A_{s_1} \times \cdots \times A_{s_n}$.

Let $u, w \in S*$, then a function $f : A\ u \rightarrow A\ w$ is *derivable (w.r.t A)* iff:

1. $u = s_1 \cdots s_n$, $w = s_i$ and f is the projection function, $f : A\ u \rightarrow A_{s_i}$.

2. $f = p_r : A_r \rightarrow A_{\mu(r)_1}$ or $f = q_r : A_r \rightarrow A_{\mu(r)_2}$.

3. $f = \kappa_v : A_{\mu(v)_1} \rightarrow A_v$, or $f = \lambda_v : A_{\mu(v)_2} \rightarrow A_v$.

4. $w = s_1 \cdots s_p$ and for each $i = 1, \ldots, p$ there is a derivable function $f_i : A\ u \rightarrow A_{s_i}$, such that $f = \langle f_1, \ldots, f_p\rangle$.

5. there exists $x \in S*$ and derivable functions $g : u \rightarrow x$ and $h : x \rightarrow w$, such that $f = g \bullet h$.

6. $u = v.x$, $v \in V$, $x \in S*$, and there are derivable functions g and h, $g : A\ \mu(v)_1.x \rightarrow A_v$ and $h : A\ \mu(v)_2.x \rightarrow A_v$ such that $f = [g, h]$.

7. $u = v.x$, $v \in V$, $x \in S*$, and there is a derivable operation, $g : A\ \mu(v)_2.x \rightarrow A\ u$, and f is the least function such that $f = [\kappa_v \times 1_x,\ g \bullet f]$.

8. Same as 7 except that $g : A\ \mu(v)_1.x \rightarrow A\ u$ and $f = [g \bullet f, \lambda_v \times 1_x]$.

9. $w = r \in R$, and there exist derivable operations g and h, $g : A\ u \rightarrow A_{\mu(r)_1}$, $h : A\ u \rightarrow A_{\mu(r)_2}$, such that $f \bullet p_r = g$, and $f \bullet q_r = h$.

□

5 Pointer Algebras

In order to be able to talk about pointers in an abstract setting we need an abstract model of computer memory. The underlying model used in this paper is the following.

Definition 5 A *Pointer presentation* specification, $\mathcal{R} = \langle G, R, V, \mathbf{1}, \mu \rangle$ together with a relation ρ on $G + R + V$.

An $\langle \mathcal{R}, \rho \rangle$*-algebra* consists of an $\mathcal{R}$-algebra A equipped with an additional operation $\sigma_{s,t} : A_s \to A_t$ for each pair $\langle s, t \rangle \in \rho$. A ρ-indexed set

$$\sigma = \langle \sigma_{s,t} : A_s \to A_t \mid \langle s, t \rangle \in \rho \rangle$$

is called a *state* for the algebra. □

The intuition for the general model is that if $\langle s, t \rangle \in \rho$ then A_s is a set of "active locations" containing values from A_t, where the value from A_t contained by "location" $a \in A_s$ in the "current state" σ is $\sigma_{s,t}(a)$. Note that "inactive locations" do not even appear in this model. When a new location or pointer is needed it is created by suitably modifying the algebra, i.e., changing it to a new algebra by freely adjoining an element to the appropriate carrier and appropriately changing σ. In the next section we introduce a very restricted version of this model that is tailored to the questions addressed in this paper. See my paper [Wag89] for other applications of this model (the formulation there is slightly different).

6 Implementing Recursive Types Using Pointers

Given a recursive type specification $\mathcal{R}$ we want to replace it by a specification for a corresponding class of pointer algebras. The desired specification is called a pointer presentation and it corresponds, intuitively, to the explicit pointer implementation of recursive types discussed in the introduction.

Definition 6 Given a pointer presentation $\mathcal{R} = \langle G, R, V, \mathbf{1}, \mu \rangle$, we define its *pointer presentation* to be $Pt(\mathcal{R}) = \langle \langle G + R + V, R, V, \mathbf{1}, \mu \sharp \rangle, \rho \rangle$ where,

1. if $x \in R \cup V$ we let $\uparrow x$ denote its image in $G \sharp = G + R + V$ under the appropriate coproduct injection $\iota_R : R \to G \sharp$ or $\iota_V : V \to G \sharp$.

2. the graph of ρ consists of exactly the pairs $\langle \uparrow x, x \rangle, x \in R \cup V$.

3. Let $S \sharp = G \sharp \cup R \cup V$, then $\mu \sharp : R \cup V \to S \sharp \times S \sharp$, where for $i = 1, 2$,

$$\mu \sharp (x)_i = \begin{cases} \mu(x)_i & \text{if } (x)_i \in G \\ \uparrow \mu(x)_i & \text{if } \mu(x)_i \in R \cup V. \end{cases}$$

Note that $\mu(x)_i$ is always in $G \sharp$, and never in $R \cup V$.

□

Example 8 Let $\mathcal{R}$ be the recursive type specification for the natural numbers given in Example 1, i.e., $\mathcal{R} = \langle G, R, V, \mathbf{1}, \mu\rangle$ where $G = \{\mathbf{1}\}$, $R = \emptyset$, $V = \{\mathbf{N}\}$, $\mu(\mathbf{N}) = \langle \mathbf{1}, \mathbf{N}\rangle$, then the corresponding pointer presentation is $\langle\langle G\sharp, R, V, \mathbf{1}, \mu\sharp\rangle, \rho\rangle$ where $G\sharp = \{\mathbf{1}, \uparrow\mathbf{N}\}, R = \emptyset, V = \{\mathbf{N}\}, \mu\sharp(\mathbf{N}) = \langle \mathbf{1}, \uparrow\mathbf{N}\rangle$, and $\rho = \{\langle \uparrow\mathbf{N}, \mathbf{N}\rangle\}$.

For each natural number n, there is a $Pt(\mathcal{R})$-algebra A_n corresponding in a natural way to the subset $\{0, 1, \ldots, n\}$ of the natural numbers, given by taking $(A_n)_{\uparrow N} = [n] = \{1, \ldots, n\}$, so that

$$(A_n)_N = \mathbf{1} + (A_n)_{\uparrow N} = \mathbf{1} + [n] \cong \{0, 1, \ldots, n\}$$

and taking $(\sigma_{\uparrow N, N})_{A_n}(i) = i - 1$ for each $i \in [n]$. This can be put in pictures as follows:

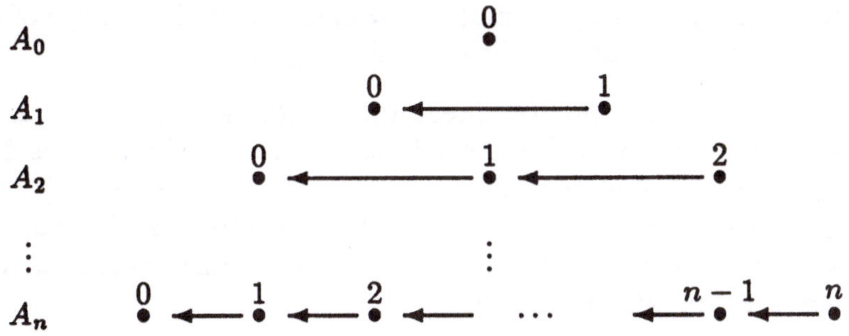

But is is also easy to see that there are $Pt(\mathcal{R})$-algebras in which a given natural number may appear more than once, e.g., the algebra corresponding to the following picture

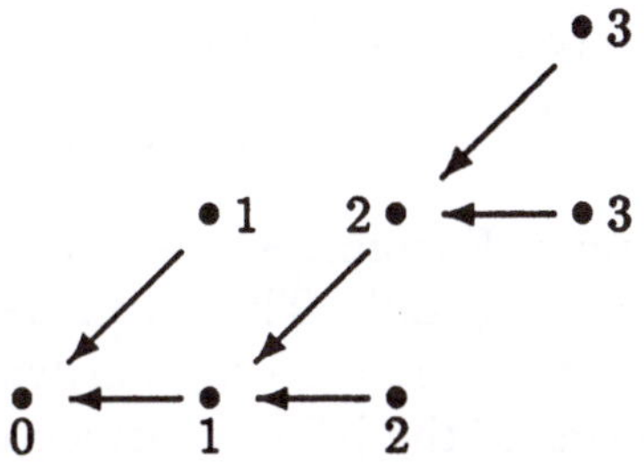

Indeed we can even have algebras that do not correspond to sets of natural numbers, for example, the algebra corresponding to the picture

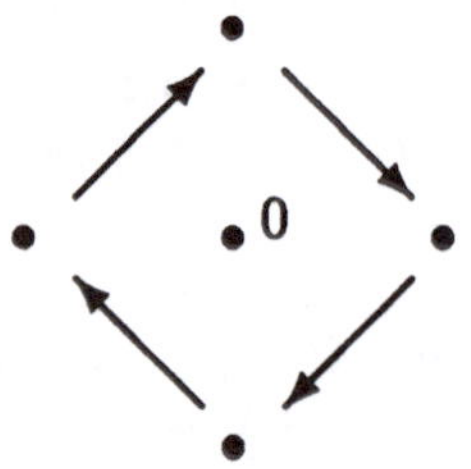

□

In this framework, the action of "creating a pointer of type x and pointing it at a given value" is captured by introducing operations for modifying the corresponding algebras by adding a new element to the carrier of sort $\uparrow x$ and appropriately modifying the operation $\sigma_{\uparrow x,x}$.

Definition 7 Given a pointer presentation $\langle\langle G+R+V, R, V, \mathbf{1}, \mu \sharp\rangle, \rho\rangle$ as above, define NEW to be the ρ-indexed family of functions NEW $= \langle NEW_{\uparrow x,x} \mid \langle\uparrow x, x\rangle \in \rho\rangle$ such that, given an $\langle\mathcal{R},\rho\rangle$-algebra A, $\langle\uparrow x, x\rangle \in \rho$, and $a \in A_x$, then $\mathrm{NEW}_{\uparrow x,x}(A,a)$ is the $\langle\mathcal{R},\rho\rangle$-algebra, B, that results from freely adjoining a new element, $\nu = \nu(A,a)$ to $A_{\uparrow x}$ and taking σ_B to be the extension of σ_A in which $(\sigma_{\uparrow x,x})_B(\nu) = a$. □

Proposition 4 *Given $\mathcal{R}$, ρ, $Pt(\mathcal{R})$-algebra A, $\langle\uparrow x, x\rangle \in \rho$, and $a \in A_x$, $NEW_{\uparrow x,x}(A,a)$ is well-defined (up to isomorphism). In particular, letting $B = NEW_{\uparrow x,x}(A,a)$, we have, $B_{\uparrow x} \cong A_{\uparrow x} + \mathbf{1}, B_y \cong A_y$ for all other $y \in G\,\sharp$, and, for every $y \in R \cup V$, there is an inclusion $\mathcal{I}_y : A_y \to B_y$. The state component $(\sigma_{\uparrow x,x})_B$ is given by the mediating morphism,*

$$(\sigma_{\uparrow x,x})_B = [(\sigma_{\uparrow x,x})_A,\ (1 \mapsto a)]$$

and for all other $\langle\uparrow y, y\rangle \in \rho$, $(\sigma_{\uparrow y,y})_B = (\sigma_{\uparrow y,y})_A \bullet \mathcal{I}_y$. □

As shown at the end of Example 8, there will generally be some $Pt(\mathcal{R})$-algebras that are "meaningless". These can be effectively eliminated by restricting our attention to the algebras constructed from from the initial $Pt(\mathcal{R})$-algebra by means of the NEW constructors. Applied to Example 8 this will still give us the algebras with multiple copies of various natural numbers, but it will eliminate the pathological examples. Let us make this precise.

Definition 8 Given $\alpha : G \to\mid \mathbf{Set} \mid$, let $I(\mathcal{R},\alpha)$ be the $Pt(\mathcal{R})$-algebra given by the assignment $\beta : (G+R+V) \to\mid \mathbf{Set} \mid$ such that $\beta(g) = \alpha(g)$, and $\beta(r) = \beta(v) = \emptyset$. Thus $I(\mathcal{R},\alpha)$ corresponds to the "memory state before any pointers have been created."

Define $\mathbf{R}(\mathcal{R},\alpha)$ then to be the set of all $Pt(\mathcal{R})$-algebras reachable from $I(\mathcal{R},\alpha)$ by repeated applications of the operations $\mathrm{NEW}_{\uparrow x,x}$, for $\langle\uparrow x, x\rangle \in \rho$. □

Notation: In order to reduce notational clutter we shall often write NEW for $\mathrm{NEW}_{\uparrow x,x}$, and σ for $\sigma_{\uparrow x,x}$.

A recursive type specification $\mathcal{R}$ together with an assignment $\alpha : G \to |\mathbf{Set}|$ has, as its semantics, a single algebra $\mathcal{R}(\alpha)$. Given $\mathcal{R} = \langle G, R, V, \mathbf{1}, \mu\rangle$, and $\alpha : G \to |\mathbf{Set}|$ the corresponding pointer presentation $Pt(\alpha)$ has as its semantics a large class $\mathbf{R}(\mathcal{R}, \alpha)$ of algebras each of which, we claim, is "a finite approximation of $\mathcal{R}(\alpha)$". We bring this into sharper focus by showing how we can construct a $\Sigma(\mathcal{R})$-algebra $\mathbf{P} = \mathbf{P}(\mathcal{R}, \alpha)$ from $\mathbf{R}(\mathcal{R}, \alpha)$ which is a homomorphic preimage of $\mathcal{R}(\alpha)$.

Definition 9 Define $\mathbf{P} = \mathbf{P}(\mathcal{R}, \alpha)$ to be the $\Sigma(\mathcal{R})$-algebra (see Definition 3) where for each $s \in (G \cup R \cup V)$, $\mathbf{P}_s = \{(A, a) \mid A \in \mathbf{R}(\mathcal{R}, \alpha),\ \textit{and}\ a \in A_s\}$; and, for any $r \in R$, or $v \in V$,

$$(p_r)_{\mathbf{P}}(A, a) = \begin{cases} (A, (p_r)_A(a)) & \text{if } \mu(r)_1 \in G \\ (A, \sigma_A((p_r)_A(a))) & \text{if } \mu(r)_1 \in R \cup V \end{cases}$$

$$(\kappa_r)_{\mathbf{P}}(A, a) = \begin{cases} (A, (\kappa_v)_A(a)) \text{ if } \mu(v)_1 \in G \\ (NEW(A, a), (\kappa_v)_{NEW(A,a)}(\nu(A, a))) \\ \qquad \text{if } \mu(v)_1 \in R \cup V \end{cases}$$

and q_r, and λ_v are similarly defined. □

Note that while $\mathbf{P}$ is a $\Sigma(\mathcal{R})$-algebra, it is not an $\mathcal{R}(\alpha)$-algebra since, for example, we do not have $\mathbf{P}_r = \mathbf{P}_{\mu(r)_1} \otimes \mathbf{P}_{\mu(r)_2}$.

Definition 10 We define an $S = (G \cup R \cup V)$-indexed family of maps $\mathbf{h} = \langle \mathbf{h}_s : \mathbf{P}_s \to \mathcal{R}(\alpha)_s \mid s \in S\rangle$, by recursion, as follows:

If $(A, a) \in \mathbf{P}_g$, $g \in G$, then $\mathbf{h}(A, a) = a$.

If $(A, a) \in \mathbf{P}_r$, $r \in R$, then $a = \langle a_1, a_2\rangle \in A_{\mu\,\sharp(r)_1} \otimes A_{\mu\,\sharp(r)_2}$ For $i = 1, 2$, either $\mu(r)_i \in G$ or $\mu\,\sharp(r)_i = \uparrow x$ for $x \in R \cup V$. In the latter case, $a_i \in A_{\uparrow x}$, and $\sigma_A(a_i) \in A_x$, by the definition of A reachable. Using this notation, define

$$\overline{a}_i = \begin{cases} a_i & \text{if } \mu(r)_i \in G \\ \sigma_A(a_i) & \text{if } \mu\,\sharp(r)_i = \uparrow x, \text{ where } x \in R \cup V \end{cases}$$

and then define

$$\mathbf{h}(A, a) = \langle \mathbf{h}(A, \overline{a}_1), \mathbf{h}(A, \overline{a}_2)\rangle$$

If $(A, a) \in \mathbf{P}_v$ for some $v \in V$, then either $a = (\kappa)_A(a_1)$ for some $a_1 \in A_{\mu\,\sharp(v)_1}$, or $a = (\lambda)_A(a_2)$ for some $a_2 \in A_{\mu\,\sharp(v)_2}$. Then, using the definition of $\overline{a}_i$ as given above, define

$$\mathbf{h}(A, a) = \begin{cases} (\kappa_v)_{\mathcal{R}(\alpha)}(\mathbf{h}(A, \overline{a}_1)) & \text{if } a = (\kappa_v)_A(a_1) \\ (\lambda_v)_{\mathcal{R}(\alpha)}(\mathbf{h}(A, \overline{a}_2)) & \text{if } a = (\lambda_v)_A(a_2). \end{cases}$$

□

Fact 2 *For any reachable* $A \in \mathbf{R}(\mathcal{R}, \alpha)$, *we have* $\mathbf{h}(NEW(A, b), a) = \mathbf{h}(A, a)$. □

Proposition 5 *The above defined family of mappings,* $\mathbf{h}$, *is a homomorphism* $\mathbf{h} : \mathbf{P} \to \mathcal{R}(\alpha)$.

Proof: We must show that

$$\mathbf{h}((p_r)_{\mathbf{P}}(A, a)) = (p_r)_{\mathcal{R}(\alpha)}(\mathbf{h}(A, a))$$

and,

$$\mathbf{h}((\kappa_v)_{\mathbf{P}}(A, a)) = (\kappa_v)_{\mathcal{R}(\alpha)}(\mathbf{h}(A, a))$$

for all $r \in R$ and $v \in V$.

$$\begin{aligned}
&\mathbf{h}((p_r)_{\mathbf{P}}(A, a)) \\
&= \begin{cases} \mathbf{h}(A, a_1) & \text{if } \mu(r)_1 \in G \\ \mathbf{h}(A, \sigma_A(a_1)) & \text{if } \mu(r)_1 \in R \cup V \end{cases} && \text{def } (p_r)_{\mathbf{P}} \\
&= \mathbf{h}(A, \overline{a}_1) && \text{def } \overline{a}_1 \\
&= (p_r)_{\mathcal{R}(\alpha)}(\langle \mathbf{h}(A, \overline{a}_1), \mathbf{h}(A, \overline{a}_2)\rangle) && \text{def } (p_r)_{\mathcal{R}(\alpha)} \\
&= (p_r)_{\mathcal{R}(\alpha)}(\mathbf{h}(A, a)) && \text{def } \mathbf{h}
\end{aligned}$$

$$\begin{aligned}
&\mathbf{h}((\kappa_v)_{\mathbf{P}}(A, a)) \\
&= \begin{cases} \mathbf{h}(a, (\kappa_v)_A(a)) & \text{if } \mu(v)_1 \in G \\ \mathbf{h}(NEW(A, a), (\kappa_v)_{NEW(A,a)}(\nu(A, a))) & \\ & \text{if } \mu(v)_1 \in R \cup V \end{cases} && \text{def } (\kappa_v)_{\mathbf{P}} \\
&= \begin{cases} (\kappa_v)_{\mathcal{R}(\alpha)}(a) & \text{if } \mu(v)_1 \in G \\ (\kappa_v)_{\mathcal{R}(\alpha)}(\mathbf{h}(NEW(A, a), \sigma_{NEW(A,a)}(\nu(A, a))) & \\ & \text{if } \mu(v)_1 \in R \cup V \end{cases} && \text{def } \mathbf{h} \\
&= \begin{cases} (\kappa_v)_{\mathcal{R}(\alpha)}(a) & \text{if } \mu(v)_1 \in G \\ (\kappa_v)_{\mathcal{R}(\alpha)}\mathbf{h}(NEW(A, a), a) & \text{if } \mu(v)_1 \in R \cup V \end{cases} && \text{def NEW} \\
&= \begin{cases} (\kappa_v)_{\mathcal{R}(\alpha)}(a) & \text{if } \mu(v)_1 \in G \\ (\kappa_v)_{\mathcal{R}(\alpha)}\mathbf{h}(A, a)) & \text{if } \mu(v)_1 \in R \cup V \end{cases} && \text{Fact 2.} \\
&= (\kappa_v)_{\mathcal{R}(\alpha)}(\mathbf{h}(A, a)).
\end{aligned}$$

□

7 Derived Operations on Pointers

The existence of **h** is still not enough to allow us to implement recursive types with pointers, but we are getting close. By further exploiting the structure underlying **P**, we can enrich **P** by adding a small number of operations (and operators), corresponding, in **P**, to operations on pointers, and, under **h** to the basic constructions used to build the derived operators in $\mathcal{R}(\alpha)$. The needed operators are given in the following definition.

Definition 11 Exploiting the structure of the algebras in $\mathbf{R}(\mathcal{R}, \alpha)$ we can define special operations in **P** as follows:

$$\begin{gathered}\gamma_r : \mathbf{P}_{\mu(r)_1} \times \mathbf{P}_{\mu(r)_2} \rightarrow \mathbf{P}_r \\ \langle (A_1, a_1), (A_2, a_2) \rangle \mapsto \\ (NEW((A_1, a_1) \oplus NEW(A_2, a_2), \langle \nu(A_1, a_1), \nu(A_2, a_2) \rangle)\end{gathered}$$

where, $A_1 \oplus A_2$ is the $\mathbf{R}(\mathcal{R}, \alpha)$-algebra A such that, for $g \in G$

$$A_g = \alpha(g)$$

for $\uparrow x \in (\uparrow R) \cup (\uparrow V)$,

$$A_{\uparrow x} = (A_1)_{\uparrow x} + (A_2)_{\uparrow x}$$

and, for all $a \in A_{\uparrow x}$,

$$\sigma(a) = \begin{cases} \sigma_1(a) & \text{if } a \in A_1 \\ \sigma_2(a) & \text{if } a \in A_2 \end{cases}$$

Secondly, for each $v \in V$, we define

$$\begin{gathered}\tau_v : \mathbf{P}_v \rightarrow \mathbf{P}_{\mu(v)_1} + \mathbf{P}_{\mu(v)_2} \\ \langle A, a \rangle \mapsto [\overline{\sigma}_1 \bullet \iota_1, \ \overline{\sigma}_2 \bullet \iota_2]_{A_v}(a)\end{gathered}$$

where, for $i = 1, 2$, ι_i is the coproduct injection,

$$\iota_i : \mathbf{P}_{\mu(v)_1} \rightarrow \mathbf{P}_{\mu(v)_1} + \mathbf{P}_{\mu(v)_2}$$

and

$$\overline{\sigma}_i : A_{\mu \sharp(v)_1} \rightarrow \mathbf{P}_{\mu(v)_i}$$

$$\overline{\sigma}_i(a) = \begin{cases} \langle A, a \rangle & \text{if } \mu \sharp(v)_i \in G \\ \langle A, \sigma(a) \rangle & \text{if } \mu \sharp(v)_i \in (\uparrow R) \cup (\uparrow V) \end{cases}$$

and $[\overline{\sigma}_1 \bullet \iota_1, \ \overline{\sigma}_2 \bullet \iota_2]_{A_v}$ is the coproduct mediator indicated in the following diagram

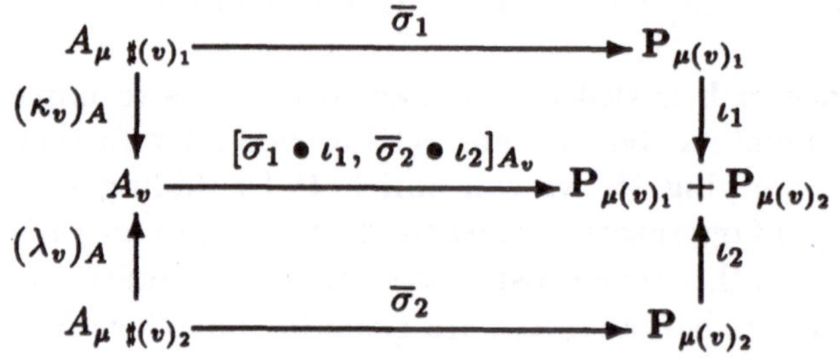

□

Definition 12 Let $u, w \in S*$, and let $f : \mathcal{R}(\alpha)\ u \to \mathcal{R}(\alpha)\ w$ be a derived operator in $\mathcal{R}(\alpha)$, we say that f has a *precursor* in $\mathbf{P}$ if there exists a derived operator $\overline{f} : \mathbf{P}\ u \to \mathbf{P}\ w$ in $\mathbf{P}$ extended with τ and γ, such that for all $\langle \overline{A}, \overline{a} \rangle = (\langle A_1, a_1 \rangle, \ldots, \langle A_{|u|}, a_{|u|} \rangle) \in \mathbf{P}\ u$,

$$\mathbf{h}(\overline{f}(\langle \overline{A}, \overline{a} \rangle)) = f(\mathbf{h}(\langle \overline{A}, \overline{a} \rangle)).$$

□

Fact 3 *We have: $(\kappa_v)_P$ is a precursor for $(\kappa_v)_{\mathcal{R}(\alpha)}$, $(\lambda_v)_P$ is a precursor for $(\lambda_v)_{\mathcal{R}(\alpha)}$, $(p_r)_P$ is a precursor for $(p_r)_{\mathcal{R}(\alpha)}$, and $(q_r)_P$ is a precursor for $(q_r)_{\mathcal{R}(\alpha)}$.*

Proof: This follows immediately from Proposition 5. □

Proposition 6 *Let $r \in R$, let $f_i : \mathcal{R}(\alpha)\ u \to \mathcal{R}(\alpha)_{\mu(r)_i}$ with precursor $\overline{f}_i$ in $\mathbf{P}$, then*

$$(\overline{f}_1, \overline{f}_2) \bullet \gamma_r : \mathbf{P}_u \to \mathbf{P}_r$$

is a precursor in $\mathbf{P}$ for product mediator

$$\langle f_1, f_2 \rangle : \mathcal{R}(\alpha)\ u \to \mathcal{R}(\alpha)_r$$

Proof: Let $\langle A, a \rangle \in \mathbf{P}\ u$, and, for $i = 1, 2$, let $\overline{f}_i(A, a) = \langle A_i, a_i \rangle$. Then we have

$$\begin{array}{lll}
\multicolumn{3}{l}{\mathbf{h}(((\overline{f}_1, \overline{f}_2) \bullet \gamma_r)(A, a))} \\
= & \mathbf{h}(\gamma_r(\langle \overline{f}_1(A, a), \overline{f}_2(A, a) \rangle)) & \text{def } (\overline{f}_1, \overline{f}_2) \\
= & \mathbf{h}(\gamma_r(\langle (A_1, a_1), (A_2, a_2) \rangle)) & f_i(A, a) = (A_i, a_i) \\
= & \mathbf{h}(A_1 \oplus A_2, \langle \nu(A_1, a_1), \nu(A_2, a_2) \rangle) & \text{def } \gamma_r \\
= & \langle \mathbf{h}(A_1, a_1), \mathbf{h}(A_2, a_2) \rangle & \text{def } \mathbf{h} \\
= & \langle \mathbf{h}(\overline{f}_1(A, a)), \mathbf{h}(\overline{f}_2(A, a)) & f_i(A, a) = (A_i, a_i) \\
= & \langle f_1(\mathbf{h}(A, a)), f_2(\mathbf{h}(A, a)) \rangle & \overline{f}_i \text{ precursor for } f_i
\end{array}$$

□

Proposition 7 *Let $v \in V$, let $f_i : (\mathcal{R}(\alpha)_{\mu(v)_i} \times \mathcal{R}(\alpha)\ w) \to \mathcal{R}(\alpha)\ u$ with precursor $\overline{f}_i$ in* **P**, *then*

$$(\tau_v \times 1_w) \bullet [\overline{f}_1, \overline{f}_2] : (\mathbf{P}_v \times \mathbf{P}\ u) \to \mathbf{P}\ u$$

is a precursor in **P** *for coproduct mediator*

$$[f_1, f_2] : (\mathcal{R}(\alpha)_v + \mathcal{R}(\alpha)\ w) \to \mathcal{R}(\alpha)\ u.$$

Proof: We must show that, for each $(A, a) \in \mathbf{P}_v$,

$$\mathbf{h}(((\tau_v \times 1_w) \bullet [\overline{f}_1, \overline{f}_2])(A, a)) = [f_1, f_2](\mathbf{h}(A, a)).$$

It is no loss of generality to assume that w is the empty string, and that $a = (\kappa_v)_A(a_1)$, in which case
$\mathbf{h}((\tau_v \bullet [\overline{f}_1, \overline{f}_2])(A, a))$

$$\begin{array}{lll} = & \mathbf{h}([\overline{f}_1, \overline{f}_2](\tau_v(A, a))) & \text{def } \bullet \\ = & \mathbf{h}([\overline{f}_1, \overline{f}_2](\iota_1(A, \overline{a}_1)) & \text{def } \tau_v \\ = & \mathbf{h}(\overline{f}_1(A, \overline{(a)}_1)) & \text{def } [\overline{f}_1, \overline{f}_2] \\ = & f_1(\mathbf{h}(\overline{a}_1)) & \overline{f}_1 \text{ precursor for } f_1 \\ = & [f_1, f_2]((\kappa_v)_{\mathcal{R}(\alpha)}(\mathbf{h}(A, \overline{a}_1))) & \text{def } [f_1, f_2] \\ = & [f_1, f_2](\mathbf{h}(A, a)) & \text{def } \mathbf{h} \end{array}$$

□

Proposition 8 *Let $f : \mathcal{R}(\alpha)\ v.w \to \mathcal{R}(\alpha)\ v.w$, $v \in V$, $w \in S*$, be defined via while-do*

$$f = [\kappa_v \times 1_w, g \bullet f],$$

where $g : \mathcal{R}(\alpha)_{\mu(v)_2} \times \mathcal{R}(\alpha)\ w \to \mathcal{R}(\alpha)\ v.w$. Let g have a precursor, $\overline{g} : \mathbf{P}_{\mu(v)_2} \times \mathbf{P}\ w \to \mathbf{P}\ v.w$. Then, where for $i = 1, 2$,

$$\iota_i : (\mathbf{P}_{\mu(v_i)} \times \mathbf{P}\ w) \to (\mathbf{P}_{\mu(v)_1} + \mathbf{P}_{\mu(v)_2}) \times \mathbf{P}\ w$$

is the indicated coproduct injection, and $F : (\mathbf{P}_{\mu(v)_1} + \mathbf{P}_{\mu(v)_2}) \times \mathbf{P}\ w \to (\mathbf{P}_{\mu(v)_1} + \mathbf{P}_{\mu(v)_2}) \times \mathbf{P}\ w$ is the least function, as given by Proposition 3, such that

$$F = [\iota_1,\ \overline{g} \bullet (\tau_v \times 1_w) \bullet F],$$

then

$$(\tau_v \times 1_w) \bullet F \bullet ([(\kappa_v)_P, (\lambda_v)_P] \times 1_w)$$

is a precursor for f. Furthermore, for $\langle \overline{A}, \overline{a} \rangle \in \mathbf{P}\ v.w$, if n is the least integer ≥ 0 such that

$$f(\mathbf{h}(\langle \overline{A}, \overline{a} \rangle)) = [\kappa_v \times 1_w, g]\ n(\mathbf{h}(\langle \overline{A}, \overline{a} \rangle)),$$

then

$$F((\tau_v \times 1_w)(\langle \overline{A}, \overline{a} \rangle)) = [\iota_1, g \bullet (\tau_v \times 1_w)]\ n((\tau_v \times 1_w)(\langle \overline{A}, \overline{a} \rangle)).$$

Proof: Without loss of generality we may assume that w is the empty string. This then gives us the diagram

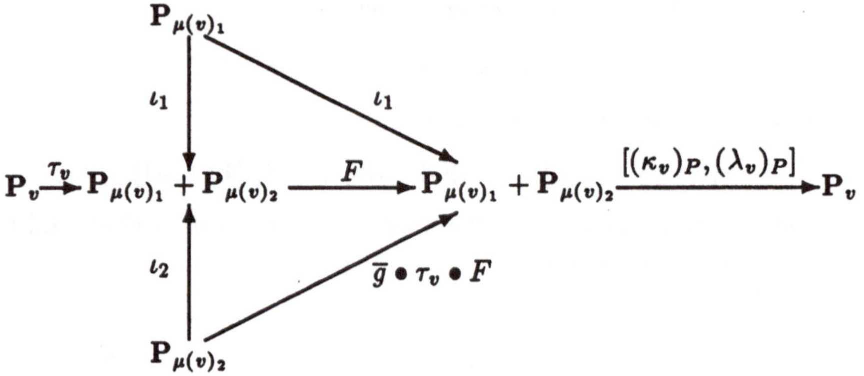

We want to show that $\tau_v \bullet F \bullet [(\kappa_v)_P, (\lambda_v)_P]$ is a precursor for

$$f = [(\kappa_v)_{\mathcal{R}(\alpha)}, g \bullet f].$$

To see this is so, let $\langle A, a \rangle \in \mathbf{P}_v$. Now either $a = (\kappa_v)_A(a_1)$, or $a = (\lambda_v)_A(a_2)$, and in either case we have

$$\begin{aligned}
&(\tau_v \bullet F \bullet [(\kappa_v)_P, (\lambda_v)_P])(\langle A, a \rangle)\\
&= (F \bullet [(\kappa_v)_P, (\lambda_v)_P])(\tau_v(\langle A, a \rangle))\\
&= (F \bullet [(\kappa_v)_P, (\lambda_v)_P])(\iota_i(\langle A, \overline{\sigma}_i(a_i) \rangle))\\
&= ([\iota_i, \overline{g} \bullet \tau_v] \bullet F \bullet [(\kappa_v)_P, (\lambda_v)_P])(\iota_i(\langle A, \overline{\sigma}_i(a_i) \rangle))\\
(*) \quad &= (\iota_i \bullet [\iota_i, \overline{g} \bullet \tau_v] \bullet F \bullet [(\kappa_v)_P, (\lambda_v)_P])(\langle A, \overline{\sigma}_i(a_i) \rangle)
\end{aligned}$$

But, if $i = 1$, then $\mathbf{h}(\langle A, a \rangle) = (\kappa_v)_{\mathcal{R}(\alpha)} \mathbf{h}(\langle A, \overline{\sigma}(a_1) \rangle)$, so

$$f(\mathbf{h}(\langle A, a \rangle)) = [(\kappa_v)_{\mathcal{R}(\alpha)}, g \bullet f](\kappa_v)_{\mathcal{R}(\alpha)}(\mathbf{h}(\langle A, \overline{\sigma}_1(a_1) \rangle)) = (\kappa_v)_{\mathcal{R}(\alpha)}(\mathbf{h}(\langle A, \overline{\sigma}_1(a_1) \rangle)).$$

Then

$$\begin{aligned}
(*) \quad &= \iota_1 \bullet F \bullet [(\kappa_v)_P, (\lambda_v)_P])(\langle A, \overline{\sigma}_i(a_i) \rangle)\\
&= \iota_1 \bullet [(\kappa_v)_P, (\lambda_v)_P])(\langle A, \overline{\sigma}_i(a_i) \rangle)\\
&= (\kappa_v)_P(\langle A, \overline{\sigma}_i(a_i) \rangle)
\end{aligned}$$

so, in this case, $\mathbf{h}(*) = f(\mathbf{h}(\langle A, a \rangle))$.
While if $i = 2$, then $\mathbf{h}(\langle A, a \rangle) = (\lambda_v)_{\mathcal{R}(\alpha)} \mathbf{h}(\langle A, \overline{\sigma}(a_1) \rangle)$ so

$$f(\mathbf{h}(\langle A, a \rangle)) = [(\kappa_v)_{\mathcal{R}(\alpha)}, g \bullet f](\lambda_v)_{\mathcal{R}(\alpha)}(\mathbf{h}(\langle A, \overline{\sigma}_1(a_1) \rangle)) = f(g(\mathbf{h}(\langle A, \overline{\sigma}_1(a_1) \rangle))).$$

Then

$$\begin{aligned}
(*) \quad &= ((\overline{g} \bullet \tau_v) \bullet F \bullet [(\kappa_v)_P, (\lambda_v)_P])(\langle A, \overline{\sigma}_i(a_i) \rangle)\\
&= \quad (\tau_v \bullet F \bullet [(\kappa_v)_P, (\lambda_v)_P])(\overline{g}(\langle A, \overline{\sigma}_i(a_i) \rangle))
\end{aligned}$$

so, since $\overline{g}$ is a precursor for g, we see that the desired result follows by induction. □

Putting the above results together, and applying them to the definitions of derived operation (Definition 4) and precursor, gives us the following result.

Theorem 1 *For any derived operation* $e(x_1, \ldots, x_n)$ *in* $\mathcal{R}(\alpha)$ *there exists in derived operation* $E(x_1, \ldots, x_n)$ *in* **P** *(extended) such that for any* $A \in$ **R**, *and* $a_1, \ldots, a_n$ *in* A

$$\mathbf{h}(E_P((A, a_1), \ldots, (A, a_n)) = e_{\mathcal{R}}(\mathbf{h}(A, a_1), \ldots, \mathbf{h}(A, a_n)).$$

□

The proof is more significant than the proof statement in the sense that it shows that "the programs are essentially the same" in both **P** and $\mathcal{R}(\alpha)$. That is we write very similar programs whether we use recursive definitions as such or as implemented by means of pointers. The only real difference is that the pointer based programs make repeated use of special "subprocedures" for creating and referencing pointers, but they are used in such a regular fashion that they can be almost forgotten.

8 References

[ADA82] *Reference Manual for the ADA Programming Language*. United States Department of Defense, Washington, DC, 1982.

[HP83] N. A. Habermann and D. E. Perry. *Ada for Experienced Programmers*. Addison-Wesley, 1983.

[Mac71] Saunders MacLane. *Categories for the Working Mathematician*. Springer-Verlag, New York, 1971.

[Van69] A. Van Wijngaarden. *Revised Report on the Algorithmic Language ALGOL 68*. 1969.

[Wag] Eric G. Wagner. Basic categorical concepts for computer scientists. Tutorial presented at Fourth Workshop on Mathematical Foundations of Programming Language Semantics, May 1988, – final written version in preparation.

[Wag87] Eric G. Wagner. Semantics of block structured languages with pointers. In *Proceedings of the 3rd Annual Workshop on the Mathematical Foundations of Programming Language Semantics*, Springer Verlag, 1987.

[Wag89] Eric G. Wagner. On declarations. In *Proceedings of the International Workshop on Categorical Methods in Computer Science*, 1989.

[WJ78] Kathleen Jenson and Niklaus Wirth. *PASCAL User Manual and Report. Lecture Notes in Computer Science, 18*, Springer-Verlag, New York, 1978.

The Abstract Galois Theory: A Survey

Isidore Fleischer[1]

In [F] the abstract Galois theory, originally developed by Krasner for automorphism groups (and subsequently endomorphism monoids) of relational structures and then extended by (him and) others for not necessarily finitary multi-argument operations, was derived by a consistent use of the index transformations under which the preserved relations are invariant (It should have been noted that this approach had previously been used for just the group case by Jurie.) The restricted length of that communication (it had been prepared as a Comptes Rendus note) precluded an explicit exhibit of a final form or a comparison with the other formulations derived by other means. It is proposed to make this up here, deriving and analyzing the other extant forms on the basis of this one.

As far as notation goes, the reader should know that a (not necessarily finitary) relation R on a set X is an arbitrary subset of X^I: its elements are functions from I to X which thus admit precomposition with selfmaps τ of I; this precomposition $p \mapsto p\tau$ is a selfmap of X^I whose inverse, also written on the right, sends subsets i.e. relations to relations. A support of R is a subset J of I such that the belonging of an element to R depends only on its restriction to J i.e. such that the characteristic function of R is factored by the projection of X^I on X^J, equivalently that R is saturated for (= a union of classes of) the kernel of this projection.

The subalgebra closure generated by a family of operations each of arity $< n$, a regular cardinal, is the union of the closures of $< n$-element subsets; if the operations are extended to act on relations (by composition with their elements) then this action commutes with the selfmaps τ of the index set: i.e. each operation induces a τ-morphism (of the same arity); symbolically, the closure satisfies $\overline{R\tau} = \overline{R}\tau$ for every relation R and every $\tau \in I^I$. Conversely, as proved in [F], if the cardinality of the index set I is not less than that of the support X of the relations, then every $< n$-ary $\{\tau\}$ commuting closure operator is so induced by a family of $< n$-ary operations on X.

A closure operator on a set may also be specified by describing the collection of closed subsets: these must be a complete $\bigcap$-subsemilattice of the power set, i.e. closed under arbitrary intersection (hence including the whole set). The τ commuting of the closure operator translates as closed-

[1]University of Windsor, Windsor, Ontario N9B 3P4, Canada

This paper is in final form and no version of it will be submitted for publication elsewhere.

ness of the collection under the action of τ and τ^{-1}. Indeed, if $C \supset \overline{C}$ then $C\tau \supset \overline{C}\tau = \overline{C\tau}$ and $C \supset \overline{C} \supset \overline{C\tau^{-1}\tau} = \overline{C\tau^{-1}}\tau$ whence $C\tau^{-1} \supset \overline{C\tau^{-1}}$; conversely if $\overline{B}\tau$ is closed then $\overline{B}\tau \supset \overline{B\tau}$ while if $\overline{B\tau}\tau^{-1}$ is closed then it $\supset \overline{B}$ hence $\overline{B\tau} \supset \overline{B}\tau$.

This condition can be made more explicit by making use of the structure of I^I as the set of all selfmaps of the index set I. Following a selfmap τ with any permutation π which inverts its image (i.e. sends each value back on a corresponding argument) converts τ to a retraction ρ (i.e. an idempotent): thus every τ has the form $\pi^{-1}\rho$ and it suffices to require closure of the closed sets under the π, ρ, and ρ^{-1}. $p\rho$ takes the values p took on $\operatorname{Im}\rho$ made constant on the classes of $\ker\rho$; hence $p'\rho^{-1}$ is non-void only if $\ker p' \supset \ker\rho$ in which case it consists of all p which agree with p' on $\operatorname{Im}\rho$. In fact, one can get this set of agreement on $\operatorname{Im}\rho$ for any p' by applying ρ^{-1} to $p'\rho$; since any non-void subset can occur as an $\operatorname{Im}\rho$, closure under the ρ^{-1} comes to (in the presence of closure under the ρ) containing the void relation and closure under passage to the elements which agree on some fixed non-void index subset with some element in a given relation.

This last passage is called "projection" or "cylindrification" by Krasner and his students. In [K,K′] it is joined with "mutation", an operation specified by a quintuple consisting of a pair of subsets I', I'' of I equipped with equivalence relations $\sim'$, $\sim''$ and a bijection $\varepsilon : I'/{\sim'} \to I''/{\sim''}$: the operation is then defined for relations with support I' which are constant on the classes of $\sim'$ and assigns every such the relation supported by I'' constant on the classes of $\sim''$, obtained by precomposing with $\varepsilon^{-1} \circ \operatorname{mod} \sim''$ (i.e. ε^{-1} composed with the quotient modulo $\sim''$) a map which sends I'' onto $I'/{\sim'}$. It may be seen from the above that this construction is unnecessarily elaborate. Indeed, "cylindrification" already having been taken care of, one only needs precomposition with an arbitrary selfmap of I: this one obtains from the special "mutations" with $I'' = I$ and $\sim' =$ identity, since the precomposition with a selfmap depends only on the restriction to its image, hence is the same for a relation and for the cylindrification which reduces its support to this image.

The term "mutation" has been taken over by Rosenberg to denote a quite different concept: the precompositon with a partial selfmap χ of I combined with "cylindrification" on the complement of the selfmap's domain: i.e. which thus assigns to a relation the set of elements whose restriction agree with the precomposite of one ot its elements with χ [R, p. 221], [R′, p. 276]. The use of partial selfmaps is a device to unify precomposition by total selfmaps with "cylindrification". Indeed, one obtains the latter by taking χ to be the identity on its domain; conversely, by extending χ in any way to be total, precomposng and cylindrifying, one gets this version of "mutation". Although not as explicit as the formulation above, it suffices (in conjunction with closure under intersection and updirected union) to generate the relational closure of invariants for a family of finite arity operations. (However [R,R′] adjoins a supplementary hypothesis stating that the total intersection shall consist of the constant elements: since the void relation is invariant, this is surely wrong.)

The formulation given by Poizat in [P] uses a different list than either of the preceding. There are first the constant relations, void and universal (the latter is subsequently obtained as the void conjunction) as well as, for every pair of indices, their "diagonal" which consists of all elements taking equal values on this pair; then there are the operations on relations of arbitrary "conjunction" (i.e. intersection), arbitrary "projection" (i.e. cylindrification) and the inverse of selfmap precomposition. Since the latter includes precomposition with premutations, only precompositon with retractions must be shown obtainable to deduce that these generate the invariant closure. But this follows by cylindrifying the complement of the image and intersecting with the diagonals for every pair in the kernel: i.e. projection + diagonals + intersection ensures precomposition by retraction; conversely, precomposing the universal relation with retractions yields the diagonals.

The setting of [BK] (see also [PK]) for finite X, and of [Ro] for infinite X, is restricted to finitary operations and relations (of arbitrary finite arity). The finitariness of the operations translates as closure of the invariant relations under updirected union; and the relations of arbitrary finite arity may be construed as relations of finite support on an infinite index set. In order to get all the invariant relations of finite support, their system should be closed under enlargement of arity by adjoining dummy arguments and under reduction by supressing arguments. These are, however, available from the "basic" list of [BK] which consists of precomposition with permutations and retractions, adjunction of "fictitious" variables and "superpositon", by which is meant relational product: the latter permits eliminating a coordinate by superposing a relation with itself at that argument; [Ro] augments these with updirected union and arbitrary intersection to take care of infinite X . Conversely, the infinitary image of a "superpositon" may be obtained from the infinitary operations: by the presence of permutations it suffices to treat finitary relations with disjoint supports and for these one can diagonalize the indices which are to be linked and then cylindrify them. Of course, reduction followed by fictitious will yield cylindrification in the system of finitary relations: to see that this finitary cylindrification in conjunction with precomposition by finitary maps suffices to generate the invariant closure in the finitary relations, one should show that the infinitary closure of finitarily closed sets of finitary relations meets all the finitary relations in the closure in no larger set. Take as generators of the closure precomposition with retractions, permutations and projection. The finitary relations in an invariant closed class are surely closed under projection and precompositon with selfmaps which leave pointwise fixed some cofinite set—since these do not lead out of the finitary relations. It will be shown that these operations already suffice to attain all the finitary relations in the invariant closure of the I-ary relations. To prove this, it must be shown that if a finite composite of generating transformations applied to some finitary relation results in a finitary relation, then the same result can be achieved with a finite composite of these special transformations. One can arrange to have the last transformation in the composite the cofinite projection onto the support of the image relation, since this leaves the image relation fixed. If the penultimate transformation is also a projection, it must be absorbed by the last one, hence can be eliminated from the composite; if the penultimate transformation is precomposition with a selfmap, then by changing this to be the identity on the complement of the support one achieves the same effect after the final projection. The result now follows by induction.

[Pl] also operates on finitary relations via infinite index sets: his "general superpositon" 2.4 p. 156 assigns, to finitary relations R_i equipped with maps π_i of their index sets m_i to a single (generally infinite) set I, a finitary relation determined by a map π of its index set m to I, by a process which may be rendered symbolically (in terms of the above symbolism) as $[\bigcap_i R_i \circ \pi_i^{-1}] \circ \pi$. (Also described in [Sz] using the pharaphanalia of predicate logic.) This obviously includes precomposition (take a single i with π_i the identity) and its inverse (take π the identity) as well as intersection (take all π_i and π the identity on the same index set) in particular, the universal relation (no π_i). To this must be added updirected union; by itself this operation will only yield the closure operator (indeed, even without intersection, i.e. by restricting to a single i) and not a characterization of the closed sets of relations on an infinite support X; for a finite support it suffices as such a characterization since it obviously includes [BK]'s "superposition" (to obtain relational product, map the two relations' index sets injectively into I so that their images overlap in just the indices to be identified and take for π a bijection to the non-overlapping indices.)

A notationally different description of the basic closure operator—but not of its resolution into explicit set theoretic operations as developed in the work described thus far—is to be found in [SS]. This starts out with a concrete category admitting direct powers; however, the category is only used to define the "dual algebra" on an object X, which is the family of all morphisms from powers of X to X, these morphisms (with their categorical composition) then functioning as the clone of operations on X. An I-ary relation on X is called "generic" if it consists of the images, of some fixed I-tuple in some Y, under all homs to X. It is clear that such a relation is clone-closed; conversely, the clone orbit of any I-tuple in X is such a relation with $Y = X^I$. This permits equating clone-closure with generic-closure.

References

[BK] Bodnarčuk (V.N), Kotov (V.N.), Kalužnin (L.A.) and Romov (B.A.)., *Galois theory for Post algebras, I-II* (Russian). Kibernetika part I: 5

(1969), 1–10, part II: **5** (1969), 1–9. English translation Cybernetics (1969), 243–252 and 531–539, MR 46 #55.

[F] Fleischer (I.), *La théorie de Galois abstraite de Krasner et Rosenberg comme théorie transformationnelle*, Bull. Sc. Math., 2e série, **102** (1978), p. 97–100.

[J′] Jurie (P.-F.). *Une extension de la théorie de Galois abstraite finitaire*, C.R., Acad. Sc. Paris, 276, Série A–B (1973), A81–A84.

[K] Krasner (M.). *Endothéorie de Galois abstraite*, Séminaire Dubreil, 22e année, 1968/1969, Exposé 6.

[K′] Krasner (M.). *Polythéorie de Galois abstraite dans le cas infini général*, Ann. Sci. Univ. Clermont Sér. Math. **13** (1976), 87–91.

[L] Leblanc (L.). "Introduction á la logique algébrique", Presse de l'Université de Montréal, 1967.

[P] Poizat (B.). *Théorie de Galois pour les algébres de post infinitaires*, Zeitshr. f. math. Logik und Grundlagen d. Math. **27** (1981), 31–44.

[PK] Pöschel (R.) and Kalužnin (L.A.). *Funktionen-und Relationenalgebren*, Veb Deutscher Verlag Der Wissenschaften, Berlin 1979.

[R] Rosenberg (I.G.). *Subalgebra systems of direct powers*, Algebra Universalis **8** (1978), 221–227.

[R′] Rosenberg (I.G.). *On a Galois connection between algebras and relations and its applications*, Proc. of the Klagenfurt Conf., May 25–28, 1978.

[SS] Sauer (N.) and Stone (M.G.) *Indigenous relations in categories of sets*, Contr. to General Algebra **5**, Proc. Salzburg Conf., Wien 1987.

[Pl] Pöschel (R.) *Closure properties for relational systems with given endomorphism structure*, Beitr zur Alg. u. Geom. **18** (1984), 153–166. Also, *Concrete representation of algebraic structures and a general Galois theory*, in contr. to Gen. Alg., Proc. Klagenfurt Conf. May 25–28, 1978, J. Heyn, Klagenfurt 1979, 249–272.

[Ro] Romov (B.A.) *Galois correspondence between iterative Post algebras and relations on infinite sets*, Kibernetika (1977), 62–64. English translation Cybernetics **13** (1977), 377–379, MR 58 #222.

[Sz] Szabo (L.) *Concrete representation of related structures of universal algebras I*, Acta. Sci. Math. (Szeged) **40** (1978), 175–184.

The Implications in Conditional Logic

Fernando Guzmán[1]

ABSTRACT Given a logical system with disjunction ($\vee$) and negation ($\neg$), we call implication, the derived operation defined by $p \Rightarrow q \equiv (\neg p) \vee q$. Conditional logic is the 3-valued logic associated with short-circuit evaluation, and up to anti-isomorphism, is the unique non-commutative regular extension of Boolean logic to 3 truth values. The two anti-isomorphic conditional logics (left to right and right to left evaluation) yield quite different implications. We study the 3-element algebras associated with each of these operations, and for each of them a complete set of laws and a recursive formula for the free spectrum is obtained.

Implication in different logical systems has been studied for some time. In [A] Abbott studies the 2-valued boolean implication. In [L-T] Lukasiewicz and Tarski study a 3-valued implication whose operation on the set $\{0, \frac{1}{2}, 1\}$ is given by the formula $xy = \min(1, 1 - x + y)$, together with the negation function. This implication has the "problem" of not being a regular (or continuous) extension of boolean implication in the sense of [K]. Later Dienes in [D] considers another 3-valued implication given by the formula $xy = \max(1 - x, y)$. This implication is a regular extension of boolean implication; in fact it is Kleene's strong implication. In [M] McCarthy proposes a 3-valued predicate calculus in order to study recursively defined predicates. In the propositional part of his calculus, the disjunction and conjunction are not commutative; they correspond to what is known as short circuit or conditional evaluation. The implication operator he proposes is the first of the two implications we consider in this paper.

In [C] Cleave considers different 3-valued implications. From philosophical and aesthetic considerations he narrows down the choices to Kleene's strong implication and the two implications we consider in this paper. At the end he chooses Kleene's strong implication on the grounds of simplicity; that is, its formula as a derived operation is simpler than those for the other two. This, of course, is a consequence of his earlier choice of Kleene's strong logic as "the" 3-valued logic, choice made on the grounds of continuity (or regularity), non-triviality and symmetry (or commutativity). However, if we drop the symmetry requirement, there are two other continuous non-trivial 3-valued logics. They are studied in [G-S] under the name of conditional logics. These two logics are anti-isomorphic, and in

[1] Dept. of Mathematical Sciences, SUNY-Binghamton, Binghamton, NY 13901
This paper is in final form and no version of it will be submitted for publication elsewhere.

them Kleene's strong implication is no longer a derived operation. Moreover, the formulas for the other two implications become very simple; in fact, they are essentially the same formula used for boolean implication.

One of the basic questions concerning logical systems is that of basic laws, that is, a collection of identities such that any other identity is a consequence of them. We call such collection a complete set of laws. A complete set of laws for 2-valued boolean implication is provided in [A]. In this paper we provide complete sets of laws for both conditional implications.

For an algebra A, the free spectrum of A is the sequence of cardinalities of the finitely generated free algebras in the variety generated by A. The question of the spectrum of finite algebras has received a great deal of attention. Berman has a "catalog" [B] of more than two hundred 3-element algebras with data about their free spectra. Conditional implication algebras appear as #058# and #053# in that catalog. This paper gives recursive formulas for the computation of the free spectrum of both conditional implications. The values of these sequences for $n \leq 5$ are listed in tables at the end of section 2 and after Theorem (3.5). They agree for $n \leq 3$ with the values computed in [B] by listing all the elements of the free algebra.

The algebra of conditional logic is the 3-element algebra $C = \{T, F, U\}$ with T and F as constants and with negation ($\neg$), disjunction ($\vee$), and conjunction ($\wedge$) as basic operations given by

	$'$
T	F
F	T
U	U

$\wedge$	T	F	U
T	T	F	U
F	F	F	F
U	U	U	U

$\vee$	T	F	U
T	T	T	T
F	T	F	U
U	U	U	U

Here we study the same 3-element set with the derived operation of implication as its basic operation. Because disjunction is non-commutative in conditional logic, we consider two (non-isomorphic) implications:

$$p \Rightarrow_1 q \equiv (\neg p) \vee q$$
$$p \Rightarrow_2 q \equiv q \vee (\neg p)$$

which correspond respectively to left to right and right to left short-circuit evaluation of implication. Thus the operations are given by the tables:

$\Rightarrow_1$	T	F	U
T	T	F	U
F	T	T	T
U	U	U	U

$\Rightarrow_2$	T	F	U
T	T	F	U
F	T	T	U
U	T	U	U

We remark that neither one of these algebras is an implication algebra in the sense of [A]. In fact, of the three laws defining implication algebras:

I1: $(ab)a = a$
I2: $(ab)b = (ba)a$
I3: $a(bc) = b(ac)$

$\Rightarrow_1$ satisfies I2 (2.1.d) but not I1 or I3

$$(T \Rightarrow_1 U) \Rightarrow_1 T = U \neq T$$
$$U \Rightarrow_1 (F \Rightarrow_1 T) = U \neq T = F \Rightarrow_1 (U \Rightarrow_1 T)$$

$\Rightarrow_2$ satisfies none of them

$$(F \Rightarrow_2 U) \Rightarrow_2 F = U \neq F$$
$$(T \Rightarrow_2 U) \Rightarrow_2 U = U \neq T = (U \Rightarrow_2 T) \Rightarrow_2 T$$
$$F \Rightarrow_2 (U \Rightarrow_2 F) = U \neq T = U \Rightarrow_2 (F \Rightarrow_2 F)$$

In Section 1 we recall some of the definitions and results about conditional logic from [G-S]. Some of the statements and definitions are taken literally from there, while some others have been modified in a way more suitable to our work. In any case, all the results in section 1 are easy consequences of [G-S] and therefore are stated without proof.

In Section 2 we study the first of the two implications ($\Rightarrow_1$). First we find a complete set of laws for this algebra (2.1),(2.12) by determining all subdirectly irreducible algebras in the variety it generates (2.11). Then we give a concrete description of the free algebras in this variety (2.22), building upon the construction in [G-S] of the free algebras of conditional logic. Finally, using this description of the free algebras, we obtain a recursion for the free spectrum.

In section 3 the second implication algebra ($\Rightarrow_2$) is studied. We show that the free algebra is the disjoint union of some principal ideals (3.5) each one of which is in a one-to-one correspondence with a free C-algebra (3.4). From this fact we compute the free spectrum (3.5), and translate the laws of C into laws for implication, to obtain a complete set of laws (3.1),(3.7).

There is a great contrast in the results obtained for these two algebras. The first one has a very simple set of laws and a very complex recursion for its free spectrum (five recursively dependent sequences, two of them doubly indexed). The second algebra has a simple recursion (when compared with the previous one) for its free spectrum but a set of laws so obscure that we do not even write down all of them explicitly.

1 General remarks

In this section we recall definitions and some results from [G-S] about conditional logic, and set up the notation common to both conditional implications.

Let $\mathcal{C}$ be the variety generated by C. Let I denote one of the implication algebras, and denote by $\mathcal{I}$ the variety generated by I. For any variety $\mathcal{V}$, any set X and any $n \in \mathbb{N}$, we denote by $F_{\mathcal{V}}(X)$ the free algebra in $\mathcal{V}$ generated by X, and by $F_{\mathcal{V}}(n)$ the free algebra in $\mathcal{V}$ on n generators. Since the operation in I is derived from the operations in C the clone $\mathrm{Clo}(I)$ is contained in $\mathrm{Clo}(C)$ and the free algebra $F_{\mathcal{I}}(X)$ can be identified with a subset of $F_{\mathcal{C}}(X)$. For both implication algebras we will determined precisely the subset, and use this result to compute the free spectrum.

Since the two implication algebras will be considered separately, we will denote their operation by concatenation; that is, in section $i+1$, xy denotes $x \Rightarrow_i y$.

Now we recall the description of $F_{\mathcal{C}}(X)$ given in [G-S]. First we need the following:

(1.1) **Definition.** *Let A be a C-algebra. If $x, p, q \in A$, define $\Gamma_x(p,q)$ by $\Gamma_x(p,q) = (x \wedge p) \vee (x' \wedge q)$.*

Observe that in particular $\Gamma_T(p,q) = p$, $\Gamma_F(p,q) = q$ and $\Gamma_U(p,q) = U$, so $\Gamma_x(p,q)$ represents the conditional expression "if x then p, else if not x then q, else U". However it is also the conditional expression "if x then p, else q" if we adopt the evaluation philosophy of conditional logic, i.e. "evaluate subexpressions only as you need them and if a subexpression evaluates to U the whole expression evaluates to U", which is what the 3 above equations say.

(1.2) **Lemma.** *Every C-algebra satisfies the laws:*
a) $\Gamma_x(p,q)' = \Gamma_x(p',q')$.
b) $\Gamma_x(p,q) \wedge r = \Gamma_x(p \wedge r, q \wedge r)$.
c) $\Gamma_x(p,q) \vee r = \Gamma_x(p \vee r, q \vee r)$.
d) $\Gamma_x(\Gamma_y(p,q), \Gamma_y(r,s)) = \Gamma_y(\Gamma_x(p,r), \Gamma_x(q,s))$.
e) $\Gamma_x(p,q) \vee \Gamma_x(r,s) = \Gamma_x(p \vee r, q \vee s)$.
f) $\Gamma_x(p,q) \wedge \Gamma_x(r,s) = \Gamma_x(p \wedge r, q \wedge s)$.
g) $\Gamma_x(T,F) = x$.

For our convenience we restate [G-S Thm.(3.14)] as follows:

(1.3) **Theorem.** *Every non-constant element of $F_{\mathcal{C}}(X)$ can be uniquely expressed (up to the congruence generated by (1.2.d)) as $\Gamma_x(p,q)$ with $x \in X$ and $p, q \in F_{\mathcal{C}}(X - \{x\})$.*

We can picture $\Gamma_x(p,q)$ as an ordered binary tree with root x and with p and q as left and right subtrees respectively. T and F are the height 0 trees.

The *evaluation endomorphism* of $F_{\mathcal{C}}(X)$, $t \mapsto t[x \leftarrow V]$ where $x \in X$ and $V \in \{T, F\}$ is determined by the two conditions:

$x[x \leftarrow V] = V$

$y[x \leftarrow V] = y$ whenever $y \in X - \{x\}$

and it has the following properties

(1.4) **Lemma.** *Let X be a set, $x, y \in X$ $x \neq y$, $p, q \in F_{\mathcal{C}}(X)$, and $V \in \{T, F\}$.*
a) $p[x \leftarrow V] \in F_{\mathcal{C}}(X - \{x\})$.
b) $\Gamma_x(p,q)[x \leftarrow T] = p[x \leftarrow T]$ and $\Gamma_x(p,q)[x \leftarrow F] = q[x \leftarrow F]$.
c) $\Gamma_x(p,q)[y \leftarrow V] = \Gamma_x(p[y \leftarrow V], q[y \leftarrow V])$.

d) $\Gamma_x(p,q) = \Gamma_x(p[x \leftarrow T], q[x \leftarrow F])$.

For $t \in F_C(X)$ *the support* $X(t)$ *of* t is the smallest subset of X such that $t \in F_C(X(t))$. It satisfies the following properties

(1.5) LEMMA. *Let* X *be a set,* $x \in X$, $p, q \in F_C(X)$, $V \in \{T, F\}$.
a) $X(T) = X(F) = \emptyset$.
b) $X(\Gamma_x(p,q)) = \{x\} \cup X(p) \cup X(q)$.
c) $X(p[x \leftarrow V]) \subseteq X(p) - \{x\}$.
d) $X(p') = X(p)$, $X(p \vee q) \subseteq X(p) \cup X(q)$ and $X(p \wedge q) \subseteq X(p) \cup X(q)$.

For $t \in F_C(X)$ *the undefining set of* t, $U(t)$ is

$$\{x \in X \mid \exists p, q \in F_C(X) \text{ with } t = \Gamma_x(p,q)\}.$$

(1.6) LEMMA. *Let* X *be a set,* $x \in X$, $p, q \in F_C(X - \{x\})$.
a) $U(T) = U(F) = \emptyset$.
b) $U(\Gamma_x(p,q)) = \{x\} \cup (U(p) \cap U(q))$.

Using (1.3),(1.6.b) and induction one can easily prove

(1.7) PROPOSITION. *Let* X *be a set,* $t \in F_C(X)$. *If* $x_1, \ldots, x_k \in U(t)$ *then there are unique* $t_1, \ldots, t_{2^k} \in F_C(X - \{x_1, \ldots, x_k\})$ *such that* $t = \Gamma_{(x_1,\ldots,x_k)}(t_1, \ldots, t_{2^k})$ *where* $\Gamma_{(x_1,\ldots,x_k)}(t_1, \ldots, t_{2^k})$ *is defined recursively by*

$$\Gamma_{(x_1,\ldots,x_k)}(t_1, \ldots, t_{2^k}) =$$
$$\Gamma_{x_1}(\Gamma_{(x_2,\ldots,x_k)}(t_1, \ldots, t_{2^{k-1}}), \Gamma_{(x_2,\ldots,x_k)}(t_{2^{k-1}+1}, \ldots, t_{2^k}))$$
$$\Gamma_{(x_1)}(t_1, t_2) = \Gamma_{x_1}(t_1, t_2)$$

Observe that from (1.4.d) we must have

$$\Gamma_{(x_2,\ldots,x_k)}(t_1, \ldots, t_{2^{k-1}}) = t[x_1 \leftarrow T],$$
$$\Gamma_{(x_2,\ldots,x_k)}(t_{2^{k-1}+1}, \ldots, t_{2^k}) = t[x_1 \leftarrow F]$$

and by induction $t_i = t[x_1 \leftarrow V_1] \cdots [x_k \leftarrow V_k]$ where $V_j \in \{T, F\}$. Moreover, the value of the V_j's for a given t_i depend on the base 2 expansion of $i-1$. If we write $i - 1 = \sum_{j=1}^{k} d_j 2^{k-j}$ with $d_j \in \{0,1\}$, then $V_j = T$ (resp. F) if $d_j = 0$ (resp. 1).

Not only is $F_{\mathcal{I}}(X)$ a subset of $F_C(X)$ but in fact $F_C(X)$ is an $\mathcal{I}$-algebra and $F_{\mathcal{I}}(X)$ a subalgebra of it. For the first implication operation in $F_C(X)$ $xy = x' \vee y = \Gamma_x(y, T)$ and for the second one $xy = y \vee x' = \Gamma_y(T, x')$. Informally they could be read as "if x then y" and "if not y then not x" respectively.

From (1.2) and (1.5) we get the following for either one of the two implication operations in $F_C(X)$.

(1.8) LEMMA. *Let* X *be a set,* $x \in X$, $p, q, r, s \in F_C(X)$.
a) $\Gamma_x(p,q)\Gamma_x(r,s) = \Gamma_x(pr, qs)$.

b) $X(pq) \subseteq X(p) \cup X(q)$

Since the operations we consider are not associative, there is the need for extensive use of parentheses. However, we can reduce it somewhat by the use of the exponent 2. Thus x^2 denotes the product xx and we will follow the usual conventions about precedence: exponentiation takes precedence over product, e.g. xy^2 is $x(yy)$.

2 Left to right implication

In this section I denotes the first implication algebra from the introduction. That is $I = \{T, F, U\}$ with a binary operation given by:

	T	F	U
T	T	F	U
F	T	T	T
U	U	U	U

2.1 The laws of I

(2.1) LEMMA. *I satisfies the following laws*
a) $x(yz) = (xy)(xz)$
b) $x^2(yz) = (x^2y)z$
c) $x^2x = x$
d) $(xy)y = (yx)x$
e) $(z(xy))x = (z(xy^2))x$
Moreover, when T is a distinguished constant
f) $Tx = x$

Proof. Check against the table.

One of our goals is to show that this is a complete set of laws for I. Let $\mathcal{V}$ be the variety of algebras with one binary operation that satisfy (2.1.a-e) and $\mathcal{V}^*$ the variety of algebras with a binary operation and a constant satisfying (2.1.a-f). Here are some consequences of the above laws.

(2.2) LEMMA. *Any algebra $A \in \mathcal{V}$ satisfies the laws*
a) $(xy)^2 = xy^2$
b) $x(xy) = xy$
c) $(x^2)^2 = x^2$

Proof. a) is a particular case of (2.1.a)
b) $x(xy) = x^2(xy) = (x^2x)y = xy$ (2.1.a,b,c)
c) $(x^2)(x^2) = (x^2x)(x^2x) = xx$ (2.1.a,c)

(2.3) LEMMA. *Any algebra $A \in \mathcal{V}$ has at most one left identity.*

Proof. Suppose $T_1, T_2 \in A$ are such that $T_1x = x = T_2x$ for all $x \in A$. Then
$$T_1 = T_1T_1 = (T_2T_1)T_1 = (T_1T_2)T_2 = T_2T_2 = T_2 \qquad (2.1.d)$$

This lemma says that the forgetful map $\mathcal{V}^* \to \mathcal{V}$ that forgets the constant is a 1-1 map, i.e. any algebra in $\mathcal{V}$ that has a left identity can be viewed in a unique way as an algebra in $\mathcal{V}^*$. If A has a left identity we will denote the left identity of A by T and will treat A as an algebra in $\mathcal{V}^*$.

For any $A \in \mathcal{V}$ and $a \in A$ we define the map
$$\lambda_a : A \to A$$
$$x \mapsto ax$$
and consider its kernel $\sim_a = \ker(\lambda_a) = \{(x,y) \in A \times A | ax = ay\}$.

(2.4) LEMMA. *Let $A \in \mathcal{V}$ and $a \in A$.*
a) λ_a is a homomorphism and $\mathrm{Im}(\lambda_a) \in \mathcal{V}^*$ with a^2 as its left identity.
b) If A has a left identity it satisfies the law $xT = x^2$.
c) $\sim_a$ is a congruence in A. If $a \neq T$ then $\sim_a$ is non-trivial.

Proof. a) λ_a is a homomorphism because of (2.1.a). Moreover,
$$a^2(ax) = (a^2a)x = ax \qquad (2.1.b,c)$$
b) If A has a left identity then xT and x^2 are both left identities in $\mathrm{Im}(\lambda_x)$, and by (2.3) they are equal.
c) If $a \neq T$ then there is an $x \in A$ such that $ax \neq x$. But by (2.2.b) $ax \sim_a x$, so $\sim_a$ is non-trivial.

We will now determine the subdirectly irreducible algebras in $\mathcal{V}$. The next proposition tells us that we only need to look into $\mathcal{V}^*$.

(2.5) PROPOSITION. *If $A \in \mathcal{V}$ is subdirectly irreducible then A has a left identity, i.e. $A \in \mathcal{V}^*$.*

Proof. Suppose A has no left identity, and let $x, y \in A$ be such that $x \neq y$. Then by (2.4.c) $\sim_x$, $\sim_y$, $\sim_{x^2}$, $\sim_{y^2}$ are non-trivial congruences in A and we claim one of them separates x and y. Otherwise we would have $x^2 = xy$, $yx = y^2$, $x^2x = x^2y$, and $y^2x = y^2y$, so
$$x = x^2x = x^2y = (xy)y = (yx)x = y^2x = y^2y = y \ (\to\leftarrow) \qquad (2.1.c,d)$$

We now need several lemmas about congruences separating elements.

(2.6) LEMMA. *Let $A \in \mathcal{V}^*$ and $x, y \in A$ be such that $x \neq y$, $x \neq T \neq y$, and $x^2 = y^2 = T$. Then one of $\sim_x$ and $\sim_y$ separates x and y.*

Proof. Suppose not. Then $xy = x^2 = T$ and $yx = y^2 = T$ so
$$x = Tx = (yx)x = (xy)y = Ty = y \ (\to\leftarrow) \qquad (2.1.f,d)$$

(2.7) LEMMA. *Let $A \in \mathcal{V}^*$ and $x, y \in A$ be such that $x \neq y$, $T \neq x, y, x^2$, and $y^2 = T$. Then if $\sim$ separates T from x^2, $\sim$ also separates x and y.*

Proof. If $x \sim y$ then $x^2 \sim y^2$, i.e. $x^2 \sim T$.

(2.8) LEMMA. *Let $A \in \mathcal{V}^*$ and $x \in A$ be such that $x \neq x^2 \neq T$. Then $\sim_{x^2}$ separates T and x.*

Proof. Assume not. Then $x^2x = x^2T$. Using (2.1.c),(2.2.c) and (2.4.b) we get $x = x^2$. $(\to\leftarrow)$

From lemmas (2.6),(2.7) and the proof of proposition (2.5) it follows that if $A \in \mathcal{V}^*$ is subdirectly irreducible, then there is an $x \in A - \{T\}$ such that no non-trivial congruence separates x and T. In particular, for any $y \in A - \{T\}$ we must have $yx = yT$, i.e. $yx = y^2$. Moreover, by lemma (2.8) either $x = x^2 \neq T$ or $x \neq x^2 = T$. In both cases we will construct a congruence in A that separates x and T, and whose quotient is isomorphic to a subalgebra of J, where J is the weak regular extension of I; i.e. $J = \{T, F, U, V\}$ with operation given by

	T	F	U	V
T	T	F	U	V
F	T	T	T	V
U	U	U	U	V
V	V	V	V	V

It will then follow that the subdirectly irreducible algebras of $\mathcal{V}$ are subalgebras of J.

(2.9) PROPOSITION. *Let $A \in \mathcal{V}^*$ and $x \in A$ be such that $x \neq x^2 = T$ and $yx = y^2$ for all $y \in A - \{T\}$. Then the equivalence relation induced by the partition consisting of the non-empty sets in*

$$\{\{T\},\ \{x\},\ \{y \in A - \{x, T\} \,|\, xy = T\},\ \{y \in A - \{x, T\} \,|\, xy \neq T\}\}$$

is a congruence in A and the quotient is isomorphic to a subalgebra of J.

Proof. We start by proving some facts. Let $y, z \in A$

Fact 1: If $yz = x$ then $y = T$ and $z = x$.

Suppose $y \neq T$. Then $y^2 = yx = y(yz) = yz = x$ (2.2.b)

so $x^2 = (y^2)^2 = y^2 = x$ $(\to\leftarrow)$ (2.2.c)

Fact 2: If $y^2 = T$ then $y = T$ or $y = x$.

Suppose $y \neq T$. Then by the hypothesis in the proposition $yx = y^2 = T$ so we get $x = Tx = (yx)x = (xy)y$ by (2.1.d,f).

But Fact 1 tells us then that $y = x$.

Fact 3: If $yz = T$ then $y = T$ or $y = x$.

$y^2 = yT = y(yz) = yz = T$ (2.4.b),(2.2.b)

Use now Fact 2.

Let $U = \{y \in A - \{x, T\} \,|\, xy = T\}$, $V = \{y \in A - \{x, T\} \,|\, xy \neq T\}$, $F = \{x\}$ and, abusing the notation, let T denote the set $\{T\}$.

We now go row by row through the table of J verifying that this partition is compatible with the operation in A, and that the operation table of the quotient is that of J (except for missing elements whenever one of U or V is empty).

Row 1: $TX = X$ for any X subset of A. (2.1.f)

Row 2: $FT = \{xT\} = \{x^2\} = FF = T$ (2.4.b)

If $U \neq \emptyset$ then $FU = T$ by the definition of U.

If $y \in V$ then $xy \neq x, T$ by Fact 1 and the definition of V.

Moreover, $x(xy) = xy \neq T$ by (2.2.b), and therefore $xy \in V$. So $FV = V$.

Row 3: Let $y \in U$. Then $yT = y^2 = yx \neq x, T$ (2.4.b), Facts 1 and 2

Moreover $x(yT) = xy^2 = (xy)^2 = T^2 = T$ (2.4.b),(2.2.a), $y \in U$, (2.1.f)

So $yT \in U$ and therefore $UT = UF \subseteq U$.

Let $y, z \in U$. Then $yz \neq x, T$ Facts 1 and 3

and $x(yz) = (xy)(xz) = T^2 = T$ (2.1.a), $y, z \in U$,(2.1.f)

so $yz \in U$ and therefore $UU \subseteq U$.

Let $y \in U$, $z \in V$. Then $yz \neq x, T$ Facts 1 and 3

and $x(yz) = (xy)(xz) = T(xz) = xz \neq T$ (2.1.a),$y \in U$,(2.1.f), $z \in V$

so $yz \in V$ and therefore $UV \subseteq V$.

Row 4: Let $y \in V$, $z \in A$. Then $yz \neq x, T$ Facts 1 and 3

and $x(yz) = (xy)(xz) \neq T$ (2.1.a), Fact 3, $y \in V$, Fact 1

so $yz \in V$ and therefore $VA \subseteq V$.

(2.10) PROPOSITION. *Let $A \in \mathcal{V}^*$ and $x \in A$ be such that $x = x^2 \neq T$ and $yx = y^2$ for all $y \in A - \{T\}$. Then there is a congruence in A that separates x and T, and whose quotient is isomorphic to a subalgebra of J.*

Proof. Since $x = x^2$ we can use (2.1.b) as an associative law but only when the first term is x.

As in the previous proposition let T denote the set $\{T\}$.

Let $F = \{y \in A - \{T\} | y^2 = T\}$

and $W = (A - F) - \{T\} = \{y \in A | y^2 \neq T\}$

Here are some facts about these sets.

Fact 1: $W = \{y \in A | xy = y\} = xA$. Moreover $WA = W$.

"$\supseteq$" Let $y \in A$ and suppose $xy = y$. If we had $y^2 = T$ then $x = x^2 = xT = xy^2 = (xy)^2 = y^2 = T$ $(\rightarrow\leftarrow)$ (2.4.b),(2.2.a)

"$\subseteq$" Let $y \in W$. Then $T \neq y, y^2$ and therefore

$yx = y^2$ and $y^2x = (y^2)^2 = y^2$ (2.2.c)

So $y^2 = y^2x = (yx)x = (xy)y = xy^2$ (2.1.b,d)

and $y = y^2y = (xy^2)y = x(y^2y) = xy$ (2.1.b,c)

The second equality follows from (2.2.b).

Finally $WA = (xA)A = x(AA) = xA = W$ (2.1.b), A has T

Fact 2: If $y, z \in F$ then $yz = T$.

$(yz)^2 = yz^2 = yT = y^2 = T$ (2.2.a),(2.4.b), $y, z \in F$

Suppose $yz \neq T$ then $(yz)x = (yz)^2 = T$ so we have

$T = (yz)x = (T(yz))x = ((yx)(yz))x$ (2.1.f)

$= (y(xz))x = (y(xz^2))x$ (2.1.a,e)

$= (y(xT))x = (yx)x = y^2x = Tx = x$ $(\rightarrow\leftarrow)$ (2.4.b)

Fact 3: If $y \in W$ and $a, b \in F$ then $ay \notin F$. Moreover $ay = T$ iff $by = T$.

If we had $ay \in F$ then $T = a(ay) = ay$ $(\rightarrow\leftarrow)$ Fact 2,(2.2b)

Now if $ay = T$ then

$T = b^2 = bT = b(ay) = (ba)(by) = T(by) = by$

Fact 2,(2.4.b),(2.2.a),(2.1.f)

It follows from Fact 3 that if $y \in W$ then either $Fy = T$ or $Fy \subseteq W$.

Let $U = \{y \in W | Fy = T\}$ and $V = \{y \in W | Fy \subseteq W\}$

Going now row by row through J's table we get:

Row 1: $TX = X$ for any $X \subseteq A$ (2.1.a)

Row 2: If $F \neq \emptyset$ then $FT \subseteq FF = T$ (2.4.b),Fact 2

$FU = T$ (Definition of U)

Let $a \in F$, $y \in V$ then $ay \in W$ (Definition of V)

if $b \in F$ then $b(ay) = (ba)(by) = T(by) = by$ (2.1.a),Fact 2,(2.1.f)

so $ay \in V$ and therefore $FV \subseteq V$.

Row 3: If $y \in U$ and $z \in A$ then by Fact 1 $yz \in W$ and for $a \in F$

$a(yz) = (ay)(az) = T(az) = az$ (2.1.a),$y \in U$,(2.1.f)

so if $z \in V$ then $a(yz) \in W$ and $yz \in V$

but if $z \notin V$ then $a(yz) = T$ and $yz \in U$

So $UV \subseteq V$, $UT \subseteq U$, $UF \subseteq U$, $UU \subseteq U$

Row 4: If $y \in V$ and $z \in A$ then again $yz \in W$ and for $a \in F$

$a(yz) = (ay)(az)$ (2.1.a)

but $ay \in W$ so by Fact 1 $a(yz) \in W$ and therefore $yz \in V$ so $VA \subseteq V$

From (2.5-10) we get the following

(2.11) THEOREM. *All subdirectly irreducible algebras in $\mathcal{V}$ (and in $\mathcal{V}^*$) are subalgebras of J.*

It is easy to check the converse: every subalgebra of J is subdirectly irreducible.

(2.12) COROLLARY. *Laws (2.1.a-e) (resp. laws (2.1.a-f)) form a complete set of laws for I (resp. I with T as a constant).*

Proof. $\{T, U\}$ is a subalgebra of I, and it is easy to check that J is the quotient of $I \times \{T, U\}$ modulo the congruence generated by $(T, U) \sim (F, U) \sim (U, U)$.

2.2 THE FREE ALGEBRAS IN $\mathcal{I}$

From (2.1.f) and (2.4.b) we can see that $|F_{\mathcal{V}^*}(X)| = |F_{\mathcal{V}}(X)| + 1$ and in fact, when we view $F_{\mathcal{V}^*}(X)$ and $F_{\mathcal{V}}(X)$ as subsets of $F_{\mathcal{C}}(X)$ we have $F_{\mathcal{V}^*}(X) = F_{\mathcal{V}}(X) \cup \{T\}$. Therefore we limit our attention to $F_{\mathcal{V}^*}(X)$ which we view as a subset of $F_{\mathcal{C}}(X)$. On the other hand $F_{\mathcal{C}}(X)$ with binary operation $pq = p' \vee q$ and constant T satisfies (2.1.a-f) since C does, and therefore, $F_C(X) \in \mathcal{V}^*$ and $F_{\mathcal{V}^*}(X)$ is a subalgebra of it.

There are two properties that will characterize the elements of $F_C(X)$ that belong to $F_{\mathcal{V}^*}(X)$. Intuitively, they are:

Left variable condition: $t \in F_C(X)$ has l.v.c. if (when pictured as a binary tree according to (1.3)) every variable that occurs in t occurs in its left-most path (from root to leaves), and maybe in other paths too.

Right false condition: $t \in F_C(X)$ has r.f.c. if there is an $x \in X$ such that every F-leaf of t is to the right of some occurrence of x in t.

One more condition will be needed in order to formalize the definitions: $t \in F_C(X)$ is said to be a *T-tree* if all its leaves are equal to T. Now the definitions.

(2.13) DEFINITION. *Let X be a set and $t \in F_C(X)$.*

a) *t has left variable condition (l.v.c.) is defined recursively by:*

T has l.v.c.

If $x \in X$, $p, q \in F_C(X-\{x\})$ are such that p has l.v.c. and $X(q) \subseteq X(p)$ then $\Gamma_x(p,q)$ has l.v.c.

b) *t is a T-tree is defined recursively by:*

T is a T-tree.

If $x \in X$, $p, q \in F_C(X-\{x\})$ are such that p, q are T-trees then $\Gamma_x(p,q)$ is a T-tree.

c) *For $x \in X$ the subset of $F_C(X)$, $\mathrm{rfc}(x)$ is defined recursively by:*

$T \in \mathrm{rfc}(x)$.

If $p, q \in F_C(X-\{x\})$ are such that p is a T-tree then $\Gamma_x(p,q) \in \mathrm{rfc}(x)$.

If $y \in X-\{x\}$ and $p, q \in F_C(X-\{y\})$ are such that $p, q \in \mathrm{rfc}(x)$ then $\Gamma_y(p,q) \in \mathrm{rfc}(x)$

d) *t has right false condition (r.f.c.) if $t = T$ or $t \in \mathrm{rfc}(x)$ for some $x \in X$.*

Note in d) that if $X \neq \emptyset$ then the second case includes the first. It is not entirely obvious but we will check that the properties defined in (2.13) do not depend on the reduced tree used to represent t (See (2.14.a) and (2.16.a,c,e)). The following lemma summarizes some properties of l.v.c.

(2.14) LEMMA. *Let X be a set, $x \in X$ and $t_1, t_2 \in F_C(X)$.*

a) *If t_1 has l.v.c. so does $t_1[x \leftarrow T]$ and $X(t_1[x \leftarrow T]) = X(t_1) - \{x\}$.*

b) *If t_1, t_2 have l.v.c so does $t_1 t_2$ and $X(t_1 t_2) = X(t_1) \cup X(t_2)$.*

Proof. a) By (1.5.c) "$\subseteq$" always holds. The proof is by induction on t_1.

If $t_1 = T$ it follows from (1.5.a).

If $t_1 = \Gamma_x(p,q)$ with $p, q \in F_C(X-\{x\})$, $X(q) \subseteq X(p)$ and p with l.v.c. then

$t_1[x \leftarrow T] = p$, $X(t_1) = X(p) \cup X(q) \cup \{x\} = X(p) \cup \{x\}$ (1.4.b),(1.5.b)

Therefore $X(t_1) - \{x\} = X(p) = X(t_1[x \leftarrow T])$.

If $t_1 = \Gamma_y(p,q)$ with $y \in X-\{x\}$, $p, q \in F_C(X-\{y\})$, $X(q) \subseteq X(p)$ and p with l.v.c. then by induction hypothesis $p[x \leftarrow T]$ has l.v.c and $X(p[x \leftarrow T]) = X(p)-\{x\}$. So $X(q[x \leftarrow T]) \subseteq X(q)-\{x\} \subseteq X(p)-\{x\} =$

$X(p[x \leftarrow T])$ and $t_1[x \leftarrow T] = \Gamma_y(p[x \leftarrow T], q[x \leftarrow T])$ has l.v.c. Moreover $X(t_1) = X(p) \cup X(q) \cup \{y\} = X(p) \cup \{y\}$ and $X(t_1[x \leftarrow T]) = X(p[x \leftarrow T]) \cup X(q[x \leftarrow T]) \cup \{y\} = X(p[x \leftarrow T]) \cup \{y\} = (X(p) - \{x\}) \cup \{y\} = X(t_1) - \{x\}$.

b) By (1.8.b) "$\subseteq$" always holds. The proof is by induction on t_1.

If $t_1 = T$ then it follows from (1.5.a) and (2.1.f).

If $t_1 = \Gamma_x(p,q)$ with $p, q \in F_C(X - \{x\})$, $X(q) \subseteq X(p)$ and p with l.v.c. then $t_1t_2 = \Gamma_x(pt_2[x \leftarrow T], qt_2[x \leftarrow F])$ (1.2.a,c),(1.4.d)
and $X(qt_2[x \leftarrow F]) \subseteq X(q) \cup X(t_2) - \{x\}$ (1.5.c),(1.8.b)
$\subseteq X(p) \cup X(t_2) - \{x\} = X(pt_2[x \leftarrow T])$ (Induction,a)
so $X(t_1t_2) = X(pt_2[x \leftarrow T]) \cup \{x\} = X(p) \cup \{x\} \cup X(t_2) = X(t_1) \cup X(t_2)$.
Moreover by induction and a) $pt_2[x \leftarrow T]$ has l.v.c so t_1t_2 has l.v.c.

(2.15) **Proposition.** *Let X be a set and $t \in F_C(X)$. If $t \in F_{\mathcal{V}^*}(X)$ then t has l.v.c.*

Proof. Follows by induction on t (as an element of $F_{\mathcal{V}^*}(X)$) using the fact that T has l.v.c., together with (1.2.g) and (2.14.b).

Let's derive now some properties about r.f.c.

(2.16) **Lemma.** *Let X be a set, $x, y \in X$, $x \neq y$, $t, t_1 \in F_C(X)$, $V \in \{T, F\}$.*

a) If t is a T-tree so is $t[x \leftarrow V]$.

b) If t is a T-tree so is t_1t.

c) If $t \in \mathrm{rfc}(x)$ then $t[x \leftarrow T]$ is a T-tree.

d) $x \in \mathrm{rfc}(x)$.

e) If $t \in \mathrm{rfc}(x)$, then $t[y \leftarrow V] \in \mathrm{rfc}(x)$.

f) If $t \in \mathrm{rfc}(x)$ then $t_1t \in \mathrm{rfc}(x)$.

g) If t has r.f.c. then either $t = T$ or there is an $x \in X(t)$ such that $t \in \mathrm{rfc}(x)$.

Proof. a) By induction on t.

If $t = \Gamma_x(p,q)$ with p, q T-trees then $t[x \leftarrow V]$ is either p or q. (1.4.b)

If $t = \Gamma_y(p,q)$ with $y \neq x$, p, q T-trees
then $t[x \leftarrow V] = \Gamma_y(p[x \leftarrow V], q[x \leftarrow V])$ (1.4.c)
so we conclude by induction.

b) By induction on t_1.

If $t_1 = T$ or F then $t_1t = t$ or T respectively.

If $t_1 = \Gamma_x(p,q)$ then $t_1t = \Gamma_x(pt[x \leftarrow T], qt[x \leftarrow F])$ so conclude using part a) and induction.

c) By induction on t.

If $t = \Gamma_x(p,q)$ with p a T-tree then $t[x \leftarrow T] = p$ (1.4.b)

If $t = \Gamma_y(p,q)$ with $y \neq x$, $p, q \in \mathrm{rfc}(x)$ then
$t[x \leftarrow T] = \Gamma_y(p[x \leftarrow T], q[x \leftarrow T])$ (1.4.c)
and by induction $p[x \leftarrow T]$ and $q[x \leftarrow T]$ are both T-trees.

d) Follows from (1.2.g)

e) By induction on t.

If $t = \Gamma_x(p,q)$ with p a T-tree then $t[y \leftarrow V] = \Gamma_x(p[y \leftarrow V], q[x \leftarrow V])$ and by part a) $p[y \leftarrow V]$ is a T-tree.

If $t = \Gamma_y(p,q)$ with $p,q \in \text{rfc}(x)$ then $t[y \leftarrow V] = p$ or q.

If $t = \Gamma_z(p,q)$ with $z \neq y$, $p,q \in \text{rfc}(x)$ then $t[y \leftarrow V] = \Gamma_z(p[y \leftarrow V], q[y \leftarrow V])$ and conclude by induction.

f) By induction on t_1.

If $t_1 = T$ or F then $t_1 t = t$ or T respectively.

If $t_1 = \Gamma_x(p,q)$ then $t_1 t = \Gamma_x(pt[x \leftarrow T], qt[x \leftarrow F])$. By parts b) and c) $pt[x \leftarrow T]$ is a T-tree and therefore $t_1 t \in \text{rfc}(x)$.

If $t_1 = \Gamma_y(p,q)$ with $y \neq x$ then $t_1 t = \Gamma_y(pt[y \leftarrow T], qt[y \leftarrow F])$ By part e) and induction $pt[y \leftarrow T], qt[y \leftarrow F] \in \text{rfc}(x)$ and therefore $t_1 t \in \text{rfc}(x)$.

g) If t has r.f.c and $t \neq T$ there is a $y \in X$ such that $t \in \text{rfc}(y)$. If $y \notin X(t)$ then $t[y \leftarrow T] = t$ so by a) t is a T-tree and for any $x \in X(t)$, $t \in \text{rfc}(x)$.

Although we don't need it here, it is worth noting that for $t \in F_{\mathcal{V}^*}(X)$ the condition "$t \in \text{rfc}(x)$ and $x \in X(t)$" is equivalent to "t is in the left ideal generated by x". (2.16.d,f) already prove one direction.

(2.17) **Proposition.** *Let X be a set, $t \in F_{\mathcal{C}}(X)$. If $t \in F_{\mathcal{V}^*}(X)$ then t has r.f.c.*

Proof. Same as proof of (2.15) using (2.16.f) instead of (2.14.b).

Next we will prove the converse of (2.15,17) combined, i.e. if $t \in F_{\mathcal{C}}(X)$ has l.v.c and r.f.c. then $t \in F_{\mathcal{V}^*}(X)$. But first a lemma and some propositions.

(2.18) **Lemma.** *Let X be a set, $x \in X$, and $t \in F_{\mathcal{C}}(X)$.*

a) t is a T-tree iff $t^2 = t$.

b) $t'((tx)t) = t$.

c) If $x \in U(t)$ then $(tx)t = t$.

Proof. a) By induction, (1.3) and (1.8.a).

b) Check that it holds for C.

c) Let $t = \Gamma_x(p,q)$ for some $p,q \in F_{\mathcal{C}}(X)$.

Then $(tx)t = \Gamma_x((pT)p, (qF)q)$, and $(pT)p = p$ (1.8.a,2.4.c,2.1.c)

$(qF)q = q$ can be checked for C.

(2.19) **Proposition.** *Let X be a set, and let $p \in F_{\mathcal{C}}(X)$ be a T-tree with l.v.c. Let $q \in F_{\mathcal{C}}(X)$.*

a) If $X(q) \subseteq X(p) \subseteq X - \{x\}$ then $\Gamma_x(p,q) \in F_{\mathcal{V}^*}(X)$.

b) If $z \in X(p)$ then there is a $k \geq 1$ and $u_1, \ldots, u_k \in F_{\mathcal{V}^*}(X)$ such that

$$u_1(u_2(\cdots(u_k(pz))\cdots)) = p.$$

Proof. The two parts are proved simultaneously by induction on p.

If $p = T$ then b) is vacuously true and for a) we have $q = T$ or F so $\Gamma_x(p,q) = x^2$ or x respectively.

If $p = \Gamma_y(r,s)$ with r a T-tree with l.v.c and s a T-tree such that

$X(s) \subseteq X(r)$ then applying a) to r we get $p \in F_{\mathcal{V}^*}(X)$. Now we consider two cases:

If $y = z$ then
$(pz)(pz) = pz^2 = \Gamma_z(r,s)\Gamma_z(T,T) = \Gamma_z(r^2,s^2) = \Gamma_z(r,s) = p$
(2.1.a,2.4.c,2.18.a)

so in b) we just take $k = 1$ and $u_1 = pz$.

If $y \neq z$ then $z \in X(r) \cup X(s) = X(r)$ so applying b) to r we find $v_1, \ldots, v_l \in F_{\mathcal{V}^*}(X)$ such that $v_1(v_2(\cdots(v_l(rz))\cdots)) = r$. Let $k = l+1$ and $u_i = yv_i = \Gamma_y(v_i, T)$ for $i = 1, \ldots, l$ and $u_k = py$. Clearly each $u_i \in F_{\mathcal{V}^*}(X)$, and $u_k(pz) = (py)(pz) = p(yz) = \Gamma_y(r,s)\Gamma_y(z,T) = \Gamma_y(rz,s)$ so $u_1(u_2(\cdots(u_k(pz))\cdots)) = \Gamma_y(v_1(v_2(\cdots(v_l(rz))\cdots)),s) = \Gamma_y(r,s) = p$. Now to prove a) let $A = \{q \in F_C(X(p)) | \Gamma_x(p,q) \in F_{\mathcal{V}^*}(X)\}$. We want to show that $A = F_C(X(p))$. First observe that $\Gamma_x(p,T) = xp$, $\Gamma_x(p,F) = (xp)x$ so $T, F \in A$. If $q \in A$ then $\Gamma_x(p,q)x = \Gamma_x(p,q')$ and therefore $q' \in A$. And if $q, r \in A$ then $\Gamma_x(p,q')\Gamma_x(p,r) = \Gamma_x(p, q \vee r)$ so $q \vee r \in A$. That is, A is a C-subalgebra of $F_C(X(p))$ and all we need to show now is that $X(p) \subseteq A$. Let $z \in X(p)$ then $(xp)z = \Gamma_x(pz,z)$ and if we set $w_i = xu_i = \Gamma_x(u_i,T)$ for $i = 1, \ldots, k$ then $w_i \in F_{\mathcal{V}^*}(X)$ and $w_1(w_2(\cdots(w_k((xp)z))\cdots)) = \Gamma_x(u_1(u_2(\cdots(u_k(pz))\cdots)),z) = \Gamma_x(p,z)$. Therefore $\Gamma_x(p,z) \in F_{\mathcal{V}^*}(X)$ and $z \in A$.

(2.20) Proposition. *Let X be a set $t \in F_C(X)$ and $x \in X(t)$. If t has l.v.c. and $t \in \mathrm{rfc}(x)$ there is a $k \geq 0$ and $u_1, \ldots, u_k \in F_{\mathcal{V}^*}(X)$ such that $u_1(u_2\cdots(u_k((tx)t))\cdots) = t$.*

Proof. If $t = \Gamma_x(p,q)$ with p a T-tree, apply (2.18.c).

If $t = \Gamma_y(p,q)$ with $y \neq x$ and $p, q \in \mathrm{rfc}(x)$ then since t has l.v.c. $X(t) = \{y\} \cup X(p)$ and therefore $x \in X(p)$. By induction there are $v_1, \ldots, v_l \in F_{\mathcal{V}^*}(X)$ such that $v_1(\cdots(v_l((px)p))\cdots) = p$. Let $k = l+1$ and $u_i = yv_i = \Gamma_y(v_i,T)$ for $i = 1, \ldots, l$ and $u_k = ty$. Clearly $u_i \in F_{\mathcal{V}^*}(X)$ for $i = 1, \ldots, l$ and $u_k = \Gamma_y(pT, qF) = \Gamma_y(p^2, q')$. By (2.18.a) p^2 is a T-tree and $X(q') = X(q) \subseteq X(p) = X(p^2)$ so by (2.19.a) $u_k \in F_{\mathcal{V}^*}(X)$.
$u_k((tx)t) = \Gamma_y(p^2,q')\Gamma_y((px)p,(qx)q) = \Gamma_y((px)p,q)$ (2.4.a,2.18.b)
so $u_1(\cdots(u_k((tx)t))\cdots) = \Gamma_y(v_1(\cdots(v_l((px)p))\cdots),q) = \Gamma_y(p,q) = t$.

(2.21) Proposition. *Let X be a set, $x \in X$, and $t \in F_C(X)$. If $t \in \mathrm{rfc}(x)$ then $(tx)t = \Gamma_x(r,s)$ with r a T-tree.*

Proof. If $t = T$ then $(tx)t = x^2 = \Gamma_x(T,T)$.
If $t = \Gamma_x(p,q)$ with p a T-tree then use (2.18.c).
If $t = \Gamma_y(p,q)$ with $y \neq x$ and $p, q \in \mathrm{rfc}(x)$ then $(tx)t = \Gamma_y((px)p,(qx)q)$.
Use induction and (1.2.d).

(2.22) Theorem. *Let X be a set and $t \in F_C(X)$. $t \in F_{\mathcal{V}^*}(X)$ if and only if t has l.v.c. and r.f.c.*

Proof. "$\Rightarrow$" (2.15), (2.17).
"$\Leftarrow$" If t has l.v.c. and r.f.c. by (2.16.g) either $t = T$ or there is an $x \in X(t)$

such that $t \in \mathrm{rfc}(x)$. In the first case there is nothing to show; in the second case, by (2.21) $(tx)t = \Gamma_x(r,s)$ with r a T-tree. By (2.14.b) $(tx)t$ has l.v.c. so by (2.19.a) $(tx)t \in F_{\mathcal{V}^*}(X)$. By (2.20) $t \in F_{\mathcal{V}^*}(X)$.

2.3 The cardinality of $F_{\mathcal{V}^*}(X)$

The condition $X(q) \subseteq X(p)$ in (2.13.a) makes difficult counting the elements of $F_{\mathcal{C}}(X)$ with l.v.c. The following definition and lemma will allow us to avoid this condition.

(2.23) **Definition.** *Let X be a set. $t \in F_{\mathcal{C}}(X)$ is said to be full if $X(t) = X$.*

The lemma characterizes full elements with l.v.c, full T-trees with l.v.c., and full elements of $\mathrm{rfc}(x)$ with l.v.c.

(2.24) **Lemma.** *Let X be a set, $t \in F_{\mathcal{C}}(X)$, $x, y \in X$, $x \neq y$, and $p, q \in F_{\mathcal{C}}(X - \{x\})$.*
a) There is a unique $Y \subseteq X$ such that t is full in $F_{\mathcal{C}}(Y)$.
b) $\Gamma_x(p,q)$ is full in $F_{\mathcal{C}}(X)$ with l.v.c if and only if p is full in $F_{\mathcal{C}}(X - \{x\})$ with l.v.c.
c) T is the only full element with l.v.c. in $F_{\mathcal{C}}(\emptyset)$.
d) $\Gamma_x(p,q)$ is a T-tree with l.v.c. full in $F_{\mathcal{C}}(X)$ iff p is a T-tree with l.v.c full in $F_{\mathcal{C}}(X - \{x\})$ and q is a T-tree.
e) $\Gamma_x(p,q)$ is full in $F_{\mathcal{C}}(X)$ with l.v.c. and in $\mathrm{rfc}(x)$ iff p is a full T-tree with l.v.c. in $F_{\mathcal{C}}(X - \{x\})$
f) $\Gamma_x(p,q)$ is full in $F_{\mathcal{C}}(X)$ with l.v.c. and in $\mathrm{rfc}(y)$ iff p is full in $F_{\mathcal{C}}(X - \{x\})$ with l.v.c. and in $\mathrm{rfc}(y)$, and $q \in \mathrm{rfc}(y)$.

Proof. For a) take $Y = X(t)$. The rest follows from (2.13),(2.14) and (2.16).

Let X be a finite set and $n = |X|$. Let $a_n = |\{t \in F_{\mathcal{C}}(X) | t \text{ is a } T\text{-tree }\}|$ and $b_n = |\{t \in F_{\mathcal{C}}(X) | t \text{ is a full } T\text{-tree with l.v.c in } F_{\mathcal{C}}(X)\}|$. In [G-S,4.2] the inclusion-exclusion principle and (1.3) are used to obtain a recursion for $f_n = |F_{\mathcal{C}}(X)|$:

$$f_0 = 2, \quad f_n = 2 + \sum_{k=1}^{n} (-1)^{k-1} \binom{n}{k} f_{n-k}^{2^k}$$

Values of f_n for $n \leq 5$ appear in the table after Theorem (3.5). Using the same technique and (2.24.d) we get

$$a_0 = 1 \ , \quad a_n = 1 + \sum_{k=1}^{n} (-1)^{k-1} \binom{n}{k} a_{n-k}^{2^k}$$

$$b_0 = 1 \ , \quad b_n = \sum_{k=1}^{n} (-1)^{k-1} \binom{n}{k} b_{n-k} a_{n-k}^{2^k - 1}$$

We now extend the definition of rfc to any subset of X.

(2.25) DEFINITION. *Let X be a set and $Y \subseteq X$. Define*

$$\mathrm{rfc}_X(Y) = \bigcap_{x \in Y} \mathrm{rfc}(x)$$

In particular, (or in addition) $\mathrm{rfc}_X(\emptyset) = F_C(X)$.

(2.26) PROPOSITION. *Let X be a finite set, $n = |X|$, $Y \subseteq X$ with $|Y| = m$. Let $C(X,Y) = \mathrm{rfc}_X(Y)$ and $c_{n,m} = |C(X,Y)|$. Then $c_{n,0} = f_n$ and for $m > 0$*

$$c_{n,m} = 1 + \sum_{k=0}^{m} \sum_{\substack{j=0 \\ k+j>0}}^{n-m} (-1)^{k+j-1} \binom{m}{k} \binom{n-m}{j} a_{n-(k+j)}^{2^{k+j}-2^j} c_{n-(k+j),m-k}^{2^j}$$

Proof. Obviously we only need to consider the case $Y \neq \emptyset$. Let Z be a subset of Y of cardinality k and W a subset of $X - Y$ of cardinality j, such that $Z \cup W \neq \emptyset$. Let $C(X,Y,Z,W) = \{t \in C(X,Y) | Z \cup W \subseteq U(t)\}$ and arrange the elements of $Z \cup W$ as in (1.7) so that the elements of Z appear first. Then any $t \in C(X,Y,Z,W)$ is uniquely determined by $t_1, \ldots, t_{2^{k+j}} \in F_C(X - (Z \cup W))$. By (2.16.c) $t_1, \ldots, t_{2^{k+j}-2^j}$ are T-trees and by (2.16.e) $t_{2^{k+j}-2^j+1}, \ldots, t_{2^{k+j}} \in C(X - (Z \cup W), Y - Z)$, so $|C(X,Y,Z,W)| = a_{n-(k+j)}^{2^{k+j}-2^j} c_{n-(k+j),m-k}^{2^j}$. From this and the inclusion-exclusion principle we get the double sum. The "1+"comes from the fact that $\{t \in C(X,Y) | U(t) = \emptyset\} = \{T\}$ whenever $Y \neq \emptyset$.

(2.27) PROPOSITION. *Let X be a finite set, $n = |X|$, $Y \subseteq X$ with $|Y| = m > 0$. Let $D(X,Y) = \{t \in \mathrm{rfc}_X(Y) | t$ is full in $F_C(X)$with l.v.c.$\}$ and let $d_{n,m} = |D(X,Y)|$. Then*

$$d_{n,m} = \sum_{k=1}^{m} \sum_{j=0}^{n-m} (-1)^{k+j-1} \binom{m}{k} \binom{n-m}{j} b_{n-(k+j)} a_{n-(k+j)}^{2^{k+j}-2^j-1} c_{n-(k+j),m-k}^{2^j}$$
$$+ \sum_{j=1}^{n-m} (-1)^{j-1} \binom{n-m}{j} d_{n-j,m} c_{n-j,m}^{2^j-1}$$

Proof. Let Z be a subset of Y of cardinality k and W a subset of $X - Y$ of cardinality j, such that $Z \cup W \neq \emptyset$. Let $D(X,Y,Z,W) = \{t \in D(X,Y) | Z \cup W \subseteq U(t)\}$ and arrange the elements of $Z \cup W$ as in (1.7) so that the elements of Z appear first. Then any $t \in D(X,Y,Z,W)$ is uniquely determined by $t_1, \ldots, t_{2^{k+j}} \in F_C(X - (Z \cup W))$. If $k > 0$ then by (2.16.c) and (2.24.d) t_1 is a T-tree with l.v.c. full in $F_C(X - (Z \cup W))$ and $t_2, \ldots, t_{2^{k+j}-2^j}$ are T-trees. By (2.16.e)

$$t_{2^{k+j}-2^j+1}, \ldots, t_{2^{k+j}} \in C(X - (Z \cup W), Y - Z),$$

so

$$|D(X,Y,Z,W)| = b_{n-(k+j)} a_{n-(k+j)}^{2^{k+j}-2^j-1} c_{n-(k+j),m-k}^{2^j}.$$

If $k = 0$ then by (2.16.e) $t_1, \ldots, t_{2^j} \in C(X - W, Y)$ and by (2.24.b) t_1 is full in $F_C(X - W)$ so $t_1 \in D(X - W, Y)$. In this case $|D(X,Y,Z,W)| =$

$d_{n-j,m}c^{2^j-1}_{n-j,m}$. As in (2.26) conclude by using the inclusion-exclusion principle. Observe that since $X \neq \emptyset$, then $U(t) \neq \emptyset$ for any $t \in D(X,Y)$.

(2.28) THEOREM. *Let X be a finite set, and $n = |X|$.*

$$|F_{\mathcal{V}^\bullet}(X)| = 1 + \sum_{l=1}^{n} \sum_{m=1}^{l} (-1)^{m-1} \binom{n}{l} \binom{l}{m} d_{l,m}$$

Proof. Once more the inclusion-exclusion principle can be used to show that there are

$$\sum_{m=1}^{l} (-1)^{m-1} \binom{l}{m} d_{l,m}$$

full elements in $F_{\mathcal{C}}(l)$ with l.v.c. and r.f.c. The outer summation comes from (2.24.a). The "1+" comes from considering the case $X(t) = \emptyset$.

The following table gives the first few values of $|F_{\mathcal{V}}(n)| = |F_{\mathcal{V}^\bullet}(n)| - 1$. (See [B,#058#]).

n	$\lvert F_{\mathcal{V}}(n)\rvert$
0	0
1	2
2	16
3	659
4	2039110
5	22275265011797

3 Right to left evaluation

In this section I denotes the second implication algebra described in the introduction. That is $I = \{T, F, U\}$ with a binary operation given by:

	T	F	U
T	T	F	U
F	T	T	U
U	T	U	U

(3.1) LEMMA. *The algebra I satisfies the following laws:*

a) $x^2x = x$

b) $((yz)x)x = y((zx)x)$

c) $((yz)x)x = ((yx)x)((zx)x)$

Proof. Check against the table.

One of our goals is to enlarge this set of laws until we get a complete set of laws for I. Let $\mathcal{L}$ be a set of laws (involving a binary operation) and let $\mathcal{V} = \mathcal{V}(\mathcal{L})$ be the variety of algebras (with one binary operation) satisfying $\mathcal{L}$.

(3.2) **Definition.** *Let $A \in \mathcal{V}$ and $x \in A$. The principal left ideal of A generated by x, denoted by A_x is defined recursively by the conditions:*

$x \in A_x$

If $t_1 \in A$ and $t_2 \in A_x$ then $t_1 t_2 \in A_x$.

Observe that since there are no constants, if X generates A then A is the union of $(A_x | x \in X)$.

(3.3) **Proposition.** *Let $A \in \mathcal{V}$ and $x, t \in A$. If $\mathcal{L}$ includes (3.1.a,b) then*

a) $t \in A_x$ iff $t = (tx)x$

b) $A_x = Ax = (Ax)x$

Proof. a) One direction is obvious; the other by induction on t.

If $t = x$ use (3.1.a).

If $t = t_1 t_2$ with $t_1 \in A$ and $t_2 \in A_x$.

By induction $t_2 = (t_2 x)x$ so $t = t_1((t_2 x)x) = ((t_1 t_2)x)x = (tx)x$ (3.1.b)

b) Follows from a).

We will now determine which subset of $F_{\mathcal{C}}(X)$ is $F_{\mathcal{I}}(X)$. Observe that as an $\mathcal{I}$-algebra $F_{\mathcal{C}}(X)$ is generated by $X \cup \{F\}$ since $tF = t'$. Consider the map

$$\begin{aligned} \psi_x : F_{\mathcal{C}}(X) &\to F_{\mathcal{C}}(X) \\ t &\mapsto (tx)x \end{aligned}$$

By (3.1.c) this is an $\mathcal{I}$-homomorphism. Moreover $\psi_x(F) = (Fx)x = x$ and therefore the image of ψ_x is contained in $F_{\mathcal{I}}(X)x$. But in fact, since $F_{\mathcal{I}}(X) \subseteq F_{\mathcal{C}}(X)$ and by (3.3.b) $\psi_x(F_{\mathcal{I}}(X)) = F_{\mathcal{I}}(X)x$ we conclude that ψ_x is onto $F_{\mathcal{I}}(X)x$. Also observe that $\psi_x(t) = \Gamma_x(T, t) = \Gamma_x(T, t[x \leftarrow F]) = \psi_x(t[x \leftarrow F])$, so the image of ψ_x does not change if we restrict the domain to $F_{\mathcal{C}}(X - \{x\})$. Finally, by (1.3) ψ_x is 1-1 when its domain is restricted to $F_{\mathcal{C}}(X - \{x\})$. So we have just proved

(3.4) **Proposition.**

$$\begin{aligned} \psi_x : F_{\mathcal{C}}(X - \{x\}) &\to F_{\mathcal{I}}(X)x \\ t &\mapsto (tx)x \end{aligned}$$

is an $\mathcal{I}$-isomorphism.

(3.5) **Theorem.** $|F_{\mathcal{I}}(n)| = n|F_{\mathcal{C}}(n-1)| = nf_{n-1}$

Proof. Let X be a finite set of cardinality n. For $t \in F_{\mathcal{C}}(X)$, $x \in X$ we have $tx = \Gamma_x(T, t')$ so $U(tx) = \{x\}$ and therefore for $x, y \in X$, with $x \neq y$ the principal left ideals $F_{\mathcal{I}}(X)x$ and $F_{\mathcal{I}}(X)y$ are disjoint. So $F_{\mathcal{I}}(X)$ is the disjoint union of $(F_{\mathcal{I}}(X)x | x \in X)$. Conclude by using (3.4).

The following table gives the values of $|F_{\mathcal{I}}(n)|$ (See [B,#053#]). and f_n for $n \leq 5$.

n	f_n	$\lvert F_{\mathcal{I}}(n)\rvert = nf_{n-1}$
0	2	0
1	6	2
2	58	12
3	6462	174
4	105783730	25848
5	39780675932043318	528918650

The bijection of (3.4) can be used to "transport" the $\mathcal{C}$-algebra structure of $F_{\mathcal{C}}(X - \{x\})$ to $F_{\mathcal{I}}(X)x$ so that ψ_x becomes a $\mathcal{C}$-algebra isomorphism. Because $F_{\mathcal{I}}(X)x$ is already a subset of the $\mathcal{C}$-algebra $F_{\mathcal{C}}(X)$ we will denote the transported $\mathcal{C}$-operations in $F_{\mathcal{I}}(X)x$ using " ". That is,

$$\text{“}T\text{”} = \psi_x(T) = x^2$$
$$\text{“}F\text{”} = \psi_x(F) = x$$

And for $p = \psi_x(u)$, $q = \psi_x(v)$ with $u, v \in F_{\mathcal{C}}(X - \{x\})$

$$\begin{aligned}
p\text{“}\vee\text{”}q &= \psi_x(u \vee v) = \Gamma_x(T, u \vee v)\\
&= \Gamma_x(T, v'u) = \Gamma_x(T, (vF)u)\\
&= (qx)p\\
p\text{“}\wedge\text{”}q &= \psi_x(u \wedge v) = \Gamma_x(T, u \wedge v)\\
&= \Gamma_x(T, (u' \vee v')') = \Gamma_x(T, (v(uF))F)\\
&= (q(px))x\\
p^{\text{“}\prime\text{”}} &= \psi(u') = \Gamma_x(T, u')\\
&= \Gamma_x(T, uF)\\
&= px
\end{aligned}$$

We can now translate the laws of $\mathcal{C}$ into laws of $\mathcal{I}$. For example the double negation law becomes $((yx)x)x = yx$ which is a consequence of (3.1.a,b). Observe that it does not translate into $(yx)x = y$ since it only holds in $F_{\mathcal{I}}(X)x$. One of DeMorgan's laws translates into $(((zx)((yx)x))x)x = (((zx)x)x)((yx)x)$ which is a consequence of (3.1.b) and the double negation law. Let $\mathcal{L}$ be the set of laws consisting of (3.1.a,b,c) and the translation of the other 5 laws of $\mathcal{C}$ (See [G-S,1.1]). Let $\mathcal{V} = \mathcal{V}(\mathcal{L})$ and let's denote by A the free algebra $F_{\mathcal{V}}(X)$ where X is a set. By (3.3.b) the principal left ideal of A generated by x is Ax. Since I satisfies $\mathcal{L}$, $F_{\mathcal{I}}(X)$ is a quotient of A. Let π be the quotient map $\pi : A \to F_{\mathcal{I}}(X)$. To show that $\mathcal{L}$ is a complete set of laws for I we need to show that π is 1-1. Since the principal left ideals $(F_{\mathcal{I}}(X)x | x \in X)$ are pairwise disjoint, so are $(Ax | x \in X)$ and π maps Ax onto $F_{\mathcal{I}}(X)x$. So we can simply check that the restriction $\pi_x : Ax \to F_{\mathcal{I}}(X)x$ is a bijection.

The choice of $\mathcal{L}$ means that Ax with operations and constants given by:

$$T = x^2$$
$$F = x$$
$$p \vee q = (qx)p$$
$$p \wedge q = (q(px))x$$
$$p' = px$$

is a $\mathcal{C}$-algebra, and $\theta_x = \psi_x^{-1} \cdot \pi_x : Ax \to F_{\mathcal{C}}(X - \{x\})$ is a surjective $\mathcal{C}$-homomorphism. Moreover $\theta_x((yx)x) = y$ for any $y \in X - \{x\}$, so all we need to show now is that as a $\mathcal{C}$-algebra Ax is generated by $\{(yx)x | y \in X - \{x\}\}$. Let B be the $\mathcal{C}$-subalgebra of Ax generated by $\{(yx)x | y \in X - \{x\}\}$.

(3.6) PROPOSITION.
a) *B is an $\mathcal{I}$-subalgebra of Ax.*
b) *If $t \in A$ then $(tx)x \in B$.*
c) *$B = Ax$*

Proof. a) Let $p, q \in B$. Then $p \in Ax$ and $(px)x = p$ (3.3.a)
so $q \vee p' = q \vee px = ((px)x)q = pq$ and therefore $pq \in B$.
b) By induction on t.
If $t = x$ then $(tx)x = x^2x = x = F \in B$ (3.1.a)
If $t = y$ for some $y \in X - \{x\}$ then $(tx)x = (yx)x \in B$
If $t = t_1t_2$ with $t_1, t_2 \in A$
then $(tx)x = ((t_1t_2)x)x = ((t_1x)x)((t_2x)x)$ (3.1.c)
and by induction $(t_1x)x, (t_2x)x \in B$ so by a) $(tx)x \in B$.
c) By b) $(Ax)x \subseteq B$ so using (3.3.b) we get $Ax = (Ax)x \subseteq B \subseteq Ax$.
From this proposition and the paragraphs preceding it we have

(3.7) THEOREM. *The set of laws consisting of (3.1.a,b,c) and the translation of the last five basic laws of $\mathcal{C}$ forms a complete set of laws for I.*

Clearly the translation of laws of $\mathcal{C}$ yields rather complicated laws for I which may be consequence of simpler ones. Such is the case with one of DeMorgan's laws as we saw before. Although we have made some simplifications along these lines, we have not found a complete set of laws for I simple enough to be worth writing down.

References

[A] Abbott, J.C., *Semi-Boolean Algebra*, Matematicki Vesnik **4** (1967), 177-198.

[B] Berman, J., *Free spectra of 3-element algebras*, in "Universal Algebra and Lattice Theory," (edited by Freese, R. and Garcia, O.), Springer Lecture Notes in Mathematics 1004, 1983.

[C] Cleave, J.P., *Some remarks on the interpretation of 3-valued logics*, Ratio **22** (1980), 52-60.

[D] Dienes, Z.P., *On an implication function on many-valued systems of logic*, Journal of Symbolic Logic **14** (1949), 95-97.

[G-S] Guzmán, F., Squier, C.C., *The algebra of conditional logic*, Algebra Universalis (To appear).

[K] Kleene, S., "Introduction to Metamathematics," Van Nostrand, New York, 1952.

[L-T] Lukasiewicz, J., Tarski A., *Untersuchungen über den Aussagenkalkül*, Comptes rendus des séances de la Société des Lettres de Varsovie Class III **23** (1930), 30-50.

[M] McCarthy, J., *Predicate calculus with "undefined" as a truth value*, Stanford Artificial Intelligence Project Memo **1** (1963).

Optimal Semantics of Data Type Extensions

Lawrence S. Moss[1]
Satish R. Thatte[2]

1 Introduction

When considering the semantics of algebraic specifications of data types, it is often natural to divide the specification into a base part and an extension part. The base defines primitive observable values, and the extension defines an abstract type whose users are expected to be interested only in its observable behavior. Final algebra semantics [Wan77,Kam83] is the formalization of this viewpoint. The hallmarks of this approach are

(1) Let A be a "correct implementation," a reachable algebra that agrees with the observable consequences of the specification. Then A will have the final algebra as a homomorphic image. The final algebra itself can be thought of as the "minimal" correct implementation.

(2) Any equation e which holds in the final algebra can be used as a program transformation rule, since any such e can be consistently imposed on any correct implementation A to yield another correct implementation B which is a quotient of A and also satisfies e.

These are very nice properties. Moreover, it is often easier and more natural to specify a data type as a final algebra rather than as an initial algebra, as we recall in a standard example below. The main drawback of final algebra semantics is its limited applicability. The existence of a final algebra follows from the condition of *sufficient completeness* first introduced by Guttag-Horning [GH78]. (Wand [Wan77] studies a related notion called Λ-*fullness*.) But this condition is rather strong. It is natural to ask if this approach can be generalized in some way to all specifications, sufficiently complete or not, in such a way that something like the two properties above continue to hold. The purpose of this paper is to present such a generalization, which we call *optimal algebra*[3] *semantics.*

---

[1]Mathematical Sciences Department, IBM Thomas J. Watson Research Center, P. O. Box 218, Yorktown Heights, NY 10598.

[2]Department of Mathematics and Computer Science, Clarkson University, Potsdam, NY 13676.

[3]The term "optimal algebra", and some of the other associated terminology, was suggested to us by Vaughan Pratt, who also independently conjectured the

For example, consider the main example from [Wan77], a specification for Nat-indexed arrays of Nat (natural numbers). We assume that basic types Nat and Bool are specified in some standard way, and we also assume that the signature contains an equality test eq defined on Nat. Moreover, we assume that Nat contains a value U for "undefined." The values of the original types Nat and Bool are taken to be visible. Nat + Bool is, in Wand's terminology, the base type. In the extension, we would like to add a new type Array, together with three new symbols:

$$\mathsf{empty} :\longrightarrow \mathsf{Array}$$
$$\mathsf{alt} : \mathsf{Nat} \times \mathsf{Nat} \times \mathsf{Array} \longrightarrow \mathsf{Array}$$
$$\mathsf{val} : \mathsf{Nat} \times \mathsf{Array} \longrightarrow \mathsf{Nat}$$

subject to the following two equations:

$$\mathsf{val}(x, \mathsf{alt}(y, z, a)) = \mathsf{if}\ \mathsf{eq}(x, y)\ \mathsf{then}\ z\ \mathsf{else}\ \mathsf{val}(x, a)$$
$$\mathsf{val}(x, \mathsf{empty}) = \mathsf{U}$$

The function alt corresponds to an update. Intuitively, only the latest modification of an array with alt for a given index can affect the outcome for val for that index. Earlier values at that index can be forgotten, and distinctions arising solely out of differences among such obsolete values need not be maintained, since such distinctions can have no observable consequences given that val is the only operation available for observing the state of an array. (These distinctions are preserved in the initial algebra, which is why it is not a satisfactory interpretation for this specification.)

Thus the natural interpretation of the extension is the set of functions from finite subsets of Nat to Nat, with the obvious interpretations of the functions. It turns out that this interpretation is the *final* object of a certain category, the category of algebras where Nat has its initial interpretation, and where every element of sort Array is the interpretation of some term in the language. In more general terminology, this is the category C of reachable algebras of the extension whose reduct to the base is initial.

Sufficient completeness is the key technical fact which allows us to conclude that C has a final object. In the example above, it amounts to showing that all terms of sort Nat are equivalent modulo the equations to a term of sort Nat in which val does not occur.

But now consider what happens if we drop the equation $\mathsf{val}(x, \mathsf{empty}) = \mathsf{U}$ from the specification. We immediately lose sufficient completeness, because terms like val(3, empty) are not equivalent, modulo the specification, to one of the original terms of sort Nat. Indeed, C has no final objects. Perhaps

existence of optimal normal models (Lemma 7). In addition, Francesco Parisi-Presicce has informed us that in unpublished notes, he too obtained the optimal semantics. Because his definition of it was in terms of optimal fixed points, he also named the semantics "optimal."

the reduced specification reflects an earlier stage in development before the possibility of anomalous expressions like val(3, empty) had been considered. However, the natural model is again the arrays of Nat. One way to formalize this intuition is to notice that in every sufficiently complete extension of the specification, we still have the fact that only the latest value at any index needs to be remembered. There are even techniques [KM86] for mechanically deducing properties of this kind for incomplete specifications.

In this paper, we show how to relax the condition of sufficient completeness for semantics based on observability. In effect, we generalize final algebra semantics by considering only those implementations which avoid all "controversial" decisions. This is made more precise in Section 3; in essence a controversial decision is one which is incompatible with some correct implementation of the specification. Sufficient completeness is equivalent to the requirement that there are no controversial decisions to make, given that correct implementations are constrained to be reachable. Indeed, one interesting consequence of the generalization is that even initial algebra semantics can be obtained as the optimal semantics, given a sufficiently general notion of correct implementation.

The rest of the paper is organized as follows: the next section gives the technical machinery and the main theorem of final algebra semantics. Section 3 defines the notions of safe and optimal algebras and proves their basic properties. Section 4 gives several examples to illustrate possible applications of optimal semantics. Section 5 shows that optimality as defined here is *not* universally applicable. The counterexample relies on constraining correct implementations to be computable in some sense. This leads to Section 6 where a modified version of optimality is applied to executable specifications and its connection to the idea of full abstraction is explained. The treatment here is necessarily sketchy, and the reason for including this material is to show that the intuitions underlying optimality have broader applicability.

2 Background on Specifications and their Semantics

Our technical machinery is based on the work of the ADJ group. An excellent tutorial introduction to this material can be found in [ADJ]. We assume familiarity with the basic notions therein.

Given an S-sorted signature Σ, we use T_Σ to denote both the initial (free) Σ-algebra, and the (many-sorted) set of all Σ-terms. Given a Σ-algebra A, the unique homomorphism from T_Σ to A evaluates terms according to their interpretation in A; it too will be denoted by A. A theory E consists of a signature Σ_E and a set (also called) E of Σ_E-equations. The equations might contain variables, but in this paper we will not consider conditional equations.

If e is a Σ-equation and the signature Σ_A of A includes Σ, then we write $A \models e$ to mean that every substitution instance of e is true in A. This notation extends to sets E of equations in the obvious way. An equation e determines a congruence $|e|$ on A; $|e|$ is the least congruence containing all substitution instances of e. The quotient $A/|e|$ then satisfies e. The same is true for sets of equations, and we write I_E for $T_\Sigma/|E|$, where $|E|$ is the smallest Σ-congruence on T_Σ including each $e \in E$. One of the basic results of algebraic semantics is that I_E is initial in the category of Σ_E-algebras which satisfy E. For all terms t and u of the appropriate signature, $I_E(t) = I_E(u)$ iff the equation "$t = u$" is deducible from E using simple equational deduction. We write $t =_E u$ as an abbreviation for $I_E(t) = I_E(u)$.

We will be concerned with **specification pairs** (BASE, EXT). Our informal intepretation of BASE is that its initial model I_{BASE} is what is "visible" or "observable" in a given situation[4]. For this reason, the set of sorts of Σ_{BASE} are called the **observable** sorts. EXT adds operations, possibly on the same sort set and possibly adding new sorts. We assume that EXT is well-formed in the sense that

1. $\Sigma_{\mathrm{EXT}} \supseteq \Sigma_{\mathrm{BASE}}$.
2. $I_{\mathrm{EXT}}|\Sigma_{\mathrm{BASE}} \cong I_{\mathrm{BASE}}$.

The reduct $A|\Sigma$ is just the algebra A considered as a Σ-algebra, forgetting everything else. Of course, this notation is used only when the signature of A includes Σ.

The second condition has two implications. It ensures that $I_{\mathrm{EXT}}|\Sigma_{\mathrm{BASE}}$ is a quotient of I_{BASE}. This is slightly weaker than requiring $\mathrm{EXT}|\Sigma_{\mathrm{BASE}} \supseteq \mathrm{BASE}$, since it allows that EXT contain a different set of axioms for I_{BASE}. The quotient relationship in the opposite direction ensures BASE-preservation, i.e., for all v_1 and v_2 from T_{BASE}, if $v_1 =_{\mathrm{EXT}} v_2$, then $v_1 =_{\mathrm{BASE}} v_2$. This is important since all elements of all carriers in I_{BASE} are considered observable, and distinctions between different observable values are also observable. It would be distressing to have a model of the I_{EXT} which destroyed the distinctions made by the I_{BASE}.

In final algebra semantics, one often considers an additional condition on the specification pair (BASE, EXT):

Definition (Sufficient Completeness) (BASE, EXT) is said to be **sufficiently complete** if for all t in T_{EXT} of observable sort, there exists some $v \in T_{\mathrm{BASE}}$ such that $t =_{\mathrm{EXT}} v$.

[4]It is possible to make weaker assumptions concerning visibility. For example, one might well want to assume some fixed subset of the base terms is visible. Our version is technically simpler, and also realistic in most practical cases.

This ensures that the carriers of I_{EXT} and I_{BASE} are isomorphic for observable sorts. There is another useful way to define the condition.

Definition (Standardness) An EXT-algebra A is said to be **standard** if for all observable sorts s, and for each $u \in A_s$, there is some term $v \in (T_{\text{BASE}})_s$ such that $u = A(v)$.

Note that sufficient completeness is a property of specification pairs, while standardness is a property of EXT-algebras. The two properties are related, since (BASE, EXT) is sufficiently complete iff I_{EXT} is standard. Now we can state the main theorem of final algebra semantics:

Theorem 1 ([Wan77]) *For every sufficiently complete specification pair* (BASE, EXT), *the category consisting of* BASE-*preserving reachable implementations of* EXT *has a final object.*

The category mentioned in this result is a natural one, since the "correct" implementations of EXT will certainly be reachable Σ_{EXT}-algebras which preserve observable values.

As its name suggests, sufficient completeness is sufficient, but not necessary, for the existence of final algebras. This can be seen with a very simple example. Let the BASE contain two symbols

$0 :\rightarrow \mathsf{Num}$
$\mathsf{succ} : \mathsf{Num} \rightarrow \mathsf{Num}$

and no equations. So the terms $\mathsf{succ}^k(0)$, for $k \geq 0$, are then distinct visible values.

We also consider the EXTENSION via a symbol $\mathsf{pred} : \mathsf{Num} \rightarrow \mathsf{Num}$ and equations

$\mathsf{succ}(\mathsf{pred}(x)) = x$
$\mathsf{pred}(\mathsf{succ}(x)) = x.$

Now every Σ_{EXT}-term is equivalent either to 0 or to a unique term of the form $\mathsf{succ}^k(0)$ or $\mathsf{pred}^k(0)$ for some $k \geq 1$. It is obvious that I_{EXT} itself is BASE-preserving, but no proper quotient of I_{EXT} is BASE-preserving. I_{EXT} is therefore the final algebra for EXT. But (BASE, EXT) is obviously not sufficiently complete, since $\mathsf{pred}(0)$ is not equivalent to any visible value.

3 Optimal Algebras

All the conditions on EXT used in Theorem 1 except sufficient completeness are well-formedness conditions. The natural generalization of the result to all extensions amounts to removing the sufficient completeness condition while preserving the spirit of the approach, which in turn is embodied in the two properties mentioned in the introduction. The loss of sufficient completeness means that all correct implementations are no longer above controversy. It is possible for two correct implementations to conflict in their visible behavior. In the incomplete specification for arrays, for instance, the value of val(4, empty) could be undefined, 1, 2, or any other natural number without violating any correctness condition. Note also that even a reachable correct implementation need no longer be standard—val(4, empty) may become a new non-standard number in its own right.

Correctness in this context follows the letter of the specification, but not necessarily its spirit. In particular, a correct implementation is not always capable of evolving with an incomplete specification to its completion. We therefore introduce a notion of safe implementations. Intuitively, a safe implementation will be one which is compatible with all correct implementations. Since the evolution of a specification to its completion in effect reduces the set of its correct implementations to a subset of the standard ones, a safe implementation is capable of accommodating any such completion. There is moreover nothing sacred about the specific definition of correctness used in Theorem 1. It is interesting to explore the consequences of varying the notion of correctness—by allowing correct implementations to enrich the signature with extra operations, for instance, or by requiring that correct implementations must be standard. We shall therefore use the specific notion of correctness as a parameter in the definition of safety.

As in the last section, we will be concerned with a system of two theories – a base theory BASE and its extension EXT. We simply fix an arbitrary pair in which EXT is well-formed relative to BASE, and preserves BASE. We begin the formalization of the ideas above with a definition of reasonable implementation. Reasonableness is simply the minimal set of conditions for correctness. Note in particular that a reasonable implementation may introduce additional operations beyond those in Σ_{EXT}.

For all the following definitions, we fix a signature Γ as the universal signature from which all signatures are drawn. Γ is based on a countably infinite set of sorts which contains a countable number of function symbols for each arity and coarity. One can think of Γ as the universe of typed identifiers, and its use ensures a certain amount of "neatness" without being restrictive.

Definition An algebra A with signature Σ_A is **reasonable** if

1. $\Sigma_{\mathrm{EXT}} \subseteq \Sigma_A \subseteq \Gamma$.
2. $A \models$ EXT.
3. $A|\Sigma_{\mathrm{BASE}} \cong I_{\mathrm{BASE}}$.

Let $\mathcal{R}$ denote the category of reasonable models.

The reasonable models comprise a category in the following way. A morphism f from A to B exists only if $\Sigma_A \subseteq \Sigma_B$, and in this case, f is a Σ_A

morphism $f : A \to B|\Sigma_A$. Identifying isomorphic algebras, $\mathcal{R}$ is a partial order category. Often we will forget the category theoretic aspects of $\mathcal{R}$ and instead emphasize the order, for instance in the definition of safety given below. In addition to being reasonable, a safe implementation will not be allowed to enrich Σ_{EXT} with new operations. Allowing a safe implementation to add new operations is practically useless since nothing safe can be done with such extra operations for most definitions of correctness, but the statements of results become unnecessarily cluttered with caveats.

Definition Let $\mathcal{Z}$ be an arbitrary nonempty full subcategory of $\mathcal{R}$. We use $\mathcal{Z}$ to fix a notion of correctness, and call the elements of $\mathcal{Z}$ **correct implementations** of EXT. An algebra $A \in \mathcal{R}$ will be called **safe** relative to $\mathcal{Z}$ iff

1. $\Sigma_A = \Sigma_{\text{EXT}}$
2. For every $B \in \mathcal{Z}$, there exists some $C \in \mathcal{Z}$ such that $A \leq C$ and $B \leq C$.

Note that A need not belong to $\mathcal{Z}$.

The safe implementations of EXT relative to $\mathcal{Z}$ form a full subcategory of $\mathcal{R}$, which we denote by $\mathbf{S}_{\mathcal{Z}}$.

The category $\mathbf{S}_{\mathcal{Z}}$ is never empty since the initial implementation I_{EXT} is always safe. This can be seen immediately from the definition above, since $I_{\text{EXT}} \leq B$ for each $B \in \mathcal{Z}$. Note the restriction that $\mathcal{Z}$ must be nonempty. If there are no correct implementations, the semantic issue is moot. Technically, the definition of a correct equation (see below) and the related results are also dependent on this assumption.

Condition 2 in the definition above formalizes a particular notion of compatibility—the ability to merge (the canonical congruences for) two algebras consistently. If we think of the two algebras as being equationally specified, the sets of equations for two compatible algebras can therefore simultaneously hold without violating BASE preservation. This turns out to be appropriate for the semantics of abstract data type specifications. In Section 6, we discuss the semantics of equational programs or executable specifications, where compatibility of algebras will be constrained by the requirements of a version of computability.

Definition An algebra A will be called **optimal** relative to $\mathcal{Z}$ iff A is a final algebra in $\mathbf{S}_{\mathcal{Z}}$.

Note that safe and optimal implementations do not have to be correct, though they do have to be reasonable. Correctness may involve stringent conditions such as being standard that are impossible to reconcile with safety for incomplete specifications.

The first property of final algebras mentioned in the introduction is just an explanation of finality, and it obviously holds for optimal algebras relative to safe ones rather than correct ones. The second property holds for all safe algebras, and therefore a fortiori for the optimal one if it exists. Any equation that holds in any safe algebra can be imposed on any correct algebra without violating correctness. Indeed, this is one way to explain the definition of safety. The precise properties involved are best expressed in terms of a notion of correct equation.

In the following, we shall drop the ubiquitous subscript EXT whenever possible without causing confusion.

Definition A ground Σ_{EXT} equation e is **correct** for $\mathcal{Z}$ if for every $\mathcal{Z}$-model B, there exists a $\mathcal{Z}$-model C such that $B \leq C$ and $C \models e$. When $\mathcal{Z}$ is clear from the context, we often do not mention it.

Proposition 2 *$A \in \mathbf{S}_{\mathcal{Z}}$ implies that every equation e which holds in A is correct.*

Proof Follows directly from the definitions of correctness and safety. ⊣

Proposition 3 *If an equation e is correct in $\mathcal{Z}$, then there is some safe algebra $A \in \mathbf{S}_{\mathcal{Z}}$ in which e holds.*

Proof Let $A = I_{\mathrm{EXT}}/|e|$. Obviously, e holds in A. Since $\mathcal{Z}$ is nonempty, there is a correct algebra in which e holds, and hence A must preserve BASE. To see that $A \in \mathbf{S}_{\mathcal{Z}}$, suppose $B \in \mathcal{Z}$ is an arbitrary correct algebra. Since e is correct in $\mathcal{Z}$, there is a $C \in \mathcal{Z}$ such that $B \leq C$ and e holds in C. Since $I_{\mathrm{EXT}} \leq C$, and e holds in C, $I_{\mathrm{EXT}}/|e| = A \leq C$ as well. ⊣

These propositions lead to a characterization of optimal algebras in terms of correct equations.

Theorem 4 *$A \in \mathbf{S}_{\mathcal{Z}}$ is optimal iff every correct equation in $\mathcal{Z}$ holds in it.*

Proof The "only if" part follows from Proposition 3 and the finality of the optimal algebra among safe ones. To see the "if" part, recall that safe algebras are reachable Σ_{EXT}-algebras. By Proposition 2, every ground Σ_{EXT}-equation which holds in a safe algebra is correct and hence holds in A. A is therefore a quotient of every safe algebra. ⊣

The characterization in Theorem 4 is not always easy to use to establish the existence of an optimal algebra. We therefore give a sufficient condition that is useful in many cases. The condition is defined in terms of maximal algebras.

Definition An algebra $A \in \mathcal{R}$ is said to be **maximal** in a subcategory

$\mathcal{Z} \subseteq \mathcal{R}$ if $A \leq B$ in $\mathcal{Z}$ implies $B \leq A$.

Proposition 5 *If C is maximal in $\mathcal{Z}$, then every correct equation in $\mathcal{Z}$ holds in C.*

Proof Suppose e is correct in $\mathcal{Z}$. By the definition of correctness, there is $D \in \mathcal{Z}$ such that $C \leq D$ and e holds in D. However, since C is maximal, $D \leq C$ and e holds in C. ⊣

For the proofs of several results below, we need to recall some of the properties of the category of all BASE-preserving quotients of I_{EXT}. We refer to these quotients as the *normal* algebras for EXT. Let

$$\mathcal{N} \quad = \quad \{A \in \mathcal{R} \mid \Sigma_A = \Sigma_{\text{EXT}}\}.$$

$\mathcal{N}$ is obviously the category we want. This category carries a natural algebraic structure that is slightly weaker than the complete lattice structure carried by all quotients of a free Σ-algebra. $\mathcal{N}$ is a complete lower semi-lattice. That is, the intersection of any family of BASE-preserving congruences on I_{EXT} is also an BASE-preserving congruence. Moreover, although least upper bounds (LUBs) of BASE-preserving quotients are not in general BASE-preserving (unless there is a final algebra for EXT in the sense of Theorem 1), LUBs of (finite and infinite) directed sets of quotients do preserve BASE. This can be seen by a simple compactness argument: any counterexample must prove the equivalence of two visible terms in a finite number of equational steps based on a finite number of congruences, but the corresponding quotients have an BASE-preserving upper bound.

Theorem 6 *If for every algebra $A \in \mathcal{Z}$ there is a $B \in \mathcal{Z}$ such that $A \leq B$ and B is maximal in $\mathcal{Z}$, then there is a final object in $\mathbf{S}_{\mathcal{Z}}$,* i.e., *an algebra that is optimal relative to $\mathcal{Z}$ exists.*

Proof The idea is to construct a safe Σ_{EXT}-algebra in which every correct equation in $\mathcal{Z}$ holds. The result then follows by Theorem 4. Recall that the equations we are concerned with are Σ_{EXT}-equations. By Proposition 5, the reduct $B|\Sigma_{\text{EXT}}$ of any maximal $B \in \mathcal{Z}$ is a (not necessarily safe) Σ_{EXT}-algebra in which every correct equation in $\mathcal{Z}$ holds. The complete lower semi-lattice structure of $\mathcal{N}$ guarantees the existence of a meet for all such reducts of maximal algebras in $\mathcal{Z}$. Let this meet be denoted by F. Obviously, every correct equation in $\mathcal{Z}$ holds in F. It is easy to see that if every algebra in $\mathcal{Z}$ has a maximal upper bound in $\mathcal{Z}$ then F is also safe. Clearly, $\Sigma_F = \Sigma_{\text{EXT}}$. Suppose $A \in \mathcal{Z}$. There is a $B \in \mathcal{Z}$ such that $A \leq B$ and B is maximal in $\mathcal{Z}$. Since F is the meet of Σ_{EXT}-reducts of all maximal algebras in $\mathcal{Z}$, $F \leq B$ as well. Both conditions for safety are satisfied and F is therefore the final object in $\mathbf{S}_{\mathcal{Z}}$. ⊣

In the next section, we consider several notions of correctness for which optimal algebras exist, and one for which they sometimes do not.

4 Variations on Correctness

We begin by identifying correctness with the category $\mathcal{N}$ of BASE-preserving quotients of I_{EXT}, which is the category of algebras normally considered in final algebra semantics. The category $\mathbf{S}_{\mathcal{N}}$ is a proper subset of $\mathcal{N}$, unless there is a final algebra in $\mathcal{N}$, in which case the two categories coincide.

Lemma 7 *There is a final algebra in* $\mathbf{S}_{\mathcal{N}}$ *which will be called the optimal normal semantics of* EXT.

Proof This result follows directly from Theorem 6, since every algebra $A \in \mathcal{N}$ has a maximal upper bound in $\mathcal{N}$. To see this, note that the set of all ground Σ_{EXT}-equations is enumerable. Let $\langle e_i : i \in \omega \rangle$ be such an enumeration. A maximal upper bound for A can be obtained as the limit of an increasing chain of quotients

$$A \quad = \quad A_0 \leq A_1 \leq A_2 \leq \cdots \leq A_i \leq \cdots$$

where $A_{i+1} = A_i$ if $A_i/|e_i|$ is not BASE-preserving, and $A_i/|e_i|$ otherwise.
The limit belongs to $\mathcal{N}$ since LUBs of directed sets of BASE-preserving quotients do preserve BASE as noted above. ⊣

Note that optimal normal semantics is applicable to all well-formed extensions. For instance, I_{EXT} is the optimal normal algebra for the succ/pred example in the introduction. The reduced specification for arrays also has an optimal normal semantics in which assertions such as

$$\mathsf{alt}(x, y, \mathsf{alt}(x, z, a)) = \mathsf{alt}(x, y, a)$$

hold, and which has a nonstandard value of sort Nat corresponding to each expression of the form $\mathsf{val}(x, \mathsf{empty})$.

As one would expect, there is a straightforward relationship between traditional final algebra semantics (in which $\mathcal{N}$ contains a final object) and optimal normal semantics. Indeed, the relationship is a little more general. Final algebra semantics based on any subset of $\mathcal{N}$ coincides with the corresponding optimal semantics when the required final algebra exists.

Lemma 8 *If* $\mathcal{Z} \subseteq \mathcal{N}$ *and* $\mathcal{Z}$ *contains a final object* F*, then* F *is final in* $\mathbf{S}_{\mathcal{Z}}$.

Proof If $\mathcal{Z}$ contains a final object F, then F is the only maximal algebra in $\mathcal{Z}$ (upto isomorphism), and F is an upper bound for every other algebra in $\mathcal{Z}$. Since $\mathcal{Z} \subseteq \mathcal{N}$, F is a Σ_{EXT}-algebra, and therefore it must be the optimal algebra by Theorem 6.

We now consider a case where a restriction of the notion of correctness is appropriate. Kapur and Musser [KM84] introduced a notion of "completable" extension of a base theory in their work on proof by consistency for ambiguous theories (their formulation is a little different, but the details are irrelevant here). For our purposes, we may define a completable extension as any extension which has a standard model. Suppose our fixed extension EXT is completable relative to BASE, and we are interested in the optimal algebra for EXT as before, but this time, we want to restrict correct implementations to those algebras in $\mathcal{N}$ which are standard. Such an optimal algebra always exists but it is not always the same as the optimal normal algebra.

Let $\mathcal{S} = \{A \in \mathcal{N} \mid A \text{ is standard}\}$. In order to ensure that $\mathcal{S}$ is nonempty, we must exclude specifications like the **succ/pred** example in the introduction. $\mathbf{S}_{\mathcal{S}}$ is usually disjoint from $\mathcal{S}$, unless EXT is sufficiently complete, in which case the two categories coincide.

Lemma 9 *There is a final object in* $\mathbf{S}_{\mathcal{S}}$ *which will be called the optimal standard semantics of* EXT.

Proof The same procedure as in the proof of Lemma 7 works here to obtain maximal correct algebras. Note that equations of observable sort can never be added to a standard algebra with any effect without loosing BASE-preservation. Only equations of nonobservable sort need therefore be considered. ⊣

Here is an example which shows the difference between the optimal normal and optimal standard algebras (it is from the paper of Kapur-Musser cited above). Let BASE be a theory of natural numbers with only

$0 :\rightarrow \mathsf{Nat}$
$\mathsf{succ} : \mathsf{Nat} \rightarrow \mathsf{Nat}$

and no equations. Now let EXT be an extension which introduces three new symbols

$\mathsf{plus} : \mathsf{Nat} \times \mathsf{Nat} \rightarrow \mathsf{Nat}$
$\mathsf{double} : \mathsf{Nat} \rightarrow \mathsf{Nat}$
$k :\rightarrow \mathsf{Nat}$

and the following four equations:

$\mathsf{plus}(0, x) = x$
$\mathsf{plus}(\mathsf{succ}(x), y) = \mathsf{succ}(\mathsf{plus}(x, y))$
$\mathsf{double}(0) = 0$
$\mathsf{double}(\mathsf{succ}(x)) = \mathsf{succ}(\mathsf{succ}(\mathsf{double}(x)))$

Now the assertion "$\mathsf{double}(x) = \mathsf{plus}(x, x)$" holds in the optimal standard algebra but not in the optimal normal algebra. The reason is that one may equate double(k) = succ(0) in an algebra in $\mathcal{N}$ but not in $\mathcal{S}$ since the identification precludes standardness. The proof techniques of Kapur-Musser [KM84] are based on the optimal standard semantics of the theories they reason about.

The optimal standard semantics is more abstract (validates more equalities) than the optimal normal semantics because it is based on a more restricted notion of correctness. One would expect the optimal semantics to be less abstract for a looser notion of correctness. The next result shows that with just a little loosening, the optimal and initial algebras coincide. The loosening allows an implementation to enrich Σ_{EXT} with new operations, but not with new sorts.

Definition $\mathcal{E} = \{A \in \mathcal{R} \mid \Sigma_A \text{ is based on the same set of sorts as } \Sigma_{\text{EXT}}\}$.

Lemma 10 I_{EXT} *is final in* $\mathbf{S}_{\mathcal{E}}$.

Proof Essentially, a new operation can be used to discriminate between any two ground terms that are not provably equal. Let t and u be arbitrary Σ_{EXT}-terms of the same sort such that $t \neq_{\text{EXT}} u$. Suppose v_1 and v_2 are two Σ_{BASE}-terms that correspond to different observable values. (Such terms exist by the assumption that all carriers of I_{BASE} are not singletons.) Now let g be a new symbol that is not in Σ_{EXT} and let $E = \{g(t) = v_1,\ g(u) = v_2\}$. Let $\Sigma_A = \Sigma_{\text{EXT}} \cup \{g\}$, $E_A = \text{EXT} \cup E$, and $A = I_{E_A}$. Clearly, $A \in \mathcal{E}$ and therefore $t = u$ is not a correct equation in $\mathcal{E}$. I_{EXT} is therefore the only safe algebra relative to $\mathcal{E}$. ⊣

Corollary 11 I_{EXT} *is final in* $\mathbf{S}_{\mathcal{R}}$.

To put these results in perspective, recall that we are interested in broadening the scope of final algebra semantics by considering categories of algebras that are compatible with all acceptable implementations. As we noted in the introduction, the choice of $\mathcal{Z}$ involves a trade-off between the desired strength of semantically valid assertions and the range of possibilities for acceptable implementations. For instance, Corollary 11 implies that if correct implementations permit the addition of arbitrary new operation symbols, then the only safe assertions are those which are valid in the initial algebra. Lemma 10 is a stronger result; it shows that no new sorts are needed. Results obtained via reasoning techniques based on final algebra semantics, such as proof by consistency ([KM86]), are not necessarily valid in implementations that introduce new symbols, even when the original specification (EXT) has a final algebra. Kapur-Musser's extension of proof by consistency [KM84] to ambiguous theories, which was originally thought to be based on optimal normal semantics, turns out to be based on optimal standard semantics because they found that they had implicitly assumed that implementations must be standard.

5 The Final Safe Model Does Not Always Exist

We show that for certain choices of BASE, EXT and $\mathcal{Z}$, the category of models which are safe for $\mathcal{Z}$ does not have a final object. In other words, there is an unsafe algebra F such that every equation which holds in F is correct.

Definition Fix a specification pair (BASE, EXT) and a signature $\Gamma \supseteq \Sigma_{\text{EXT}}$. A normal model A is **finitary** iff there is a signature Σ_A and a *finite* set E of Σ_A-equations such that $A \cong I_E|\text{EXT}$. Let $\mathcal{F}$ be the collection of finitary reasonable models.

This condition is a natural one for some purposes, since implementations must after all be computable, and in this context equational computation seems to be the natural choice. Of course finite axiomatization only guarantees semi-computability for the word problem, but allows all semantically implied observable results to be computed, which is what one really needs in an implementation.

Our negative result rests on the fact that there are algebras which can be specified as final algebras of finite specifications, but for which the word problem is not semi-computable; such an algebra is not finitary. More precisely,

Lemma 12 *There is a specification pair* (BASE, EXT) *for which the category of normal models has a final object which is not semicomputable and hence not finitary.*

We know of no simple or natural examples illustrating this lemma. The interested reader can find details in [MMG].

Lemma 13 *There exists a specification* (BASE, EXT) *for which the category of models safe for* $\mathcal{F}$ *does not have a final object.*

Proof Let (BASE, EXT) and $\mathcal{Z}$ be as in Lemma 12, and let F be the final normal model. We first show that F is not safe for $\mathcal{F}$. If it were, there would be some $C \in \mathcal{F}$ such that $I_{\text{EXT}} \leq C$ and $F \leq C$. But since F is final among normal algebras and C is normal, $F \cong C$. Thus F is finitary, and this contradicts Lemma 12.

We now claim that there is no final safe-for-finitary model. Suppose that G is any model which is safe for $\mathcal{F}$. We know that $G \neq F$, and by finality, $G \leq F$. Let e be an equation that does not hold in G but holds in F. Let $H = G/|e|$. We claim that H too is safe for $\mathcal{F}$. For this, consider a finitary B. We know that there is a finitary C such that $B \leq C$ and $G \leq C$. Now let E be the finite axiomatization of C. $C/|e|$ preserves the BASE since $F \models e$. Moreover, $C/|e|$ is finitary as $E \cup \{e\}$ is a finite axiomatization of it. We therefore have $B \leq C \leq C/|e|$ and $H \leq C/|e|$. This shows that H is safe. Since G is not a quotient of it, G is not final. ⊣

This proof exploits the fact that the partial order of $\mathcal{F}$ is usually not chain complete. In conjunction with the result of Lemma 12, we were able to turn the situation into a non-existence proof. It should be noted, however, that chain completeness of a class $\mathcal{Z}$ is not a necessary condition for the optimal model to be safe. For example, consider, in contrast to the situation of Lemma 13, the case where Γ is infinite, and has infinitely many function symbols of each arity and co-arity. Let $\mathcal{Z}$ be the collection of finitary reasonable models. $\mathcal{Z}$ is not chain complete. But the proof of Lemma 10 goes through exactly as before. So we know that the optimal $\mathcal{Z}$ model is I_{EXT}. This model is trivially safe, for any $\mathcal{Z}$.

It can be shown that Lemma 13 also holds if "finitary" is replaced by "computability of the word problem" (that is, "recursiveness"). The proof is much more complicated, and beyond our scope.

Computability is in some sense an unnatural restriction for correctness in the present context since other aspects of optimality, such as safety, do not take it into consideration. In the next section, we consider a modification of the notion of optimality in which a notion of computability plays the central role.

6 The Semantics of Executable Specifications

There is a class of programming languages which use equational theories as programs. Examples include Standard ML [Mil84], Miranda [Tur85] and the equational language of Hoffmann and O'Donnell [HO82]. Operationally, equational computation can be understood as term rewriting. The denotational properties of rewriting have been studied by Nivat [Niv75] using the notion of initiality. Raoult and Vuillemin [RV80] have used a more complex approach to establish a correspondence between the denotational and operational properties of their language along the lines of full abstraction [Mil77]. Intuitively, initiality formalizes a view of equational computation as a completely transparent visible process, while full abstraction views an equational program as a black box with both visible and invisible aspects; the visible values being typically those involved in input/output. Full abstraction is clearly akin to finality. The results in this section make the connection precise for equational languages.

Our goal is to consider equational programs as extensions of a base theory which describes the input/output values, and arrive at a variation of optimal semantics which captures full abstraction. Consider first the notion of correctness. Since the setting is that of a programming language, meanings must be provided for theories in all stages of completion. Intended implementations may therefore require enrichment of the signature with both new symbols and new sorts. On the other hand, implementations must be "computable" in some sense. We shall use the following minimally restrictive variation of computability which simply ensures that all semantically implied observable results can be computed by a finite program.

Definition An algebra $A \in \mathcal{R}$ is **observable** iff there is a finite set E_A of Σ_A-equations such that $t \in T_{\Sigma,A}$, $v \in T_{\mathbf{BASE}}$ and $A(t) = A(v)$ imply $t =_{E_A} v$. In this case, A is said to be observable with E_A.

To put it another way, an algebra is observable if the word problem for terms denoting observable values is solvable with simple equational deduction using a finite axiomatization. The connection between equational deduction and "real" computation is that for programs with the Church-Rosser property, equational deduction can be implemented with rewriting. This notion of computability is natural for the semantics of programs (cf. the "termination lemma" in [RV80]) because the ultimate means of observation in a program is the output of actual execution. It would be strange to claim in such an environment that an expression is semantically equal to 1, say, when upon evaluation it does not yield 1 as output. Note that observability is a weaker notion than finitariness.

There is one further natural constraint on correctness that is required for full abstraction. A subcategory $\mathcal{Z} \subseteq \mathcal{R}$ of correct algebras corresponds to all observable models of programs extending EXT in a particular language. For our purposes, an algebra A is a model of program E iff $\Sigma_A = \Sigma_E$ and A is observable with E. The restriction we impose is that if $\mathcal{Z}$ contains one observable model of a program, then it must contain all observable models of that program—the rationale being that when such a model is present the program can be assumed to be in the language.

Definition Any $\mathcal{Z} \subseteq \mathcal{R}$ will be said to be a **language model** if

1. Every algebra in $\mathcal{Z}$ is observable.
2. Whenever $A \in \mathcal{Z}$ is observable with a theory E, any $B \in \mathcal{R}$ such that B is observable with E is also in $\mathcal{Z}$.
3. $I_{\mathbf{EXT}} \in \mathcal{Z}$.

The last restriction merely says that EXT itself must be in the language concerned. For some applications, it may be more appropriate to weaken constraint (2) in this definition by introducing a notion of language as a collection of theories, and requiring that all observable models of theories in the language must be in $\mathcal{Z}$. The notion of observable safety below would also need to be modified accordingly. Observability will also be the new ingredient in safety. For an algebra to be safe, not only must it be possible to merge it with any correct algebra, but the result of the merger must be an observable model for the same program as before.

Definition Given a category $\mathcal{Z}$ of observable algebras, an algebra A is

observably safe iff for every $B \in \mathcal{Z}$, there is a $C \in \mathcal{Z}$ such that $B \leq C$ and $A \leq C$ and there is a set E of equations such that B and C are both observable with E. The category of observably safe algebras relative to $\mathcal{Z}$ will be denoted by $\mathbf{CS}_{\mathcal{Z}}$.

The main result we wish to prove is that there is always a final object in $\mathbf{CS}_{\mathcal{Z}}$ whenever $\mathcal{Z}$ is a language model. This final object is in fact the fully abstract semantics of EXT relative to the language of programs which serve as observers for algebras in $\mathcal{Z}$. The complete details explicating this observation are beyond the scope of this paper, but we shall define the fully abstract semantics and show that it is the final algebra in $\mathbf{CS}_{\mathcal{Z}}$.

A fully abstract model semantically distinguishes between two expressions only when they are are observably distinct in some program context. Traditionally, two expressions are considered separable in this sense when there is a context in which insertion of each leads to two distinct observable results. This definition is too strong for implementations of programming languages with non-strict functions because it does not take into account distinctions between unsolvable and solvable expressions. For instance, suppose g is defined by the equation "$g = g$". Clearly, g is unsolvable and does not provide any information about its value. In any of the languages mentioned at the beginning of this section, g will not be separable from any other term because any context in which the insertion of g leads to an observable result must essentially ignore the inserted term, which means that the insertion of any other term also leads to exactly the same observable result. A more realistic notion of observable distinction is obtained if separability is weakened by not requiring the insertion of both expressions to produce observable results. Wadsworth [Wad76] used a similar weakening of separability which he called semi-separability in studying the correspondence between operational and denotational distinctions in the λ-calculus. In order to give a version for equational languages, it is necessary to introduce the notion of a context, which needs a few syntactic notions about terms.

Syntactically, Σ-terms are understood as trees labeled with a function symbol from Σ at each node. The symbols at leaves may also be variables. A path p is a (possibly empty) string of positive integers. Given a Σ-term t and a path, we assume familiarity with what is meant by the subterm **reached by p in t**; we use t/p as a notation for this term. The expression $t[p \leftarrow w]$ denotes the term obtained by replacing t/p at p by w. A **context** in Σ is a pair (c, q) such that $c \in T_\Sigma$ and q is a path in c. A context in which $q = \Lambda$ is said to be **empty**. Following traditional notation, a context will be denoted by $C[\cdot]$, and the term obtained by inserting a term t in $C[\cdot]$ will be written as $C[t]$. If $C[\cdot]$ is the pair (c, q), then $C[t]$ is the term $c[q \leftarrow t]$.

Definition Let $\mathcal{Z}$ be a category of algebras. Two ground terms t and u in T_{EXT} are said to be **inseparable in $\mathcal{Z}$** if for all $B \in \mathcal{Z}$, all Σ_B-contexts

$C[\cdot]$, and all $v \in T_{\text{BASE}}$,

$$B(C[t]) = B(v) \quad \text{iff} \quad B(C[u]) = B(v),$$

and vice–versa. We write $t \approx_{\mathcal{Z}} u$ to mean that t and u are inseparable in $\mathcal{Z}$. Whenever possible, we suppress the class $\mathcal{Z}$ from the notation and just write $t \approx u$.

Our definition of inseparability might also be called non-semi-separability.

It is important to note that throughout this discussion, we permit the contexts to be empty. This is not always done in discussions of inseparability, but it is quite natural here. One consequence of allowing empty contexts is that if a term t is inseparable from a visible term v in an algebra A, then indeed $A(v) = A(t)$.

The following lemma is quite general, and does not assume that $\mathcal{Z}$ is a language model.

Lemma 14 *Let $A \in \mathcal{Z}$. Then the relation $\approx$ is a congruence on A, and the quotient $A/\!\approx$ is a* BASE*-preserving quotient of A. Furthermore, if A is computable by E, then $A/\!\approx$ is computable by E as well.*

Proof The relation $\approx$ is obviously an equivalence. It is a congruence due to the fact that applying a function symbol to a sequence of congruent terms merely creates a partial context for them. Placing such an application in a larger context corresponds to placing the congruent subterms in a context in which they are already known to be inseparable. It is obvious that $t =_{\text{EXT}} u$ implies $t \approx u$. As we noted above, the empty context shows that if $v \approx w$ are visible terms, then already $A(v) = A(w)$. This shows that the quotient $A/\!\approx$ preserves the BASE, and also that any set of equations which observes A also observes $A/\!\approx$. ⊣

$T_{\text{EXT}}/\approx$ is meant to be the fully abstract model for EXT. It is not hard to show that $T_{\text{EXT}}/\approx$ is observable with EXT and final in $\mathbf{CS}_{\mathcal{Z}}$ whenever $\mathcal{Z}$ is a language model.

Proposition 15 *For any language model $\mathcal{Z}$, $T_{\text{EXT}}/\approx$ belongs to $\mathbf{CS}_{\mathcal{Z}}$.*

Proof Suppose $A \in \mathcal{Z}$, and A is observable with E_A. Let $t \approx u$. By the definition of inseparability, $B = A/|t = u|$ is also observable with E_A and hence $B \in \mathcal{Z}$. One can therefore add all such equations $t = u$ to A along the lines of the proof of Lemma 10, and the result must still be in $\mathcal{Z}$. ⊣

Theorem 16 *For any language model $\mathcal{Z}$, $T_{\text{EXT}}/\approx$ is final in $\mathbf{CS}_{\mathcal{Z}}$.*

Proof Suppose for the sake of contradiction that it is not. Let $A \in \mathbf{CS}_{\mathcal{Z}}$ be such that for some $t, u \in T_{\text{EXT}}$, $A(t) = A(u)$ but $t \not\approx u$. Therefore there must be $B \in \mathcal{Z}$ observable with E_B, and $C[\cdot] \in \Sigma_B$, such that $B(C[t]) = B(v)$ and $B(C[u]) = /B(v)$ for some $v \in T_{\text{BASE}}$. However, since A is observably safe, there must be a $D \in \mathcal{Z}$ such that $D(t) = D(u)$, $B \leq D$, and D is observable with the same theory E_B. However, $D(t) = D(u)$ implies $D(C[t]) = D(v)$ iff $D(C[u]) = D(v)$. Therefore D cannot be observable by E_B. ⊣

This version of fully abstract semantics is the "right" semantics for equational programming because for most real languages it achieves the right degree of abstraction. A denotational semantics is too abstract if the operational semantics of the language cannot realize it even in the limit. The technical notion involved in making this statement precise was introduced by Wadsworth [Wad76], and is called limiting completeness. It turns out that a number of typical equational languages can be shown to possess the limiting completeness property relative to the fully abstract semantics we have defined. The interested reader is referred to [Tha89] for further details – the formulation there is completely different but the definition of fully abstract semantics is equivalent.

7 Conclusions

The generalization of final algebra semantics described in this paper is based on two key ideas – the separation of safety from correctness and the relativization of the category of interest with respect to correctness as a parameter.

The idea of safety is motivated by the fact that in reasoning about a specification, one is interested in those assertions which hold without regard for the incidental or arbitrary decisions which may be reflected in correct implementations. The defining property of safe algebras is that they capture only such assertions. In this framework, one can say that the applicability of traditional final algebra semantics is limited because it is applicable only in situations where all correct algebras happen to be safe.

The relativization of correctness is a necessity from a practical viewpoint. Finality is perhaps the simplest embodiment of a kind of intensionality that contrasts with the extensionality of initial algebra semantics. However, as we show in Section 4, finality may be reduced to initiality if correct implementations are given sufficient liberty to enrich the signature. If correctness must therefore be restricted in order to allow finality to offer something new, it seems unlikely that some fixed absolute restriction will be satisfactory in all situations. Relativization allows a kind of "semantic engineering" to adjust finality to a variety of contexts, as we have illustrated with several examples.

The connections of finality with the fully abstract semantics of languages of equational specifications and programs are yet to be properly explored. The material in Section 6 makes a beginning in this direction. These connections are likely to be interesting and fruitful.

8 References

[ADJ] J. A. Goguen, J. W. Thatcher, E. G. Wagner, and J.B. Wright (ADJ). An initial algebra approach to the specification, correctness, and implementation of abstract data types. In R. T. Yeh, editor, *Current Trends in Programming Methodology IV*. Prentice-Hall, 1978.

[GH78] J. V. Guttag and J. J. Horning. The algebraic specification of abstract data types. *Acta Informatica*, 10(1):27–52, 1978.

[HO82] M. C. Hoffman and M. J. O'Donnell. Programming with equations. *ACM TOPLAS*, 4(1):83–112, 1982.

[Kam83] S. Kamin. Final data types and their specifications. *ACM TOPLAS*, 5(1):97–121, 1983.

[KM84] D. Kapur and D. R. Musser. Proof by consistency. *General Electric Corporate Research and Development Report 84GEN008*, 1984.

[KM86] D. Kapur and D. R. Musser. Inductive reasoning with incomplete specifications. In *Proceedings of the Symposium on Logic in Computer Science*, pages 367–377, 1986.

[Mil77] R. Milner. Fully abstract models of typed λ–calculi. *Theoretical Computer Science*, 4:1–22, 1977.

[Mil84] R. Milner. A proposal for standard ML. In *Proc. 1984 Symp. on LISP and Functional Programming*, pages 184–197. ACM, 1984.

[MMG] L. S. Moss, J. Meseguer, and J. A. Goguen. Final algebras, cosemicomputable algebras, and degrees of unsolvability. *Theoretical Computer Science*, To Appear.

[Niv75] M. Nivat. On the interpretation of recursive polyadic program schemes. *Symposia Mathematica*, 15:255–281, 1975.

[RV80] J. C. Raoult and J. C. Vuillemin. Operational and semantic equivalence between recursive programs. *J. Assoc. Comp. Mach.*, 27(4):772–776, 1980.

[Tha89] S. R. Thatte. Full abstraction and limiting completeness in equational languages. *Theoretical Computer Science*, 65(1), 1989.

[Tur85] D. A. Turner. Miranda: A non-strict functional language with polymorphic types. In *LNCS 201*. Springer-Verlag, 1985.

[Wad76] C. Wadsworth. The relation between computational and denotational properties for Scott's D_∞ models of the λ–calculus. *SIAM Journal of Computing*, 5(3):488–520, 1976.

[Wan77] M. Wand. Final algebra semantics and data type extensions. *Journal of Computer and System Sciences*, 19(1):27–44, 1977.

OTHER LOGICS FOR (EQUATIONAL) THEORIES

G. C. Nelson
The University of Iowa

We consider finding logical systems in which one can give conceptually simple formal proofs for certain kinds of formulas from nonlogical axioms which are usually identities. Birkhoff in [3] developed an equational logic adequate for proving equations from a set of equations and Selman in [12] developed a logical system adequate for proving quasi–identities from a set of quasi–identities. Our approach is different in that we do not require the syntactic form of the desired theorems to be the same as that of the nonlogical axioms or we allow a restricted class of models to be considered. In general both problems that we study here fit into the following pattern: Given a set of nonlogical axioms Σ, a set Γ of formulas, and a class $\mathscr{K}$ of structures all in a common language L describe a notion of proof from Σ such that a formula φ in Γ is provable from Σ iff φ is true in all elements of $\mathscr{K}$. Of course the predicate calculus answers this question for first order languages L with $\mathscr{K}$ equal to the class of all models of Σ and Γ equal to the set of all formulas, but one would hope that when Σ or Γ possess special properties that one can place more restrictions on what form a proof may take, as was done in [3] or [12]. Of course when one varies $\mathscr{K}$ a multitude of difficulties can arise, but the above problem can be even more interesting from a mathematical point of view when $\mathscr{K}$ consists of a single model of Σ instead of all models of Σ. In fact in both sections of this paper our results can be viewed as taking $\mathscr{K}$ to be a special one–element set.

In section 1 we develop a notion of positive proof such that a positive formula has a positive proof from a set Σ of identities iff it is true in all models of Σ. It turns out that an analogous result holds for sets Σ of nonlogical axioms which are universal Horn sentences [5] or [4]. This latter result can be applied to logic programming languages such as Eqlog [7] or Prolog [8] where one is restricted to a

---

This paper is in final form and no version of it will be submitted for publication elsewhere.

set Σ of universal Horn nonlogical axioms (for one's program) but where one still may be interested in what positive sentences are provable from Σ. The basic fact which our results depend upon is that there is a free model of Σ on infinitely many generators and that this free model has a syntactic construction which is analogous to the fact used in [7] or [15] that a free model on zero generators exists.

In section 2 we develop a notion of formal proof from a set of identities Σ whose validity in an algebra $\mathbf{A}$ depends upon that algebra having a finite cardinality. Using this notion of k+1–proof, it is shown that the identities true in a k element algebra $\mathbf{A}$ in a finite algebraic language L are all k+1–provable from a finite set Σ of identities of $\mathbf{A}$ where Σ is easily constructed from the free algebra on k generators in the variety generated by $\mathbf{A}$. This demonstrates that the notions of evaluation and derivation can be merged to a single notion of derivation in this setting, namely that of k+1–provability for an appropriate k and axioms Σ.

§1. Logic for Positive Sentences

We assume for exposition that we are given an algebraic first order language L consisting entirely of a set $\mathcal{F}$ of function symbols each of which has a finite arity together with equality $\approx$. Our basic definitions and notation can be found in [4] or [6]. At the end of this section we will indicate how these results apply to other languages such as heterogeneous relational languages as described in [7]. We point out that consequently our results give a proof theory for the positive formulas in logic programming languages such as Eqlog [7] as well as Prolog [9].

Let $X = \{x_1,x_2,\dots,x_n,\dots\}$ be the list of formal variable symbols of L and we denote by T(X) the set of terms in these variables while $T(X_n)$ is the subset of T(X) consisting of all terms in at most the variables in $X_n = \{x_1,x_2,\dots,x_n\}$. An *atomic formula* or *equation* is a finite string of symbols of the form $t_1 \approx t_2$ where t_1 and t_2 are in T(X). Finally the set P of *positive formulas* of L is the smallest set of formulas of L containing the atomic formulas and closed under the formula building operations using only the logical symbols &, $\vee$, $\forall x$, and $\exists x$. Recall that Lyndon [9] has proven that an elementary sentence is preserved under the formation of homomorphic images iff it is logically equivalent to a positive sentence. We use the standard definition of homomorphism for relational languages found in [4] or [5] which is different from that in [6], and we only need the fact that the positive formulas are preserved by surjective homomorphisms, which is quite easy to prove.

Let Σ denote a set of atomic formulas of L. We let $\mathcal{V}(\Sigma)$ denote the variety determined by the identities in Σ and denote by $\mathbf{F}_{\Sigma}(\overline{X})$ the free algebra for $\mathcal{V}(\Sigma)$ on the generators $\overline{X} = \{\overline{x}_1,\overline{x}_2,\ldots,\overline{x}_n,\ldots\}$. We know that $\mathbf{F}_{\Sigma}(\overline{X}) \in \mathcal{V}(\Sigma)$ [4] (Theorem 10.12). Moreover $T(X)$ is the ***absolutely free algebra*** for L so that given $s : X \longrightarrow A$ for any structure $\mathbf{A}$ of L there is a unique homomorphism $\overline{s} : T(X) \longrightarrow \mathbf{A}$ called ***evaluation*** and we denote by $\mathbf{A} \vDash \varphi[s]$ the assertion that s satisfies the formula φ in $\mathbf{A}$ [6]. In particular recall that for $id : X \longrightarrow \overline{X}$ given by $id(x_i) = \overline{x_i}$ for $i = 1,2,\ldots$ that $\overline{id} : T(X) \longrightarrow \mathbf{F}_{\Sigma}(\overline{X})$ is onto. We write $\varphi(x_1,\ldots,x_n)$ to denote a formula all of whose free variables occur in X_n.

Lemma 1.1. For a positive formula $\varphi(x_1,\ldots,x_n)$, φ is true in $\mathcal{V}(\Sigma)$ iff $\mathbf{F}_{\Sigma}(\overline{X}) \vDash \varphi[id]$.

Proof: One direction is clear since $\mathbf{F}_{\Sigma}(\overline{X}) \in \mathcal{V}(\Sigma)$. Suppose now $\mathbf{F}_{\Sigma}(\overline{X}) \vDash \varphi[id]$ and suppose $\mathbf{A} \in \mathcal{V}(\Sigma)$. We assume momentarily that $\mathbf{A}$ is countable and $a_1,\ldots,a_n \in A$; then there exists a homomorphism h of $\mathbf{F}_{\Sigma}(\overline{X})$ onto $\mathbf{A}$ such that $h(\overline{x}_i) = a_i$ for $i = 1,\ldots,n$. Thus, $\mathbf{A} \vDash \varphi[h \circ id]$ since φ is positive. If $\mathbf{A}$ were uncountable, then by extending $\overline{X}$ to a set $\overline{Y}$ of cardinality $|A|$ the same argument as above using $F_{\Sigma}(\overline{Y})$ works since $F_{\Sigma}(\overline{Y})$ is an elementary extension of $F_{\Sigma}(\overline{X})$ [13]. q.e.d.

Corollary 1.2. If $\varphi(x_1,\ldots,x_n)$ is positive, then $\mathbf{F}_{\Sigma}(\overline{X}) \vDash \varphi[id]$ iff $\mathbf{F}_{\Sigma}(\overline{X}) \vDash \forall x_1 \ldots \forall x_n \varphi$.

We now extend the equational logic of Birkhoff [3] or [4] to one which is adequate for proving positive formulas from Σ. We refer the reader to [6] or another source for the definitions of when a term t is ***substitutable*** or ***free*** for a variable x in a formula φ and what it means for θ' to be an ***alphabetic variant*** of θ. Our rules of inference are the following which are just a few special cases of those found in first order logic:

(a) From $\emptyset$ infer $t \approx t$ for any term t.

(b) From $t_1 \approx t_2$ infer $t_2 \approx t_1$ for any terms t_1 and t_2.

(c) From $t_1 \approx t_2, t_2 \approx t_3$ infer $t_1 \approx t_3$ for any terms t_1, t_2 and t_3.

(d) From $t_1 \approx t_2$ infer $\bar{s}(t_1) \approx \bar{s}(t_2)$ for any $s : X \longrightarrow T(X)$.

(e) From $t_1 \approx t_2$ infer $t \approx t'$ where t_1, t_2, t, and t' are terms and t' is the result of replacing one or more occurrences of t_1 in t by t_2.

We note that (a) through (e) give the usual equational logic due to Birkhoff [3] which is adequate for proving all identities from Σ which are true in all models of Σ, i.e., in $\mathscr{V}(\Sigma)$.

(f) From φ and ψ infer $(\varphi \,\&\, \psi)$ for φ and ψ both positive formulas.
(g) From φ infer $\forall x(\varphi)$ for φ a positive formula.
(h) From φ infer $(\varphi \vee \psi)$ for φ and ψ both positive formulas.
(i) From φ infer $(\psi \vee \varphi)$ for φ and ψ both positive formulas.
(k) From θ'^{x}_{t} infer $\exists x(\theta)$, for θ positive, θ' an alphabetic variant of θ, t is a term substitutable for x in θ', and θ'^{x}_{t} is the result of replacing all free occurrences of x in θ' by t.

Definition 1.3. A positive formula φ has a *positive proof from* Σ, denoted by $\Sigma \vdash_p \varphi$, if there is a finite sequence $\theta_1,\dots,\theta_n$ of positive formulas such that $\theta_n = \varphi$ and for each $1 \leq i \leq n$, θ_i is either in Σ or can be inferred using one of the rules (a)–(k) above from formulas which occur prior to θ_i in this sequence.

Next we have a completeness theorem for this setting.

Theorem 1.4. For any positive formula φ, $\Sigma \vdash_p \varphi$ iff φ is true in $\mathscr{V}(\Sigma)$.

Proof: In view of Lemma 1.1 we are able to replace the condition φ is true in $\mathscr{V}(\Sigma)$ by $\mathbf{F}_{\Sigma}(X) \models \varphi[\mathrm{id}]$. To prove the implication from left to right it is sufficient to show that our rules of inference preserve this satisfaction. Suppose φ is positive and $\mathbf{F}_{\Sigma}(X) \models \varphi[\mathrm{id}]$, then $\mathbf{F}_{\Sigma}(X) \models \forall x\, \varphi[\mathrm{id}]$ by Corollary 1.2; thus rule (g) preserves this satisfaction. Suppose we have that θ'^{x}_{t} is as in the premise of rule (k) and suppose $\mathbf{F}_{\Sigma}(X) \models \theta'^{x}_{t}[\mathrm{id}]$. By the Substitution Lemma [6] (p. 127),

$\mathbf{F}_\Sigma(X) \vDash \theta'[\mathrm{id}(x/\overline{\mathrm{id}}(t))]$ but then $\mathbf{F}_\Sigma(X) \vDash \theta[\mathrm{id}(x/\overline{\mathrm{id}}(t))]$ since θ' is an alphabetic variant of θ. Thus $\mathbf{F}_\Sigma(X) \vDash \exists x\,(\theta)[\mathrm{id}]$ and rule [k] preserves satisfaction in $\mathbf{F}_\Sigma(X)$ by id. The other rules of inference are easy to check.

For the converse we argue that $\mathbf{F}_\Sigma(X) \vDash \varphi[\mathrm{id}]$ implies $\Sigma \vdash_p \varphi$ by induction on the number of symbols of the form &, $\vee$, $\forall$, and $\exists$ occurring in the positive formula φ. If φ has none of these symbols in it, then φ is an atomic formula and $\Sigma \vdash_p \varphi$ follows from the fact that rules of inference (a)–(e) are adequate for identities. Suppose now that φ is $\forall x(\theta)$ and that the result holds for all positive formulas with fewer of those symbols than φ. Suppose $\mathbf{F}_\Sigma(X) \vDash \forall x(\theta)[\mathrm{id}]$, then $\mathbf{F}_\Sigma(X) \vDash \theta[\mathrm{id}]$ and by hypothesis $\Sigma \vdash_p \theta$, but then $\Sigma \vdash_p \forall x(\theta)$ by (g). Suppose $\mathbf{F}_\Sigma(X) \vDash \exists x(\theta)[\mathrm{id}]$, then for some element $a \in F_\Sigma(X)$, $\mathbf{F}_\Sigma(X) \vDash \theta[\mathrm{id}(x/a)]$. The map $\overline{\mathrm{id}} : T(X) \longrightarrow \mathbf{F}_\Sigma(X)$ is onto and let t be a term such that $\overline{\mathrm{id}}(t) = a$ and let θ' be an alphabetic variant of θ such that t is substitutable for x in θ'. By the Substitution Lemma [6] (p.127), $\mathbf{F}_\Sigma(X) \vDash \theta'^{x}_{t}[\mathrm{id}]$ and by our inductive hypothesis $\Sigma \vdash_p \theta'^{x}_{t}$. Thus, $\Sigma \vdash_p \exists x(\theta)$ by rule (k). The other connectives are easy to check. q.e.d.

Corollary 1.5. If $\varphi(y,x_1,\dots,x_n)$ is a positive formula and $\forall x_1 \dots \forall x_n\, \exists y\, \varphi$ is true in all models of Σ, then there is a term t in $T(X_n)$ such that $\forall x_1 \dots \forall x_n\, \varphi(t,x_1,\dots,x_n)$ is true in all models of Σ.

One can extend this result in several directions by changing Σ or changing L. First let Σ be a set of quasi–identities in an algebraic language L and let $\mathrm{Mod}(\Sigma) = \{\mathbf{A} : \text{all formulas in } \Sigma \text{ are true in } \mathbf{A}\}$. One has that $F_\Sigma(X)$ exists and belongs to $\mathrm{Mod}(\Sigma)$ by [11]. A version of Lemma 1.1 holds with $\mathrm{Mod}(\Sigma)$ replacing $\mathcal{V}(\Sigma)$. There are axioms (infer from empty set of premises) and rules of inference adequate for proving from Σ precisely those quasi–identities true in $\mathrm{Mod}(\Sigma)$ in [12]. If one adjoins these to rules (f)–(k) listed above, then one is able to establish a completeness theorem for positive theorems in this setting by imitating the proof of Theorem 1.4. One can extend the results of this paragraph to relational first order languages L which has relation symbols as well as function symbols where Σ is a consistent set of universal Horn sentences, but this will be subsumed in a discussion below.

One may wonder why in logic programming languages ([15], [8], or [7]) universal Horn sentences play such a fundamental role. After all, in logic one characterization of these sentences is that they are precisely the sentences preserved under substructures as well as products [5] or [4]. Actually one can for a set of axioms restrict oneself to universal closures of basic Horn formulas (sometimes called McKinsey formulas) which by definition have the form $(\beta_1 \vee \beta_2 \vee \cdots \vee \beta_n)$ where each β_i is atomic or negated atomic with at most one β_i being atomic. But in any case a set Σ of universal Horn sentences has the property that the class $\mathrm{Mod}(\Sigma)$ of all models of Σ is closed under direct products and the formation of substructures. Among the fundamental objects of the theory of logic programming are the minimal Herbrand models of Σ obtainable as the intersection over all Herbrand models of Σ, see [15] for how the assumption of universal Horn formulas in Σ gets used. However one can obtain essentially the same model by taking the direct product of all Herbrand models of Σ and next taking the substructure of this product generated by the closed terms, this turns out to be a Herbrand model of Σ using the above preservation properties of universal Horn sentences. An analogous result is valid for any set of Herbrand models of Σ. The minimal Herbrand model can be obtained syntactically as in [15] using what is called there fixed point semantics. It would seem that the fundamental property being utilized here is that the minimal Herbrand model (defined model theoretically) and the syntactic model (defined using formal proofs) are isomorphic. Of course it is a more general version of this connection involving variables that we have used to obtain our results characterizing formal proofs of positive formulas.

We now describe how to obtain a result like Theorem 1.4 in the setting of a countable heterogeneous relational language L with equalities $\approx_i$ for each sort $i \in I$ as described in [7] with I nonempty. We assume that we have a denumerable set of formal variables $X^i = \{x_1^i, x_2^i, \ldots, x_n^i, \ldots\}$ for each sort $i \in I$. One has function symbols and relation symbols of sorted arities. One defines the set of terms of sort i, as usual, and denote it by $T^i(X)$ where $X = \cup_{i \in I} X^i$ and $T(X) = \cup_{i \in I} T^i(X)$. A structure $\mathbf{B}$ for L consists of a collection of nonempty sets B^i for each $i \in I$ which interpret the sort i (on B^i, $\approx_i$ is by definition $=$) as well as appropriate functions and relations defined on cartesian products of the correct B^i's. Given $s : X^i \longrightarrow B^i$ for $i \in I$, there is a unique extension $\bar{s} : T(X) \longrightarrow B = \cup_{i \in I} B^i$. $T(X)$ can be viewed as a structure for the algebraic part of L. We assume that the set Σ of nonlogical axioms are all basic Horn formulas, i.e., they all have the form $(\beta_1 \vee \cdots \vee \beta_k)$ where each β_i is a negated atomic or atomic formula with at most one β_i being atomic (note because the universal quantifiers have been suppressed that these are quantifier–free formulas). Our rules of inference are the following. For any $i \in I$:

(a^i) From $\emptyset$ infer $t \approx_i t$ for any term t of sort i ($t \in T^i(X)$).

(b^i) From $t_1 \approx_i t_2$ and $t_2 \approx_i t_3$ infer $t_1 \approx_i t_3$ for any t_1, t_2, t_3 in $T^i(X)$.

(c^i) From $t_1 \approx_i t_2$ infer $t_2 \approx_i t_1$ for any t_1, t_2 in $T^i(X)$.

(d) For each $\alpha \in \Sigma$ and each $s : X^i \longrightarrow T^i(X)$ for $i \in I$ infer $\bar{s}(\alpha)$. (For example, $\bar{s}(t_1 \approx_i t_2) = \bar{s}(t_1) \approx_i \bar{s}(t_2)$ and $\bar{s}(R(t_1,t_2)) = R(\bar{s}(t_1), \bar{s}(t_2))$.)

(e^i) From $t_1 \approx_i t_2$ and α atomic infer α' where α' is the result of replacing zero or more occurrences of t_1 by t_2 in α.

(M.P.) From $\neg\alpha \vee \beta$ and α infer β.

These rules of inference apply for now just to quantifier–free formulas. A proof of α from Σ is a finite sequence of formulas ending with α such that each formula is an instance of (a^i) or (d) or follows by a rule of inference from previous formulas. For any s, s_1 as in (d), it is clear that for all t, $(\bar{s}_1 \circ \bar{s})(t) = \overline{\bar{s}_1 \circ s}(t)$. From this it follows that $\Sigma \vdash \alpha$ implies $\Sigma \vdash \bar{s}(\alpha)$ for any s as in (d). Also it is clear that $\Sigma \vdash \alpha$ implies α is true in any model of Σ. Assuming Σ is consistent, i.e., for all α, not $- \Sigma \vdash \alpha$ or not $- \Sigma \vdash \neg\alpha$, we can construct a syntactic model $\mathbf{M}^*$ of Σ by defining $t_1 \equiv_i t_2$ for t_1, t_2 in $T^i(X)$ to mean $\Sigma \vdash t_1 \approx_i t_2$ and M^i be the equivalence classes of $T^i(X)$ determined by $\equiv_i$. One defines $f^{\mathbf{M}^*}([t_1], [t_2]) = [f(t_1,t_2)]$ which is well–defined by use of (e^i) and one defines $([t_1], [t_2]) \in R^{\mathbf{M}^*}$ iff $\Sigma \vdash R(t_1,t_2)$ which again is well–defined by (e^i). One can argue that each element of Σ is true in $\mathbf{M}^*$. Suppose $\neg R(t_1,t_2) \vee \neg V(t_3)$ is an element in Σ (here R, V are relation symbols) and suppose for some $s : X^i \longrightarrow M^i$, $\mathbf{M}^* \models R(t_1,t_2)[s]$. We claim $\mathbf{M} \models \neg V(t_3)[s]$. For each $i \in I$ and $x \in X^i$ choose $s_x \in \bar{s}(x)$ and define $s'(x) = s_x$; clearly $\overline{s'}(t) \in \bar{s}(t)$ for all t. But $\Sigma \vdash R(\overline{s'}(t_1), \overline{s'}(t_2))$ as well as $\Sigma \vdash \neg R(\overline{s'}(t_1), \overline{s'}(t_2)) \vee \neg V(\overline{s'}(t_3))$ by (d). Hence by (M.P.), $\Sigma \vdash \neg V(\overline{s'}(t_3))$

and by consistency, not $- \Sigma \vdash V(\overline{s}^T(t_3))$, i.e., not $- \mathbf{M}^* \models V(t_3)[s]$. The other form of basic Horn axioms in Σ can also be verified to hold in $\mathbf{M}^*$ in a similar manner. It is straightforward to verify that $\mathbf{M}^*$ is the free model on ω–generators in each sort $i \in I$; namely that given any $\mathbf{B} \models \Sigma$ and $s^{\mathbf{B}} : X^i \longrightarrow B^i$ then there is a unique homomorphism φ from $\mathbf{M}^*$ into $\mathbf{B}$ such that $\varphi([x_j^i]) = s^{\mathbf{B}}(x_j^i)$. Clearly an analogous result to Lemma 1.1 also holds for $\mathbf{M}^*$. One can also construct $\mathbf{F}_{\overline{X}^i}(\Sigma) \in \mathrm{Mod}(\Sigma)$ as done in [2] as a substructure of the direct product of "all" (one needs ω–copies of each one from each countable isomorphism type) models of Σ. Again $\mathbf{M}^*$ and $\mathbf{F}_{\overline{X}^i}(\Sigma)$ are isomorphic. Finally one can expand our rules of inference given above by adding versions of (f), (g), (h), (i) and (k) as additional rules of inference to obtain $\Sigma \vdash_{p(I)} \varphi$. Consequently we obtain the following version of Theorem 1.4.

<u>Theorem 1.6</u>. For any positive formula φ, $\mathbf{M}^* \models \varphi[\mathrm{id}]$ iff $\Sigma \vdash_{p(I)} \varphi$.

Finally we remark that one could formulate and carry out in quite a general setting the results of the last paragraph. In this way one obtains a general universal algebraic result that for most "languages" L any consistent set Σ of universal Horn axioms possesses syntactic models $\mathcal{M}^*$ as well as free models $F_{\omega(I)}(\Sigma)$ which are isomorphic to each other. Moreover "contained" in these objects are the positive sentences of L true in all models of Σ.

§2. Logic for Finite Algebras

In this section we assume $\mathbf{A}$ is a finite algebra with k elements in A in a finite algebraic language L. We wish to produce a logical system which when given an appropriate finite set Σ of identities true in $\mathbf{A}$ will prove from it exactly all identities true in $\mathbf{A}$. This, in turn, leads to an effective procedure for determining whether or not an equation is or is not an identity of $\mathbf{A}$. Of course there is an easy algorithm which can be used to answer whether or not an equation is an identity of $\mathbf{A}$; namely, just evaluate both sides of the given equation at all possible inputs (finitely many) and check to see whether or not the result on both sides is always the same. This method although effective is not very interesting. On the other hand, it is well known that using Birkhoff's equational logic alone that for certain finite algebras, called non–finitely based algebras [4] or [14], it is impossible to find a finite set Σ of identities from which one can prove all other identities. It turns out that there is a simple remedy to this situation utilizing the finiteness of the algebra **A**. We will expand the usual equational logic to the *logic for* k–*element algebras* by adding new rules of inference R_n with $n > k$. R_n, described below, will have the property that if its premises are identities of **A** and **A** has fewer than n elements, then the conclusion of R_n is an identity of **A**.

The rule R_n with $n \geq 2$ has as premises $\binom{n}{2}$ identities which are the following. Let t_1 and t_2 be terms in at most the variables in $X_n = \{x_1,\dots,x_n\}$. For $i < j$ let $t_{x_i}^{x_j}$ be the result of replacing x_j everyplace by x_i in t. Then rule R_n is

$$\frac{\{t_1{}_{x_i}^{x_j} = t_2{}_{x_i}^{x_j} : 1 \leq i < j \leq n\}}{t_1 \approx t_2}$$

which means one can infer $t_1 \approx t_2$ from the premises consisting of all the elements in the set above the line.

Lemma 2.1. If **A** has k elements and $n > k$, then R_n is valid in **A**.

Proof: Suppose **A** has k elements and let t_1 and t_2 be terms in X_n with $n > k$. Suppose for a contradiction that R_n is not valid in **A**, i.e., $t_1 \approx t_2$ is not an identity in **A** but $t_1{}_{x_i}^{x_j} = t_2{}_{x_j}^{x_i}$ are identities in **A** for $1 \leq i < j \leq n$. Thus, $s : X_n \longrightarrow A$ exists such that $\bar{s}(t_1) \neq \bar{s}(t_2)$ but for some i,j, $1 \leq i < j \leq n$ it must be that $s(x_i) = s(x_j)$ since $n > k$. But then $\bar{s}\left[t_1{}_{x_i}^{x_j}\right] = \bar{s}(t_1)$ and $\bar{s}\left[t_2{}_{x_i}^{x_j}\right] = \bar{s}(t_2)$, which is a contradiction. q.e.d.

The next result is proven in [4] (Theorem V.4.2) and is attributed to Birkhoff. Its proof can be thought of as writing down the operation tables of $\mathbf{F}_n(\mathbf{A})$, i.e., the free algebra on n generators in the variety generated by $\mathbf{A}$.

Lemma 2.2. If $\mathbf{A}$ is finite in a finite language L, then the identities in X_n true in $\mathbf{A}$ are finitely based.

The next result is well known also.

Lemma 2.3. If $\mathbf{A}$ is finite and generated by n or fewer elements, then $\mathbf{F}_n(\mathbf{A})$ and $\mathbf{A}$ have exactly the same identities.

Definition 2.4. We say that an equation $t_1 \approx t_2$ is m–provable from a set of equations Σ and denote this by $\Sigma \vdash_m t_1 \approx t_2$ if there is a finite sequence of equations $\sigma_1,\ldots,\sigma_q$ (called an m–proof) such that σ_q is $t_1 \approx t_2$ and each equation in the sequence is either in Σ or follows from previous equations in the sequence using rules of inference (a)–(e) or R_n with $n \geq m$.

We finally have the following completeness theorem for this setting. Again we assume the language here is finite.

Theorem 2.5. If $\mathbf{A}$ has k–elements, then there is a finite set of identities Σ such that for any equation $t_1 \approx t_2$, $\Sigma \vdash_{k+1} t_1 \approx t_2$ iff $t_1 \approx t_2$ is true in $\mathbf{A}$.

Proof: let Σ be a finite set of equations in X_k guaranteed by Lemma 2.2. We know then that any identity $t_1 \approx t_2$ of $\mathbf{A}$ in k or fewer variables is provable from Σ using rules of inference (a)–(e) and so certainly $\Sigma \vdash_{k+1} t_1 \approx t_2$. Moreover, since $\mathbf{A}$ has k elements it follows by induction on the lengths of k+1–proofs that $\Sigma \vdash_{k+1} t \approx t_2$ implies $t_1 \approx t_2$ is true in $\mathbf{A}$.

Conversely, suppose $t_1 \approx t_2$ is identity of $\mathbf{A}$. We prove by induction on the number of variables occurring in $t_1 \approx t_2$ that $\Sigma \vdash_{k+1} t_1 \approx t_2$. Suppose this result holds for all identities of $\mathbf{A}$ in n or fewer variables with $n \geq k$. Let $t_1 \approx t_2$ be an equation in $n+1$ variables which is true in $\mathbf{A}$. We can assume the variables in $t_1 \approx t_2$ are $x_1,\dots,x_{n+1}$. For $1 \le i < j \le n+1$, we have that $t_1 {}^{x_j}_{x_i} = t_2 {}^{x_j}_{x_i}$ is an identity of $\mathbf{A}$ and by inductive hypothesis $\Sigma \vdash_{k+1} t_1 {}^{x_j}_{x_i} = t_2 {}^{x_j}_{x_i}$. Thus, $\Sigma \vdash_{k+1} t_1 \approx t_2$ using rule R_{n+1} since $n+1 \ge k+1$. q.e.d.

We remark that in the context of Theorem 2.5 a $k+1$–proof is very close to calculating in $\mathbf{F}_k(\mathbf{A})$ the value of all possible substitutions in the equation $t_1 \approx t_2$. So for this proof system the two separate operations of determining if $t_1 \approx t_2$ is an identity and looking for an appropriate proof from other identities have merged into the single operation of finding a $k+1$–proof of $t_1 \approx t_2$. Moreover Theorem 2.5 is constructive, i.e., given an equation $t_1 \approx t_2$ and the operation tables for $\mathbf{F}_k(\mathbf{A})$ one obtains effectively either a $k+1$-proof of $t_1 \approx t_2$ or a calculation of the fact that some evaluations of t_1 and t_2 are different in $\mathbf{F}_k(\mathbf{A})$ and hence in $\mathbf{A}$ by Lemma 2.3.

Example 2.6. Let $\mathbf{A} = 2$, the two–element Boolean algebra. The question as to whether or not $t_1 \approx t_2$ is an identity in $\mathbf{A}$ is equivalent to the question as to whether or not $t_1 \longleftrightarrow t_2$ is a tautology. Thus Theorem 2.5 together with a table of values of operations on $\mathbf{F}_2(\mathbf{A})$ and the notion of 3–proof gives a new way of developing the propositional logic in infinitely many sentence symbols. Consider trying to prove the identity $x_1+(x_2+x_3) \approx (x_1+x_2)+x_3$ where $x_1+x_2 = (x_1 \wedge x_2') \vee (x_2 \wedge x_1')$. Using R_3 it suffices to see that $x_1+(x_1+x_2) \approx (x_1+x_1)+x_2$, $x_1+(x_2+x_1) \approx (x_1+x_2)+x_1$, and $x_1+(x_2+x_2) \approx (x_1+x_2)+x_2$ are identities. But the middle one follows immediately from $x_1+x_2 \approx x_2+x_1$ and the first and last are essentially the same once one does a change of variables in view of $x_1+x_2 \approx x_2+x_1$. It is easy to check the first identity. Even if one did not want to choose the notion of 3–proof for proof here, it may be advantageous to view R_n with $n \ge 3$ as an additional meta–rule in the traditional development of propositional logic. However, in view of the usual proof of the completeness theorem for propositional logic in [10], 3–proof becomes more appealing.

We remark that our development here is quite straightforward. However if one were to start with $\mathbf{F}_k(\mathbf{A})$ and attempt to show that its identities all had k+1–proofs from Σ used in Theorem 2.5 for the identities of $\mathbf{A}$ which we know to be clear on the basis of Lemma 2.3 and Theorem 2.5, it would seem to be a more difficult task.

One can view Theorem 2.5 as asserting that any finite algebra's identities has a finite basis in an expanded logic for that algebra's identities. This expanded logic has the pleasant feature that it just depends upon the cardinality of the finite algebra. Observing that there are finite algebras without a finite basis and that rule R_n only permits one to prove identities in n or fewer variables one obtains.

Corollary 2.7. There exists finite algebras $\mathbf{A}$ with k elements such that in order to give k+1–proofs for all its identities from a finite Σ requires rule R_n for infinitely many $n \geq k+1$.

For $\mathbf{A}$ finite with k–elements it is well known that $|F_n(\mathbf{A})| \leq k^{(k^n)}$. The following result gives in some cases a better way of estimating these bounds in certain instances, see [1] for examples.

Corollary 2.8. For $\mathbf{A}$ finite with k elements and $n+1 > k$, then $|F_{n+1}(\mathbf{A})| \leq |F_n(\mathbf{A})|^{\binom{n+1}{2}}$.

Proof: Rule R_{n+1} asserts that the function defined by the term t in variables in X_{n+1} is completely determined by the functions determined by $t\frac{x_j}{x_i}$, $1 \leq i < j \leq n+1$. There are $|F_n(\mathbf{A})|$ such functions in n variables. q.e.d.

Another way of viewing an instance of the rule of inference R_n would be the sentence $\forall x_1 \ldots \forall x_n\, (\wedge\{t_1\frac{x_j}{x_i} \approx t_2\frac{x_j}{x_i} : 1 \leq i < j \leq n\}) \longrightarrow \forall x_1 \ldots \forall x_n (t_1 \approx t_2)$ with t_1 and t_2 terms in $T(X_n)$; the validity of this instance of R_n being equivalent in a model to the truth of this sentence in that model. This sentence, in turn, is a logical consequence of the sentence $\forall x_1 \ldots \forall x_n\, ((\wedge\{t_1{}^{x_j}_{x_i} \approx t_2{}^{x_j}_{x_i} : 1 \le i < j \le n\})$ $\longrightarrow t_1 \approx t_2)$ which moreover is a quasi–identity which we denote by $r(n, t_1, t_2)$. The proof of Lemma 2.1 verifies that each of the quasi–identities $r(n, t_1, t_2)$ are true in $\mathbf{A}$, if $n > k$ with A having k elements. Let $\Sigma_{k+1} = \{r(n, t_1, t_2) : t_1, t_2$ are terms in $T(X_n)$ with $n \ge k+1\}$. $Mod(\Sigma_{k+1})$ are closed under direct products and substructures (as well as reduced products or bounded Boolean powers). Thus, it follows that for $j \ge 1$, $\mathbf{F}_j(\mathbf{A}) \in Mod(\Sigma_{k+1})$. Since each instance of rule R_n with $n > k$ is valid in each element of $Mod(\Sigma_{k+1})$, by imitating the proof of Theorem 2.5 we obtain the next result.

Proposition 2.9. If $\mathbf{M} \in Mod(\Sigma_{k+1})$, then any identity of $\mathbf{M}$ has a k+1–proof from the set of identities in k–variables true in $\mathbf{M}$.

Using Proposition 2.9 and Lemma 2.2 it follows that for $j \ge 1$, every identity of $\mathbf{F}_j(\mathbf{A})$ has a k+1–proof from a finite set of identities. Thus, in this property, $\mathbf{A}$ and $\mathbf{F}_j(\mathbf{A})$ are alike.

REFERENCES

[1] J. Berman, Free spectra of 3–element algebras, in: R.S. Freese et al., *Universal Algebra and Lattice Theory*, LNM 1004, Springer–Verlag: Berlin–New York, 1983.

[2] G. Birkhoff and J.D. Lipson, Heterogeneous algebras, *J. Combinatorial Theory*, **8**(1970), 115–133.

[3] G. Birkhoff, Universal algebra, *Proceedings of the First Canadian Mathematical Congress*, The University of Toronto Press, Toronto, 1946, 310–325.

[4] S. Burris and H.P. Sankappanavar, *A Course in Universal Algebra*, Springer–Verlag: Berlin–New York, 1981.

[5] C.C. Chang and H.J. Keister, *Model Theory*, North–Holland: Amsterdam, 1973.

[6] H.B. Enderton, *A Mathematical Introduction to Logic*, Academic Press: New York, 1972.

[7] J.A. Goguen and J. Meseguer, Equality, types, modules, and (why not?) generics for logic programming, *J. Logic Programming* **2**(1984), 179–210.

[8] J.W. Lloyd, *Foundations of Logic Programming*, Springer–Verlag: Berlin–New York, 1984.

[9] R.C. Lyndon, Properties preserved under homomorphisms, *Pacific J. Math.* **9**(1959), 143–154.

[10] E. Mendelson, *Introduction to Mathematical Logic*, Van Nostrand: Princeton, 1964.

[11] W. Peremans, Some theorems on free algebra and on direct products of algebras, *Simon Stevin* **29**(1952), 51–59.

[12] A. Selman, Completeness of calculi for axiomatically defined classes of algebras, *Alg. Univ.* **2**(1972), 30–32.

[13] A. Tarski and R.L. Vaught, Arithmetical extensions of relational systems, *Compositio Math.* **13**(1958), 81–102.

[14] W. Taylor, Equational logic, *Houston J. Math*, Survey (1979).

[15] M.H. van Emden and R.A. Kowalski, The semantics of predicate logic as a programming language, *J. Assoc. Comput. Math.* **23**(1976), 733–742.

MAL'CEV ALGEBRAS FOR UNIVERSAL ALGEBRA TERMS

Ivo G. Rosenberg
Département de mathématiques et statistique
Université de Montréal
C.P. 6128, Succ. A, Montréal H3C 3J7 Canada

The paper is in final form and no version of it will be submitted for publication elsewhere.

Abstract.*We discuss formal treatments of composition and terms in universal algebra, propositional logics etc. which may serve as an indispensable base for computer programs capable of term building and term comparison, an important problem in theoretical computer science. After a survey we discuss various Mal'cev algebras introduced for this purpose: preiterative, preiterative with identity, iterative and postiterative. These algebras seem to be simple to implement. We then show how these algebras allow to bring certain universal algebra concepts (as varieties and subvarieties, interpretation and hyperidentities) one conceptual level down. We conclude with a list of properties of preiterative algebras.*

O. In this note we discuss formal treatment of composition (substitution or superposition) in universal algebra. The implicit motivation is the automated term generation and term comparison. The construction of terms has been formally described in several ways but usually it is treated quite informally. A correct formal treatment seems to be indispensable for a computer program capable of building and comparing terms. A.I. Mal'cev introduced a formal treatment based on a purely algebraic "superstructure" on the set 0_A of all finitary operations (on a fixed universe A). This structure, a monoid $*$ on 0_A with three unary operations ζ, τ and Δ, does not explicitly depend upon the arities of elements of 0_A and so avoids problems inherent in the other approaches (e.g. unnecessary restrictions or countably many partial operations). The fact that at most 6 relatively simple operation symbols are used could be crucial for a term writing program. In Mal'cev algebra the terms are coded as words built from $*$, ζ , τ , Δ and the given operations of an algebra on A (but without variables). This coding may look complicated and usually non-transparent (e.g. for f binary, the associative law becomes $f * f = \zeta(\tau f * f))$ but in a computer program the relative simplicity of basic operations probably outweighs the lack of transparency.

After a brief introduction and a survey we introduce four variants of Mal'cev's algebras: preiterative, unitary preiterative, iterative and postiterative (the third uses an additional unary operation and the second and fourth an additional nullary operation). Then we show how they allow to bring concepts one level down. Thus a variety Var ($\underline{\underline{A}}$), generated by an algebra $\underline{\underline{A}} = \langle A ; F \rangle$, may be identified with the class of homomorphisms from the iterative algebra $\underline{\underline{I}}$, generated by F , onto iterative algebras. Similarly, the subvarieties of Var $\underline{\underline{A}}$ may be identified with the congruences θ of $\underline{\underline{I}}$ such that $\underline{\underline{I}}/\theta$ is isomorphic to an iterative algebra, hypervarieties and hyperidentities generated by the terms of $\underline{\underline{A}}$ with Var $\underline{\underline{I}}$ and the identities of $\underline{\underline{I}}$ etc.

The four Mal'cev algebras are of interest by themselves. We conclude with a list of properties satisfied by every preiterative algebra. They consist of one existential rule and an infinite family of quasiidentities. Although preiterative algebras were designed to avoid the formal arity problems, it seems that every description of the class of preiterative algebras must cope with them in one way or another.

For most algebras term creation and comparison is a hard problem and anyway the set a nontrivial terms is infinite. Perhaps Mal'cev algebras could help in this task for some low arity terms.

The partial financial assistance provided by NSERC Canada operating grant A-5407 and FCAR Québerc subvention d'équipe Eq-0539 is gratefully acknowledged.

1. In the sequel the universe is a fixed non-empty set A . For a positive integer n denote by $0_A^{(n)}$ (or shortly by $0^{(n)}$), the set of all n-ary operations on A (i.e. maps from A^n to A) and put $0_A := \bigcup_{n=1}^{\infty} 0_A^{(n)}$ (or shortly by 0) . The use of nullary operations (or zero operations which are essentially elements of A) slightly complicates the formal treatment and so for simplicity's sake we replace each nullary operation by the constant unary operation with the same value. However, it will be convenient to use nullary operations in the concrete cases of unitary preiterative and postiterative algebras. Put $\partial f := n$ for every $f \in 0^{(n)}$.

For a subset F of 0 the pair $\underline{\underline{A}} = \langle A ; F \rangle$ is called a (non-indexed) *universal algebra*. Universal algebras appeared in 1847 in the propositional calculus of logics, starting from the 1930's became important as generalizations of classical algebras (groups, rings, lattices,etc.) and switching theory and more recently started to play a role in theoretical computer science. In all these contexts new operations, called terms (or, earlier, polynomials) of $\underline{\underline{A}}$ are formed from F via *composition* (also referred to as substitution or superposition). This roughly means that we are allowed to perform four constructs: 1) We can replace a variable of an already constructed term by an operation from F. For example, if F contains two binary operations $f(x_1x_2) = x_1+x_2$ and $g(x_1, x_2) = x_1 \circ x_2$ we can replace the second variable of g by $f(x_2, x_3)$ to form a ternary operation $x_1(x_2+x_3)$. Clearly this is a multivariable analog of the familiar composition "$\circ$" of selfmaps of A. Note that we should specify what are the variables of the composite operation (cf. sections 2-3 below). 2) All the above applications more or less explicitly allow free exchange of variables. Thus if f is a binary term of $\underline{\underline{A}}$ so is $g \in 0^{(2)}$ defined by $g(x_1, x_2) \approx f(x_2, x_1)$ (here and in the sequel $\approx$ stands for the equality of the two operations i.e. $g(a_1, a_2) = f(a_2, a_1)$ holds for all $a_1, a_2 \in A$). This fact is e.g. implicit in the commutative law $f \approx g$ equating two terms of $\underline{\underline{A}}$. 3) In the above applications new terms are obtained through identification (or fusion) of variables. For example, if f is a binary term of $\underline{\underline{A}}$ so is $g \in 0^{(1)}$, defined by $g(x_1) \approx f(x_1, x_1)$. This is implicit e.g. in the semilattice idempotent law $x_1 \vee x_1 \approx x_1$.4) Sometimes, but not always, it is convenient to use *projections* (also called selectors or trivial operations) e_i^n defined by setting $e_i^n(x_1,\ldots,x_n) \approx x_i$ for all $1 \le i \le n$. The set $Q := \{e_i^n : 1 \le i \le n < \omega\}$ plays the role of the multivariable analog of the identity selfmap id_A in the symmetric semigroup $< 0^{(1)} ; \circ >$.

Finally the set $T(\underline{\underline{A}})$ of all terms of $\underline{\underline{A}}$ is the least subset of 0 containing $F \cup Q$ and closed under the constructs 1-3. The subsets of 0 of the form $T(\underline{\underline{A}})$ (for some algebra $\underline{\underline{A}}$ on A) are the *clones* on A.

2. The above description is not fully precise. Note that in most cases in the literature a precise definition is either omitted or passed over by a kind of hand-waving (to quote from [4, p. 33] "the systematic treatment of polynomial symbols...turns out to be..., surprisingly, one of the topics most neglected in literature"). Of course, there is the inductive definition (cf. [5]): Put $T_0 := F \cup Q$ and if T_i has been defined let T_{i+1} consist of all $h \in 0^{(n)}$ such that

$$h(x_1,\ldots,x_n) \approx f(g_1(x_{i_{11}},\ldots,x_{i_{1n(1)}}),\ldots,g_m(x_{i_{m1}},\ldots,x_{i_{m,n(m)}})) \quad (1)$$

where $f \in T_i$ is m-ary, $g_i \in T_0$ is n(i)-ary ($i = 1,\ldots,m$) and $1 \leq i_{11},\ldots,i_{m,n(m)} \leq n$ are fixed. Finally put $T(\underline{\underline{A}}) := \bigcup_{i<\omega} T_i$. This definition, although formally correct, seems to be unwielding and could hardly serve as a formal basis for a computer term generating program.

A remedy was attempted by restricting the set 0 (of all operations on A) to the set $0^{(n)}$ of all n-ary operations. This led to the n-ary *Menger algebras* $\underline{\underline{M}}_A^{(n)} = \underline{\underline{M}}^{(n)}$ $< 0^{(n)}; c^n, e_1^n,\ldots,e_n^n >$. Here $e_1^n,\ldots,e_n^n$ are nullary operations on $0^{(n)}$ and the (n+1)-ary operation c^n on $0^{(n)}$ is defined by (1) in the particular case $m = n(1) = \ldots = n(m) = n$ and $i_{j^*} = j$ for all $1 \leq i, j \leq n$. Thus c^n assigns to $f, g_1,\ldots,g_n \in 0^{(n)}$ the operation $h \in 0^{(n)}$, usually denoted $f[g_1,\ldots,g_n]$, defined by

(2) $$h(x_1,\ldots,x_n) \approx f(g_1(x_1,\ldots,x_n),\ldots,g_n(x_1,\ldots,x_n)).$$

Note that $\underline{\underline{M}}^{(n)}$ is a universal algebra on 0 of type $< n+1, 0,\ldots,0 >$. The class of algebras isomorphic to subalgebras of $M_{\underline{\underline{A}}}^{(n)}$ is a variety (or equational class) defined by a superassociative law (cf. [9-11,13-15,17])). In spite of this neat characterization, the arity restriction is so drastic and $M^{(n)}$ so removed from the needs of universal algebra that the interest in Menger algebras mostly faded away (e.g. it seems that even a complete description of the subalgebras of $M_{(0,1)}^{(2)}$ has not yet been published).

A standard definition of composition is based on a modification of (2). Define a partial (n+1)-ary operation

(3) $$h(x_1,\ldots,x_m) \approx f(g_1(x_1,\ldots,x_m),\ldots,g_n(x_1,\ldots,x_m))$$

whenever $f \in 0^{(n)}$ and $g_1,\ldots,g_n \in 0^{(m)}$ for some $0 < m < \omega$. Note that c^n maps $\bigcup_{m=1}^{\infty} 0^{(n)} \times (0^{(m)})^n$ into 0. For an algebra $\underline{\underline{A}} = < A; F >$ the subuniverse (i.e. the carrier of a subalgebra) of $\underline{\underline{C}} := < 0; \{c^n : 0 < n < \omega) \cup Q >$ generated by F (i.e. the least subuniverse containing F) is exactly the set $T(\underline{\underline{A}})$ of terms of $\underline{\underline{A}}$ or, equivalently, the clone generated by F. (Note that often c^n is replaced by the family $\{c_m^n : 0 < m < \omega\}$ where the heterogeneous operation $c_m^n : 0^{(n)} \times (0^{(m)})^n\, n \to 0^{(m)}$ is a restriction of c^n). While this provides an algebraic description of $T(\underline{\underline{A}})$, unfortunately it involves a countably infinite family of partial operations. Properties of partial algebras are less transparent and known that those of full algebras and $\underline{\underline{C}}$ is probably not the best formal approach to a term generating computer program.

3. An approach which is purely algebraic and based on a minimum number of basic operations has been proposed by A.I. Mal'cev [8] more than 20 years ago. The handling of arities, one of the main sources of difficulties, is avoided by a neat trick. It may seem that its presentation takes up more space than the other definitions. However, a properly done definition takes up approximately the same space and almost invariably the attempts to save space bring in either a more complex structure or formal difficulties later on. While Mal'cev's approach is quoted in the East European literature, it is not widely used elsewhere. Mal'cev formulated it in terms of a monoid $*$ on 0 together with three unary operations ζ, τ and Δ on 0. We first define $*$ as follows. For $f \in 0^{(n)}$ and $g \in 0^{(m)}$ put $r := m+n-1$ and define an r-ary operation $h := f * g$ by setting

$$(4) \qquad h((x_1,\ldots,x_r) \approx f(g(x_1,\ldots,x_m), x_{m+1},\ldots,x_r).$$

Thus the construct (1) from 1 is restricted to the replacement of the first variable of f by g while keeping the variables as distinct as possible. It is easy to verify (cf. the rule (viii) in Proposition 11 below) that $*$ is associative and that e_1^1 (= id_A or 1_A) is its neutral element (i.e. $e_1^1 * f = f * e_1^1$ holds for all $f \in 0$). Thinking in terms of switching circuits we may interpret $f \in 0^{(n)}$ as a gate with n inputs and a single output assigning the output $f(a_1,\ldots,a_n)$ to signals a_i on the i-th input ($i = 1,\ldots,n$). Now $f * g$ may be seen as the circuit with r inputs and a single output obtained by joining the output of g to the first input of f (cf. Fig. 1).

The unary operations ζ and τ on 0 (i.e. selfmaps of 0) are defined as follows: (i) for $f \in 0^{(1)}$ set $\zeta f = \tau f := f$, and (ii) for $f \in 0^{(n)}$ ($n > 1$) define $\zeta f, \tau f \in 0^{(n)}$ by setting

$$(\zeta f)\ (x_1,\ldots,x_n) \approx f(x_2,\ldots,x_n, x_1),$$
$$(\tau f)\ (x_1,\ldots,x_n) \approx f(x_2, x_1, x_3,\ldots,x_n).$$

Thus ζf and τf are obtained from f by a cyclic exchange of variables and the transposition (or switch) of the first two variables. In terms of gates this represents two types of shuffles of inputs (cf. fig 2a-b). It is known that the cycle $z = (1,\ldots,n)$ and the transposition $t = (1,2)$ generate the symmetric group S_n [2,p.63], and so we are able to obtain every permutation of variables by repeatedly applying ζ and τ. This takes care of the construct 2 in section 1.

The operation Δ is defined as follows: (i) For $f \in 0^{(1)}$ put $\Delta f := f$, and (ii) for $f \in 0^{(n)}$ ($n > 1$) define $\Delta f \in 0^{(n-1)}$ by setting

$$(\Delta f)\ (x_1,\ldots,x_{n-1}) \approx f(x_1, x_1, x_2,\ldots,x_{n-1}).$$

(Note that repeated applications of ζ, τ and Δ can produce every permutation and identification of variables.)

Call $P'_A = \underline{\underline{P}}'$: = < 0 ; $*$, ζ , τ , Δ > (of type <2, 1, 1, 1 >) the (Mal'cev) *full preiterative algebra on* A , its subalgebras the *preiterative algebras* and its subuniverses the *preiterative sets*. Thus a preiterative set is a subsemigroup C of < 0 ; $*$ > such that $\Delta(C) \subseteq \zeta(C) = \tau(C) = C$. The preiterative sets are usually called closed classes which is not specific enough and may be misleading. Preiterative sets are slightly more general than clones (cf. Sec. 1) because they need not contain the projections. (However, in most applications the adjunction of projections to a preiterative set does no harm.)

The proof that the preiterative algebras do what they are supposed to do is straightforward [8] (the key is that we may replace other variables than the first one in f by g; indeed this is achieved through the possibility of free exchange and fusion of variables gained through appropriate successive applications of ζ, τ and Δ to f).

Preiterative sets and algebras containing e_1^1 are *unitary*. Obviously the unitary preiterative sets are exactly the subuniverses of < σ ; $*$, τ , A , e_1^1 > (where e_1^1 is a nullary operation).

4. A preiterative set may be closed with respect to the adjunction of fictitious (dummy or non-essential) variables. Mal'cev formalized this as follows. Let ∇ be the unary operation on 0 assigning to each n-ary operation f the (n+1)-ary operation ∇f defined by setting

$$(\nabla f)(x_1,\ldots,x_{n+1}) \approx f(x_2,\ldots,x_{n+1}).$$

Note that the first variable of ∇f is indeed fictitious in the sense that a change in the first argument alone has no impact on the value. The algebra

$$\underline{\underline{P}}_A = \underline{\underline{P}} := <\sigma ; *, \zeta, \tau, \Delta, \nabla>$$

is the (Mal'cev) *full iterative algebra* and its subuniverses are the *iterative sets*.

The clones (i.e. preiterative sets containing all projections) are the subuniverses of the (Mal'cev) *full postiterative algebra*

$$\underline{\underline{P}}_A^0 = \underline{\underline{P}}^0 := <0 ; *, \tau, \Delta, e_1^2>$$

(of type < 2, 1, 1, 1, 0 > where $e_1^2 \in 0^{(2)}$ is nullary; in [16] a full postiterative algebra is called a centered preiterativealgebra). A *postiterative algebra* is a subalgebra of $\underline{\underline{P}}_A^0$. In view of $e_1^{n+1} = e_1^n * e_1^2$ (n = 1, 2,...) the clones are indeed the subuniverses of $\underline{\underline{P}}_A^0$. Noting that for all $f \in 0$

$$\nabla f = f * e_2^2 \ , \ e_2^2 = \nabla e_1^1$$

we see that clones are exactly the unitary iterative sets.

The above Mal'cev algebras may be easily adapted to other algebraic structures; e.g. for partial algebras this was already done in [8].

5. An identity (axiom or law) $t_1 \approx t_2$ between two terms of an algebra $\underline{\underline{A}}$ may be interpreted as an equality of two words in $\underline{\underline{P}}^0$. For example, the idempotent, commutative and associative laws for $f \in 0^{(2)}$ are

(5) $$\Delta f = e_1^1 \,,\; \tau f = f \,,\; f * f = \zeta(\tau f * f) \,.$$

This and the sections 8-10 below illustrate the fact that the use of Mal'cev algebras may bring certain problems conceptually one level down. We start with the following fact. Let φ be a map from a subset C of 0_A into 0_B. Call φ *arity non-increasing (preserving)* if $\partial \varphi f \le \partial f \; (\partial \varphi f = \partial)$ for all $f \varepsilon C$.

6. Proposition *Every preiterative (iterative) homomorphism is arity non-increasing (preserving).*

Proof. Let $f \in 0$. Then f has arity n if and only if n is the least positive integer such that $\Delta^{n-1} f = \Delta^n f$ (where, as usual, $\Delta^0 f := f$). From this it follows that a preiterative homomorphism φ is arity non-increasing. As noted in [16], f has arity n if and only if $\nabla f * f = \nabla^n f$ and therefore an iterative homomorphism is arity preserving.

Example. Let $\underline{\underline{I}}$ be a preiterative algebra and let e denote the only selfmap of $\{0\}$ (defined by $e(0) := 0$). Then the map φ from I onto $\{0\}$ is a homomorphism from $\underline{\underline{I}}$ onto $< \{e\}; *, \zeta, \tau, \Delta >$ which, in general, is not arity preserving.

7. We recall two natural and well-known maps. Let $\underline{\underline{A}} = < A ; F >$ be an algebra and X a set. Following [1, III.2] denote by $W_F(X)$ the set of F-words in X (i.e. the set of finite sequences built from $F \cup X$ and parentheses so that the arities are respected or, equivalently, the absolutely free F-algebra with the free generating family X, cf. [4, Ch. 4]; note that for simplicity we do not use the Polish parenthesis-free notation here). For example, for binary $f_1, f_2 \in F$ and $x_1, x_2, x_3 \in X$ the word $f_1(f_2(x_2,x_1), f_1(x_3, x_2))$ belongs to $W_F(X)$.

Let $\Omega := \{*, \zeta, \tau, \Delta, \nabla\}$, let $\underline{\underline{I}} = < I ; \Omega >$ denote the iterative algebra generated by F and let $X = \{x_i : 0 < i < \omega\}$. Consider $W_\Omega(F)$ and $W_F(X)$. By induction on the length we construct a map $w \to w^0$ from $W_\Omega(F)$ into $W_F(X)$. 1) For each one-symbol word f with $f \in F$ n-ary, put $f^0 := f(x_1,\ldots,x_n)$. 2) Suppose u^0 and v^0 have been defined for $u, v \in W_\Omega(F)$ and let n and m denote the largest integer such that x_n and x_m occur in u and v. Define $(u * v)^0$ as the word obtained from u^0 if we replace each x_1 in u^0 by v^0 and each x_i in u^0 by x_{i+m-1} $(i = 1,\ldots,n)$. 3) $(\zeta u)^0$,

$(\tau u)^0$, $(\Delta u)^0$ and $(\nabla u)^0$ are defined in a similar fashion. For example, if $w = f_1 * \tau f_2$ where $f_1, f_2 \in F$ are binary, then $w^0 = f_1(f_2(x_2, x_1), x_3)$. The map $w \to w^0$ may be seen as a "decoding" of w. Let $w \in W_\Omega(F)$. Denote by w^A the operation on A calculated according to w^0 (e.g. for the above $w = f_1 * \tau f_2$, the ternary operation w^A is defined by setting $w^A(a_1, a_2, a_3) := f_1(f_2(a_2, a_1), a_3)$ for all $a_1, a_2, a_3 \in A$; note that here f_1 and f_2 are concrete operations on A (which could have been denoted f_1^A and f_2^A while in w and w^0 they serve as more operation symbols). We have:

8. Proposition. *Let* $\underline{\underline{A}} = \langle A ; F \rangle$ *be an algebra,* $\underline{\underline{I}}$ *the iterative algera generated by* F *and* Var $\underline{\underline{A}}$ *the variety generated by* $\underline{\underline{A}}$. *There is a bijection between* Var $\underline{\underline{A}}$ *and the class of iterative homomorphisms from* $\underline{\underline{I}}$ *onto iterative algebras.*

Proof: Let B be a universe and let $\varphi : f \to f'$ be an arity preserving map from F onto a subset G of O_B. Note that φ provides an indexing of the operations of $\underline{\underline{B}} := \langle B ; G \rangle$ by F which is sometimes called an interpretation (together with the notation $f^A := f$ and $f^B := f'$ for all $f \in F$).

Denote by Σ the set of all identities of Var $\underline{\underline{A}}$. Such an identity may be written as $w_1^0 = w_2^0$ for some $w_1, w_2 \in W_\Omega(F)$. Obviously this is equivalent to $w_1 = w_2$ and so we interpret Σ in this way. Note that $w_1 = w_2$ iff $w_1^A \approx w_2^A$. We mention a minor point. An identity of Var $\underline{\underline{A}}$ may be of the form $w = e_i^n$ for some $1 \le i \le n$ (e.g. the idempotency of a binary f is expressed as $\nabla f = e_1^1$), while in principle I need not contain e_1^1. However, if $w = e_i^n$ is a valid identity, then w^A is the projection e_i^n on A and so $e_i^n \in I$. In other words, the use of iterative algebras instead of the postiterative ones does no harm here.

To $w \in W_\Omega(F)$ assign $w' \in W_\Omega(G)$ obtained from w by replacing each symbol $f \in F$ occuring in w by f'.

Finally let Σ' denote the set of identities $w_1' = w_2'$ such that $w_1 = w_2$ is an identity from Σ. Clearly $\underline{\underline{B}}$ belongs to Var $\underline{\underline{A}}$ if and only if $\underline{\underline{B}}$ satisfies Σ'. It is well known and easy to see that $\underline{\underline{B}}$ satisfies Σ' if and only if the map $\varphi : f \to f'$ from F onto G extends to a homomorphism from the iterative algebra $\underline{\underline{I}}$ onto the iterative algebra $\underline{\underline{J}}$ generated by G.

Conversely, let B be a universe and let $\varphi : f \to f'$ be a homomorphism from $\underline{\underline{I}}$

onto an iterative subalgebra $\underline{\underline{J}}$ of O_B. Put $G := \text{im}\,\varphi$ and $\underline{\underline{B}} := \langle B ; G \rangle$. Since F generates the iterative algebra $\underline{\underline{I}}$, clearly $\underline{\underline{J}}$ is the iterative algebra generated by G. Now if $w_1 = w_2$ is an identity in $\underline{\underline{A}}$, then clearly $w_1' = w_2'$ is an identity $\underline{\underline{B}}$ and hence $\underline{\underline{B}} \in \text{Var}\,\underline{\underline{A}}$.

9. Remarks. (1) Let $\underline{\underline{A}} = \langle A ; F \rangle$ be an algebra and let $\underline{\underline{I}}$ be the iterative algebra generated by F. A subvariety of Var $\underline{\underline{A}}$ is determined (up to iterative isomorphisms) by an iterative homomorphism φ from $\underline{\underline{I}}$ onto an iterative algebra (i.e. it may be identified with the class of all $\varphi \circ \psi$ where ψ is an iterative isomorphism. By the homomorphism theorem the subvarieties of Var ($\underline{\underline{A}}$) are determined by the congruences θ of $\underline{\underline{I}}$ such that I/θ is an iterative algebra. Among such congruences are trivially the least congruence (but not the greatest congruence) and the following "arity" congruence k_a defined by $(f, g) \in k_a$ whenever $f, g \in I$ are of the same arity (note that I/k_a may be identified with $\underline{\underline{P}}_A'$ with $|A| = 1$ whereby the corresponding subvariety is the trivial one [8]). It is known that k_a is covered by the greatest congruence. The congruences of $\underline{\underline{I}}$ contained in k_a (incomparable to k_a) are termed congruences of the *first (second) kind.* If I is rich enough k_a may be the only non-trivial congruence of $\underline{\underline{I}}$, i.e. Var $\underline{\underline{A}}$ may only have trivial subvarieties ([8]; for more information cf [7, Part III, pp. 177-199]).

(2) A slight modification of Proposition 8 yields: *Let* $\underline{\underline{A}} = \langle A ; F \rangle$ *and* $\underline{\underline{B}} = \langle B ; G \rangle$ *be algebras and* $\underline{\underline{I}}$ and $\underline{\underline{J}}$ *the iterative algebras generated by* F *and* G. *Then the variety is interpretable in* Var $\underline{\underline{B}}$ (cf. [3]) *if and only if there is a homomorphism from* $\underline{\underline{I}}$ *into* $\underline{\underline{J}}$.

(3) Let $\underline{\underline{A}} = \langle A ; F \rangle$ be an algebra, $\Omega := \{*, \zeta, \tau, A)$ and $\underline{\underline{I}} = \langle I ; \Omega \rangle$ the preiterative algebra generated by F. Suppose I is not closed under ∇. Call an identity $w_1 = w_2$ (for $w_1, w_2 \in W_\Omega(F)$) *regular* if for all $0 < i < \omega$ the symbol x_i appears in w_1^0 exactly if it appears in w_2^0 (note that for $w \in W_\Omega(F)$ the symbols x_i appearing in w^0 are exactly $x_1, \ldots, x_n$ for some $0 < n < \omega$). Some of the identities of Var $\underline{\underline{A}}$ may be non-regular. Indeed, if $u, v \in W_\Omega(F)$ are such that $u^A = f$ and $v^A = g$, where f is unary and g is binary, and it happens that $f(a_2) = g(a_1, a_2)$ holds for all $a_1, a_2 \in A$ then $u^0(x_2) = v^0(x_1, x_2)$ is a valid identity of Var $\underline{\underline{A}}$ which cannot be written as $u = v$ (as f and g have different arities; in an iterative algebra we get around this difficulty by using $\nabla u = v$). Inspired by Proposition 8 we may consider the class $V(\underline{\underline{A}})$ of preiterative homomorphisms from $\underline{\underline{I}}$ onto preiterative algebras. The subclass $V_r(\underline{\underline{A}})$ of arity preserving maps from $V(\underline{\underline{A}})$ may be identified with the variety $\text{Var}_r\,\underline{\underline{A}}$ generated by the regular identities of $\underline{\underline{A}}$.

10. The preiterative, unitary preiterative, iterative and postiterative algebras are concrete algebras (semigroups with 3 or 4 unary operations and possibly one nullary operation) and so it is natural to ask about their properties. The algebras were constructed so that preiterative sets, preiterative sets with e_1^1 , iterative sets and clones are exactly their subuniverses. It is known that their structure may be complex; e.g., for A finite, $|A| > 2$ there are $2^{\aleph_0}$ clones on A [6] . Already in [8] Mal'cev asked about their congruences and automorphisms. He partially answered these questions by showing that in many cases these structures are quite simple (cf. 9.1 above). For other algebras this was carried further (cf. [7] ch. III). Let $\underline{\underline{A}} = < A ; F >$ be an algebra and $\underline{\underline{I}}$ the postiterative algebra generated by F . It is natural to ask about the variety Var $\underline{\underline{I}}$ generated by $\underline{\underline{I}}$. The identities of Var $\underline{\underline{I}}$ (with the translation $w \rightarrow w^0$) are the hyperidentities of $\underline{\underline{A}}$. If $\underline{\underline{I}}$ is non-trivial then Var $\underline{\underline{I}}$ contains non-iterative algebras. Indeed, we have seen in the section 6 an example of non-iterative homomorphic image. Moreover, the countable power $\underline{\underline{I}}^{\omega}$ contains $f = < e_1^1 , e_1^2 ,... >$. From $\nabla e_1^i = e_2^{i+1}$ and $e_2^{i+1} * e_1^i = e_2^{2i}$ for $0 < i < \omega$ (in $\underline{\underline{I}}$) we obtain $\nabla f * f = < e_2^2 , e_2^4 ,... > \neq \nabla^n f$ for all n and consequently f has no arity.

As mentioned before, the various Mal'cev algebras were cleverly designed to avoid the bothersome arities in their formal description. However, it seems that each characterization of the classes of Mal'cev algebras must reintroduce arities in some way or another.

The class of postiterative algebras was characterized in [16] Thm.1. The existence of an arity is postualted in Axiom 1.1. The other Axioms 1.2-1.5, 2.1 - 2.8, 3.1-3.5 and 4 are classes of quasiidentities (e.g. if f has arity n then $\zeta^n f = f$). The same paper contains an abstract characterization of the full postiterative algebras (based on the description of the constant unary operations). The proof relies on the presence of projection-like elements and so it falls already for iterative algebras. The "axioms" 3.1-3.5 are tailored for the proof and so they are not quite transparent.

We conclude with a list of properties satisfied by all preiterative algebras. Presently, we have not proved that each algebra $< C ; * , \zeta , \tau , \Delta >$ (of type (2 , 1 , 1 , 1)) satisfying the properties is isomorphic to a (concrete) preiterative algebra (this was erroneously claimed in [12]).

Let $< C ; * , \zeta , \tau , \Delta >$ be a universal algebra of type (2 , 1 , 1 , 1) . As in section 6 define the arity ∂c of $c \in C$ as the least positive integer n such that $\Delta^n c = \Delta^{n-1} c$ (provided it exists). Thus the arity c is defined if and only if the sequence $c = \Delta^0 c , \Delta^1 , \Delta^2 c$, becomes stationary and ∂c equals to the first exponent n such that $\Delta^{n-1} c = \Delta^n c = ...$

In the following proposition we use the Lukasiewicz' or Polish parenthesis-free notation and write $*cd$ instead of $c*d$, e.g.

$$\Delta\zeta*\zeta*c_0*c_1c_2c_3$$

stands for

$$\Delta(\zeta(\zeta(c_0*(c_1*c_2))*c_3)) .$$

In our case this notation does not seem to be too cumbersome because we have only one non-unary operation $*$. We also use non-negative powers of unary operations (defined by ψ^0 = identity map and $\psi^{i+1} = \psi\circ\psi^i$ for $i \geq 0$). The product of a and b is denoted $*ab$ instead of $a*b$ which e.g. hides the fact that (viii) is the ordinary associative law (i.e. $(a*b)*c = a*(b*c)$ in the customary notation).

11. Proposition. Let $< C ; * , \zeta , \tau , A >$ *be a preiterative algebra. Then*

(i) ∂c *is defined for every* $c \in C$

and for all $a , b , c \in C$ with $\partial a = \alpha , \partial b = \beta , \partial c = \gamma$

(ii) $\partial\zeta a = \partial\tau a = \alpha$, $\partial*ab = \alpha+\beta-1$

(iii) $\tau^2 a = a$

(iv) $\zeta^\alpha a = (\zeta\tau)^{\alpha-1}a$

(v) $(\zeta^\upsilon\tau v^{\alpha-\nu}\tau)^2 a = a$ for $1 < \nu \leq \frac{1}{2}\alpha$

(vi) $\Delta\tau a = \Delta a$

(vii) $\Delta^\upsilon\zeta^{\upsilon-1}\tau\zeta^{\alpha+1-\nu}a = \Delta^\nu a$ for $1 < \nu \leq \alpha$

(viii) $**abc = *a*bc$

(ix) $*\zeta^\nu*abc = \zeta^{\gamma+\nu-1}*\zeta^{\alpha-\nu}*\zeta^\nu acb$ for $1 \leq \nu < \alpha$

(x) $*\zeta^\nu*abc = (\zeta^{\gamma+\nu-\alpha-1}(\zeta\tau)^{2\alpha+\beta-\nu-3})^{\alpha-1}\zeta^2*a*\zeta^{\nu-\alpha+1}bc$ for $\alpha \leq \nu \leq \alpha+\beta-2$

(xi) $*a\zeta b = \zeta^{\alpha+\beta-2}(\zeta\tau)^{\beta-1}\zeta^{\alpha+1}*ab$

(xii) $*\Delta ab = \zeta^{\alpha+\beta-3}\prod_{0\leq i<\alpha}(\zeta^{\beta-i}\Delta(\zeta\tau)^{\alpha+\beta-i-3})\zeta^{\beta+1}*\zeta^{\alpha-1}*abb$ for $\alpha > 1$

(xiii) $*\zeta^\nu\Delta ab = \zeta^{\alpha+\nu-1}\Delta\zeta^{\alpha-\nu}*\zeta^\nu ab$ for $1 \leq \nu \leq \alpha-2$

(xiv) $\Delta*ab = *a\Delta b$ for $\beta > 1$

(xv) $\Delta\zeta^\nu*ab = \zeta^{\nu-1}*\zeta^{\alpha-\nu}\Delta\zeta^\nu ab$ for $2 \leq \nu < \alpha$

(xvi) $*\tau ab = \zeta^\beta(\tau\zeta)^{\alpha-2}\zeta*\zeta^{\alpha-1}ab$ for $\alpha > 1$

(xvii) $*\zeta^\nu\tau ab = \zeta^{\beta+\nu-1}\tau\zeta^{\alpha-\nu}*\zeta^\nu ab$ for $1 \leq \nu \leq \alpha-2$

(xviii) $*\zeta^{\alpha-1}\tau ab = \zeta^{\alpha+\beta-2}(\zeta^{\alpha+\beta-2}\tau)^{\alpha-2}\zeta^{\alpha-1}*ab$ for $\alpha > 1$

(xix) $*a\tau b = \tau*ab$ for $\beta > 1$.

12. Remarks. (1) The straightforward but sometimes cumbersome proofs are omitted.

(2) The complicated form of some of the rules is due to the fact that we prefer the rules to be explicit. For example the rule (xvi) can be expressed in the following form: $* \tau ab = \sigma * \zeta^{\alpha-1} ab$ for $\alpha > 1$ where σ is a sequence of ζ's and τ's with the property that the sequence obtained from σ by replacing ζ by $z_{\alpha+\beta-2}$ (cyclic shift $(1,\ldots,\alpha+\beta-2)$) and τ by $t_{\alpha+\beta-2}$ (transposition (1,2)) yields the cyclic permutation $(\beta, \beta+1,\ldots,\alpha+\beta-2)$. The first half of (ii) and (iii) - (v) guarantee that ζ and τ act on the variables of an n-ary a in the same way as the cyclic permutation $\{1,\ldots,n\}$ and the transposition (1, 2) in the symmetric group of all permutations of $\{1,\ldots,n\}$ (cf. [2]). Similarly, the above rules together with (vi) and (vii) are designed so that ζ, τ and Δ act on the variables of an n-ary a in the same way as the above permutations and the map d corresponding to Δ (defined by $d(1) := 1$ and $d(i) := i-1$ for all $i < i \leq n$) act on the symmetric semigroup on $\{1,\ldots,n\}$.

(3) From the formal point of view the rule (i) is existential, rules (ii), (iv), (v), (vii) and (ix) - (xix) quasiidentities and only the three rules (iii), (vi) and (viii) are identities (axioms).

BIBLIOGRAPHY

[1] Cohn, B.M., *Unicversal Algebra,* Harper and Row, New York, 1965; second edition: D. Reidel Publ. Co. Dordrecht, Boston, London 1981.

[2] Coxeter, H.S.M.; Moser, W.O.J., *Generators and relations for discrete groups.* Ergeb. d. Math. u. ihrer Grenzgebiete 14, Springer Verlag 1972.

[3] Garcia, D.C.; Taylor, W.F., The lattice of interpretability types of varieties. *Mem. Amer. Math. Soc.* (50) 1984, no. 305, v+125 pp.

[4] Grätzer, G., *Universal algebra.* D. van Nostrand Co., Princeton N.J. 1968. 2nd edition Birkhauser Verlag Basel 1979.

[5] Jablonskiï , S.V., Functional constructions in a k-valued logic (Russian). *Trudg Mat. Inst. Steklov* 51 (1958) 5-142 .

[6] Janov, Ju.I.; Mu¨cnik, A.A., Existence of k-valued closed classes without a finite basis (Russian). *Dokl . Akad . Nauk SSSR* 127 (1959) 44-46.

[7] Lau, D., Function algebras over finite sets (German) *Dissertation,* Wilhelm-Pieck-Universität Rostock, 1984, 214 pp.

[8] Mal'cev, A.I., Iterative algebras and Post's varieties (Russian). *Algebra i logika (Sem.)* 5 (1966) No.2, pp. 5-24. English translation in *The metamathematics of algebraic systems,* Collected papers 1936-67. Studies in Logics and Fondations of Mathematics, vol. 66, North-Holland 1971.

[9] Menger, K., Function algebra and propositional calculus. *Self-organizing systems* 1962. Washington: Spartan Books 1962.

[10] Menger, K., On substitutive algebra and its syntax. *Z. Math. Logik Grundl. Math.* 10 (1964) 81-104.

[11] Menger, K. ; Whitlock, H.I., Two theorems on the generation of systems of functions. *Fund.Math.* 58 (1966) 229-240.

[12] Rosenberg, I.G., *Characterization of Mal'cev's preiterative algebra* (prelim. announc.) Preprint CRM-594, Université de Montréal 1976, 15 pp.

[13] Schweitzer, B.; Sklar, A., A mapping algebra with infinitely many operations. *Colloq. Math. 9* (1962) 33-38.

[14] Skala, H.L., Grouplike Menger Algebras. *Fund. Math. 79* (1973) 199-207.

[15] Trokhimenko, V.S., Ordered algebras of multiplace functions (Russian). *Izv.Vyš̈s Učebn Zaved. Matematika* 104 (1971), 90-98.

[16] Trokhimenko, S.V., Characteristic of some algebras of functions of many-valued logic (Russian, English summary). *Kibernetika Kiev* 1987, no. 3, 63-67, 80, 135.

[17] Whitlock, H.I., A composition algebra for multiplace functions. *Math. Ann.* 157 (1964) 167-178.

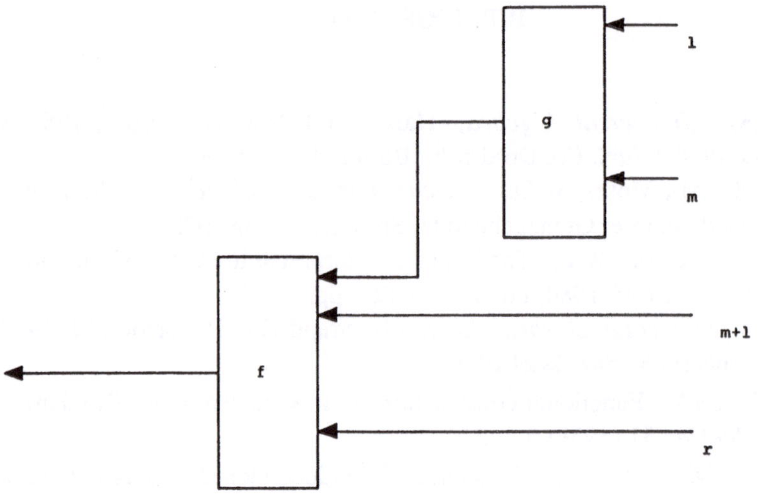

FIG. 1

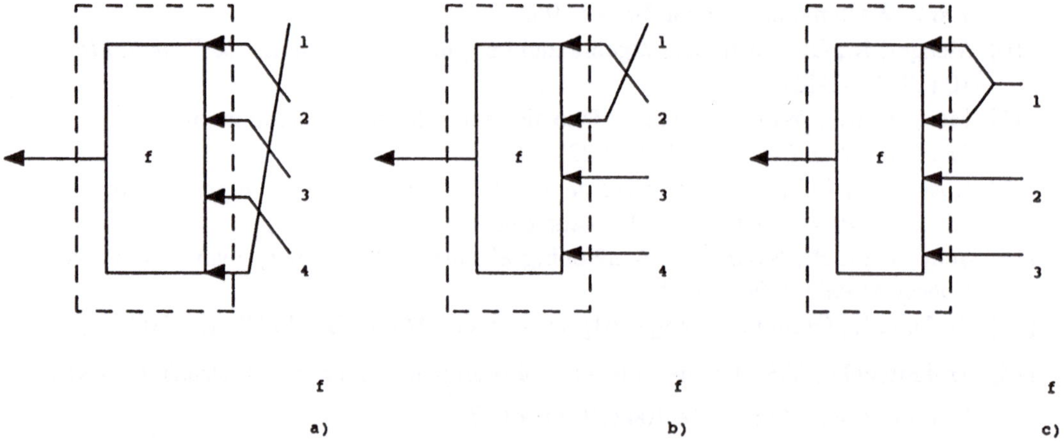

FIG. 2

Beth's and Craig's Properties via Epimorphisms and Amalgamation in Algebraic Logic

Ildikó Sain[1]

Here we investigate algebras associated with algebraizations of (variants of) first-order logic and of multi-modal logics. We will use the notation of the fundamental textbook Henkin-Monk-Tarski [10] (Parts I, II) of algebraic logic, but we will try to keep the present note self-contained. In a way, this work is a continuation of [31], but we do not assume familiarity with that work.

For a class K of algebras we understand the *amalgamation property* (AP), *embedding property* (EP), and *surjectiveness of epimorphisms* (ES) in the standard algebraic sense, see e.g. Kiss et al. [15], Bergman-McKenzie [4], and Pigozzi [30]. In particular, let $\mathfrak{A}, \mathfrak{B} \in K$ and let $f : \mathfrak{A} \to \mathfrak{B}$ be a homomorphism. f is an *epimorphism* (*epi* for short) of K iff for all $\mathfrak{C} \in K$ and all $h, g \in Hom(\mathfrak{B}, \mathfrak{C})$ we have $(h \circ f = g \circ f \Longrightarrow h = g)$. In other words, f is an epi iff it is right cancellative. K has ES iff every epimorphism f as above is onto $\mathfrak{B}$. The rest of the above properties (e.g. AP) will be recalled in Definition 0 below.

CONNECTIONS WITH LOGIC:

A class $Alg(\mathcal{L})$ of algebras can be correlated with a logic $\mathcal{L}$ in a natural way ([10, pp. 255–258 of §5.6]). Under this correlation various important metalogical properties of $\mathcal{L}$ are transformed into natural algebraic properties of $Alg(\mathcal{L})$. For instance, $\mathcal{L}$ has Beth's definability property ([3, p. 32 and p. 689]) iff $Alg(\mathcal{L})$ has ES. The equivalence of Beth property with ES is proved as Thm.5.6.10 in [10]. Similar connections between Craig's interpolation property for $\mathcal{L}$ and AP for $Alg(\mathcal{L})$ were established by Pigozzi [30], Maksimova [19], Wroński [33], and Ono [29]. (The latter used the abstract model theoretic framework of [3].) In this connection we note that it would be nice to see some of the problems left open in [30] solved because then one could connect the chart of abstract model theoretic properties (like Beth, weak Beth, Craig) on p. 715 (Table 7) of Barwise-Feferman [3] ([18]) with the chart of algebraic properties (like ES, AP) in Kiss et al. [15].

Here we will investigate the above mentioned algebraic properties (like

[1]Mathematical Institute of the Hungarian Academy of Sciences, Budapest, Pf.127, H-1364, Hungary

AP) for classes of algebras playing a central rôle in algebraic logic. By the above outlined connections between algebra and logic, every statement about ES or AP below implies a parallel statement about Beth's or Craig's property for one of the following logics: Keisler's logic [14], the Henkin-Tarski logic(s) discussed in §4.3 of [10], first-order logic with finitely many variables (L_n), the logics investigated in Tarski-Givant [32], or multi-modal logic(s). We will look into the so called *weak* Beth property (in the sense of [3], recalled below) in items 7(iii), 10, and 11.

For completeness we recall the Beth properties of logics. Since the logics listed above (as the counterparts of the classes of algebras investigated herein) are all extensions or fragments of classical first-order logic $L_{\omega\omega}$, we will restrict ourselves to these special logics. (An example of a fragment of $L_{\omega\omega}$ is first-order logic L_n with n variables for $n \in \omega$. Monadic first-order logic is another example.)

Let $\mathcal{L}$ be an extension or a fragment of $L_{\omega\omega}$. Let Λ be a similarity type or "language" of[2] $\mathcal{L}$, and let Fm_Λ be the set of $\mathcal{L}$-formulas of similarity type Λ. For some cardinal κ, let $R_i, i < \kappa$ be new ranked relation symbols not occurring in Λ such that $\Lambda^+ \stackrel{\text{def}}{=} \Lambda + \{R_i : i < \kappa\}$ is again a similarity type of $\mathcal{L}$.

Let $Th \subseteq Fm_\Lambda$ and $\Sigma(R_i)_{i<\kappa} = \Sigma(R_0 \ldots R_i \ldots)_{i<\kappa}$ be a subset of Fm_{Λ^+}. Then

(I) $Th + \Sigma(R_i)_{i<\kappa}$ is a *weak definition*[3] of $R_0, \ldots R_i, \ldots$ iff each model $\mathfrak{M}$ of Th has *at most one* expansion $\langle \mathfrak{M}, R_i^+ \rangle_{i<\kappa}$ which is a model of $Th + \Sigma(R_i)_{i<\kappa}$.

(II) $Th + \Sigma(R_i)_{i<\kappa}$ is a *strong definition* of $R_0, \ldots R_i, \ldots$ iff each model $\mathfrak{M}$ of Th has *exactly one* expansion $\langle \mathfrak{M}, R_i^+ \rangle_{i<\kappa}$ which is a model of $Th + \Sigma(R_i)_{i<\kappa}$.

Summing up, $(\forall \mathfrak{M} \in Mod(Th))(\exists! R_0^+, \ldots R_i^+, \ldots)\langle \mathfrak{M}, R_i^+ \rangle_{i<\kappa} \models (Th + \Sigma)$ means that $(Th+\Sigma)$ is a strong definition. As a contrast, weak definition requires only uniqueness of the defined $R_0^+, \ldots R_i^+, \ldots$ without requiring their existence.

(III) $\mathcal{L}$ has the *Beth property* iff for every choice of $\Lambda, \kappa, \langle R_i : i < \kappa \rangle$, and for every weak definition $Th + \Sigma(R_i)_{i<\kappa}$, there are formulas $\varphi_i(\bar{x}) \in Fm_\Lambda$, $i < \kappa$, such that

(1) $$Th + \Sigma(R_i)_{i<\kappa} \models_{\mathcal{L}} R_i(\bar{x}) \leftrightarrow \varphi_i(\bar{x}), \quad \text{for all } i < \kappa.$$

Here $\models_{\mathcal{L}}$ is the semantic consequence relation of the logic $\mathcal{L}$.

(IV) $\mathcal{L}$ has the *weak Beth property* iff for every Λ and $\langle R_i : i < \kappa \rangle$, and for every strong definition $Th + \Sigma(R_i)_{i<\kappa}$, statement (1) above holds.

---

[2]Then Λ is a set of ranked relation symbols such that their ranks are subject to the restrictions posed by $\mathcal{L}$. Namely the choice of $\mathcal{L}$ will permit or prohibit certain cardinals to occur as ranks.

[3]In standard model theory this is called an "implicit definition".

We note that in the case of Keisler's logic or the Henkin-Tarski logics, $R_0, \ldots R_i \ldots$ may have infinite ranks, while in propositional multi-modal logics they all are of rank zero.

We note that by a theorem in Németi [23], the cases of $\kappa < \omega$ and $\kappa = 1$ are equivalent. That is, if $\mathcal{L}$ has the Beth property for $\kappa = 1$ then it has the Beth property for each $\kappa < \omega$.

* * *

DEFINITION 0. *Throughout $f : \mathfrak{A} \rightarrowtail \mathfrak{B}$ means that f is an embedding of $\mathfrak{A}$ into $\mathfrak{B}$. Let K be a class of algebras.* $\mathbf{I}K$ *denotes the class of all isomorphic copies of members of K.*

(i) *K has* EP *if for any $\mathfrak{A}, \mathfrak{B}, \mathfrak{C} \in \mathbf{I}K$ with $\mathfrak{B} \supseteq \mathfrak{A} \subseteq \mathfrak{C}$, there are $\mathfrak{N} \in K$ and $\mathfrak{B} \overset{f}{\rightarrowtail} \mathfrak{N} \overset{h}{\leftarrowtail} \mathfrak{C}$.*

(ii) *K has* AP *if* (i) *can be strengthened*[4] *by requiring $f \restriction A = h \restriction A$.*

(iii) *K has* strong AP (SAP) *if* (ii) *can be strengthened by requiring that $f(B \setminus A) \cap h(C \setminus A) = 0$.*

(iv) *K has* strong EP *if* (i) *can be strengthened by requiring $f(B \setminus A) \cap h(C \setminus A) = 0$.* ■

Throughout α is an ordinal. Recall, from [10] Thm.5.1.31 p. 191 that $\mathbf{S}Dr_\alpha$ is the class of multi-modal (S5) algebras with α many modalities. That is, $\mathfrak{A} \in \mathbf{S}Dr_\alpha$ iff $\mathfrak{A} = \langle \mathfrak{B}, c_i \rangle_{i \in \alpha}$ where $\mathfrak{B}$ is a Boolean algebra and $c_i : B \to B$ is an S5 operator on $\mathfrak{B}$, i.e., c_i is a closure operator distributing over joins, and the complements of c_i-closed elements are c_i-closed (for all i). So, $\mathfrak{A}$ is a Boolean algebra (BA) with α many S5 modalities c_i ($= \Diamond_i$).

PROPOSITION 1. *$\mathbf{S}Dr_\alpha$ has the* SAP *(hence it has* ES*).*

To prove Proposition 1, we will use Lemmas 1.2 and 1.3 below. Recall from Henkin-Monk-Tarski [10] that Bo_α is the variety of Boolean algebras with operators and that an equation (of Bo_α's) is positive if "$-$" does not occur in it. "Positive in the wider sense" was also introduced in [10]. Our reason for mentioning this is that equations positive in the wider sense satisfy the conditions we need in the proofs of our lemmas, namely they are preserved under the so called canonical embedding algebras (i.e. complex algebras of Kripke models) by item 2.7.16 of [10].

Notation 1.1. *Let $\mathfrak{A} = \langle \mathfrak{B}, c_i \rangle_{i \in \alpha}$ with $\mathfrak{B}$ a Boolean algebra (BA) and $c_i : B \longrightarrow B$.*

(i) *$Kripke(\mathfrak{A})$ is a structure $\langle U, T_i \rangle_{i \in \alpha}$ where U is the set of ultrafilters of $\mathfrak{B}$ and $T_i \subseteq U \times U$ is defined such that*

$$F_1 T_i F_2 \quad \text{iff} \quad c_i(F_1) \subseteq F_2,$$

for all $F_1, F_2 \in U$.

---

[4]We use this to abbreviate that if we strengthen (i) by adding the requirement $f \restriction A = h \restriction A$ to the conclusion, this stronger version of (i) will still hold in K.

(ii) $\mathfrak{Cm}(\mathfrak{A})$ *is the complex algebra (algebra of complexes, see* [10]*)*[5] *of* $Kripke(\mathfrak{A})$*. The atom structure* $\mathfrak{At}(\mathfrak{Cm}(\mathfrak{A}))$ *of* $\mathfrak{Cm}(\mathfrak{A})$ *is the original* $Kripke(\mathfrak{A})$*. (So intuitively,* $\mathfrak{At}$ *is the inverse of the formation of complex algebras.)*

(iii) *For* $V \subseteq Bo_\alpha$ *we let* $Kripke(V) = \{Kripke(\mathfrak{A}) : \mathfrak{A} \in V\}$. ■

In recalling $\mathfrak{Cm}$ and $\mathfrak{At}$ from [10] and in introducing $Kripke(\dots)$, we disregard the distinguished constants of Bo_α's, called diagonal elements in [10], because they can be considered as special cases of the unary c_i's.

Intuitively, $Kripke(\mathfrak{A})$ is the "atom structure" suitable for representing $\mathfrak{A}$ in the weaker, nongeometric sense of §2.7 of [10]. Using the multi-modal intuition, if we think of $\mathfrak{A}$ as a multi-modal theory in algebraic form, then $Kripke(\mathfrak{A})$ is the Kripke frame providing a semantics for this theory.

Let $\mathfrak{M} = \langle M, R_i\rangle_{i\in I}$ be a relational structure. If $N \subseteq M$ then

$$\mathfrak{M} \upharpoonright N \overset{\text{def}}{=} \langle N, (R_i \cap {}^{\rho(i)}N)\rangle_{i\in I}$$

where $\rho(i)$ is the rank of R_i. We call $\mathfrak{M} \upharpoonright N$ a ***relative substructure*** of $\mathfrak{M}$.

Lemma 1.2. *Let* $V \subseteq Bo_\alpha$ *be a variety definable by a set of positive (in the wider sense) equations. Assume that* $Kripke(V)$ *is closed under relative substructures and finite direct products up to isomorphisms; this amounts to assuming that for all* $\mathfrak{M}, \mathfrak{N} \in Kripke(V)$ *and* $H \subseteq M$*, both* $\mathfrak{M} \upharpoonright H$ *and* $\mathfrak{M} \times \mathfrak{N}$ *are isomorphic to members of* $Kripke(V)$.

Then V *has the* SAP.

Proof: Since V is defined by positive equations, $\mathfrak{Cm}(\mathfrak{A}) \in V$ by 2.7.16 of [10]. Thus the natural embedding $em : \mathfrak{A} \rightarrowtail \mathfrak{Cm}(\mathfrak{A})$ for any $\mathfrak{A} \in V$ can be applied to show that V satisfies the conditions of Lemma 3 in [24] as follows: By 2.7.8(iii) and 2.7.9 of [10] p. 434 (or equivalently by the proof of Lemma 4 in [24]), em is a compact and saturating representation (in the sense of [24]) of $\mathfrak{A}$. This implies the conditions of Lemma 3. ■

By a $Df\,Bo_\alpha$ we understand a diagonal free reduct of a Bo_α. I.e. the only "extra Boolean" operations of a $Df\,Bo_\alpha$ are the c_i's ($i \in \alpha$), so in particular, it has no extra Boolean constants.

Lemma 1.3. *Let* $V \subseteq Df\,Bo_\alpha$ *be a variety definable by a set of positive (in the wider sense) equations. Assume that* $Kripke(V)$ *is closed under finite direct products. Assume* $V \models \{c_i0,\ c_i1\} \subseteq \{0,1\}$ *for* $i \in \alpha$.

Then V *has the* strong EP.

Proof: Notice that the only difference between the assumptions of Lemmas 1.2 and 1.3 is that here we omit the assumption of closure under relative substructures. So, repeating the proof of Lemma 1.2, we obtain all the conditions of Lemma 3 of [24] except the closure condition for relative substructures. A careful reading of the proof of Lemma 3 in [24]

[5] The complex algebra of a structure $\langle M, R_i\rangle_{i\in I}$ consists of the powerset BA of M together with the operations induced on this powerset by the R_i's.

reveals that this closure condition was not used in the strong EP part of the proof except for checking that the constructed embeddings preserve the constants, i.e., the elements of the minimal subalgebras. Dropping the d_{ij}'s and assuming $\{c_i0,\ c_i1\} \subseteq \{0,1\}$ ensure that the minimal subalgebras are trivial. Thus the proof for strong EP goes through without the omitted closure condition. ■

Problem 1.4. *The condition of closure under relative substructures in Lemma 1.2 above seems to be unnecessarily strong, preventing some desirable applications. E.g.* SAP *for Maddux's weakly associative relation algebras (WA's), cf. Theorem 8 below, was proved in* [28] *by observing that though this closure condition fails for them, those special relative substructures that are really needed in the proof of Lemma 3 of* [24] *are in Kripke(WA). So the problem is to find a stronger (but natural) version of Lemma 1.2 which would apply to WA as well.* ■

Proof of Proposition 1: $\mathbf{S}Dr_\alpha$ can be defined by adding positive equations to those defining Bo_α; cf. [10] Chapter 5. $Kripke(\mathbf{S}Dr_\alpha)$ is closed under relative substructures and products, e.g. by (the appropriate parts of) 2.7.40 of [10]. Thus Lemma 1.2 completes the proof. ■

Further generalizations of the results in [10] 2.7.16 saying that positive equations are preserved under taking canonical embedding algebras (i.e. under $\mathfrak{Em}$), hence implicitly also generalizations of our Lemmas 1.2 and 1.3, can be found in Goldblatt [8].

In passing we note that, by a result of Andréka, $\mathbf{S}Dr_\alpha = Dr_\alpha$.

Recall from [10] that Df_α is the subvariety of $\mathbf{S}Dr_\alpha$ in which the modalities commute. I.e.,

$$Df_\alpha = \{\,\mathfrak{A} \in \mathbf{S}Dr_\alpha : \mathfrak{A} \models c_ic_jx = c_jc_ix \quad \text{for all } i,j \in \alpha\,\}.$$

RDf_α is the class of representable members of Df_α in the geometric- or relation-algebraic sense introduced in Ch.5 of [10]: The elements of an RDf_α are α-ary relations, and $c_i(R)$ for some relation R is obtained from R by abstracting away from its i-th argument. E.g. if $\alpha = 2$ then the greatest element of the algebra is of the form $U_0 \times U_1$ and $c_1R = Dom(R) \times U_1$ while $c_0R = U_0 \times Rng(R)$. Intuitively, when $\alpha = 1+\beta$ and the greatest element of our RDf_α is ${}^\alpha U$, then by rewriting $R \subseteq {}^\alpha U$ in the form $R \subseteq U \times ({}^\beta U)$ we again can write $c_0(R) = U \times Rng(R)$, and similarly for c_i with $i < \alpha$ (note that since $Rng(R) \subseteq {}^\beta U$, $c_0(R)$ can be considered as an α-ary relation over U).

The AP was shown to fail in Df_α and RDf_α for $1 < \alpha < \omega$ in Comer [5]. The Df_α-part of Theorem 2(ii) below solves a problem on p. 315 of [5] (where the (ordinary) EP for RDf_α was already announced). Theorem 2 also solves problems on pages 346 and 348 of Pigozzi [30]. (The second problem is implicit; see lines 13 and 12 from the bottom of p. 348.)

THEOREM 2.

(i) ES *fails in* Df_α *and* RDf_α, *for* $\alpha > 1$.

(ii) *Df_α and RDf_α have the* strong EP.

Outline of proof:

The Df_α case of (ii) is immediate by showing that the conditions of Lemma 1.3 are satisfied by Df_α. The RDf_α case of (ii) uses ideas in the proof of Theorem 3 below; namely we start by blowing up the base sets involved to arbitrarily large cardinalities. We omit the details.

To prove (i), let $R \subseteq 4 \times 4$ be

$$\begin{array}{ccc} 1 & \rightarrow & 2 \\ \uparrow & \nearrow & \downarrow \\ 0 & \leftarrow & 3 \end{array}.$$

Let $V = 4 \times 4 \times 1 \times \cdots \times 1 \times \ldots (\subseteq {}^\alpha 4)$, $x = \{q \in V : q_0 \; R \; q_1\}$, and $d = \{q \in V : q_0 = q_1\}$. Let $\mathfrak{A}$ be the full diagonal-free cylindric set algebra with greatest element V, and let $\mathfrak{B} \subseteq \mathfrak{A}$ be generated by $\{x, d\}$. We claim that $Id_B : \mathfrak{B} \rightarrowtail \mathfrak{A}$ is the desired nonsurjective epimorphism of both Df_α and RDf_α. Indeed, $B \neq A$ is easy to check. To see that Id_B is an epimorphism, we first note that $\mathfrak{A}$ is simple. Hence in any $\mathfrak{B} \rightarrowtail \mathfrak{A} \overset{h}{\underset{g}{\rightrightarrows}} \mathfrak{C}$, h and g have to be embeddings.

First we will prove that $Rng(h) = Rng(g)$. From this we will conclude $h = g$ which in turn implies that Id_B is an epi. Since our purpose is proving that Id_B is an epi and since $d \in B$, we may assume that $h(d) = g(d)$ and that $\mathfrak{C}$ is subdirectly irreducible. Let $At(d) = \{x \in At\mathfrak{A} : x \leq d\}$. Clearly, $At(d)$ generates $\mathfrak{A}$. Assume

$$h(At(d)) \neq g(At(d)). \tag{2}$$

Then there are $a, b, c \in At(d)$ with $b \neq c$, $h(a) \cap g(b) > 0 < h(a) \cap g(c)$, and $\Delta a = \Delta b = \Delta c = \{0, 1\}$. This is so because $h(At(d))$ and $g(At(d))$ are two different partitions of the same "set" $h(d)$. Let $x \overset{\text{def}}{=} c_1(h(a) \cap g(b))$ and $y \overset{\text{def}}{=} c_0(h(a) \cap g(c))$. Then $\Delta x = 1$ and $\Delta y = \{1\}$; hence $\Delta x \cap \Delta y = 0$. In any subdirectly irreducible Df_α, if two nonzero, finite dimensional elements have disjoint dimension sets, then they are not disjoint. Therefore $x \cap y > 0$. But $c_1 g(b) \cap c_0 g(c) \geq x \cap y$ and $c_1(b) \cap c_0(c) \leq -d$ imply $x \cap y \leq -g(d) = -h(d)$. So $c_1 h(a) \cap c_0 h(a) \geq x \cap y \nleq h(d)$ contradicting the fact that $c_1 a \cap c_0 a \leq d$ holds in $\mathfrak{A}$. This proves that (2) is false. But then, since $At(d)$ generates $\mathfrak{A}$, we may assume that $\mathfrak{C} \cong \mathfrak{A}$, and hence that $\mathfrak{A} = \mathfrak{C}$. Now $h = g$ follows relatively easily. (Hint: Since $\mathfrak{A}$ is full, by the proof of 3.1.47 of [10], h and g are base-automorphisms. So $h \neq g$ would imply the existence of a base-automorphism $\tilde{f}$ of $\mathfrak{A}$ leaving B fixed but moving some of the atoms. But a glance at the picture defining R reveals that $f(0) = 0$ must hold. This implies $f(1) = 1$ etc.) This proves (i). ■

Proposition 1 and Theorem 2 imply that the behavior of multi-modal logic(s) change drastically if we require modalities to commute ($\Diamond_i \Diamond_j \varphi \leftrightarrow$

$\Diamond_j \Diamond_i \varphi$). If the modalities are not required to commute, the logic has both Beth's definability and Craig's interpolation properties, but it has neither if they are required to commute. (This is analogous to the fact that this logic is decidable by [25] but not if modalities commute, in case $\alpha > 2$, by Thm.5.1.66 of [10]. Necessity of $\alpha > 2$ was shown by Dana Scott.) Actually it was conjectured in Németi [26], [27] that set theory might be finitely axiomatizable in Df_3 or Df_4, i.e. in commutative multi-modal (S5) logic, in the manner of Tarski-Givant [32]. This would imply a very strong version of undecidability, but the conjecture is still open.

* * *

As in [10], *PA* abbreviates "polyadic algebra", the letter *R* (in *RPA*, *RDf* etc.) means "representable", and the index α means that the elements of RPA_α's are relations of *rank* α. Quasi *PA*'s (*QPA*'s) are Boolean algebras with operators where the operators are c_i, p_{ij}, s^i_j for $i, j \in \alpha$. The intuitive meanings of the latter are described below (p_{ij} is the "substitution" interchanging i and j while s^i_j is the one sending i to j and leaving the rest fixed). *QPA*'s are discussed in §V.7 (p. 120ff) of Halmos [9] and e.g. in [10]. *QPEA*'s are *QPA*'s with equality (and *PEA* is obtained similarly from *PA*).

(3) An $RQPEA_\alpha$ is an $\mathfrak{A} = \langle \mathfrak{C}, d, p_{ij} \rangle_{i,j \in \alpha}$ where $\mathfrak{C}$ is an RDf_α with greatest element ${}^\alpha U (= U \times U \times \ldots)$, a constant $d = \{q \in {}^\alpha U \ : q_0 = q_1\}$ ($= Identityrelation \times {}^{(\alpha-2)}U$), and unary operations p_{ij} such that for any α-ary relation $R \subseteq {}^\alpha U$,

$$p_{01}(R) = \{\langle q_1, q_0, q_2, q_3, \ldots \rangle : \langle q_0, q_1, q_2, q_3, \ldots \rangle \in R\} \qquad \text{and}$$

$$p_{12}(R) = \{\langle q_0, q_2, q_1, q_3, \ldots \rangle : \langle q_0, q_1, q_2, q_3, \ldots \rangle \in R\} \qquad \text{etc.}$$

To be precise, $RQPEA_\alpha$'s are subdirect products of algebras $\mathfrak{A}$ of the above kind. (They form a variety.)

(4) $RQPA_\alpha$'s are obtained from $\mathfrak{A}$ in (3) above by replacing d with the unary operator s^0_1 where

$$s^0_1(R) = \{\langle q_0, q_1, q_2, \ldots \rangle \in {}^\alpha U \quad : \quad \langle q_1, q_1, q_2, q_3, \ldots \rangle \in R\}$$

(i.e. $s^0_1 x = c_0(d \cap x)$). In the case of $RQPA_\alpha$'s we do not need to consider other algebras than $\mathfrak{A}$ above (i.e., there is no need for subdirect products):

THEOREM 3. *Let $\alpha \geq \omega$. Then $RQPA_\alpha$'s of the form* (4) *above (i.e. with greatest elements of the form ${}^\alpha U$) form a variety (up to isomorphism).*

Outline of proof: By Lemma 6.4 (and by the construction preceding it) on p. 226 of Halmos [9] one first proves that $RQPA_\alpha$ coincides with ${}_\infty RQPA_\alpha$ (for ${}_\infty K$ see Part II of [10]; ${}_\infty K$ is the class of those members of K which are representable over an infinite base). Then by adopting the proof method of 3.1.106 of [10], one proves that Cs-like $RQPA$'s are closed under **SP** (we use the fact that we need consider only ${}_{\infty}Cs$-like $RQPA$'s by our present first step). The proof of closure under **H** is similar to the proof of 3.1.103 of [10]. ■

Theorem 4(ii) below solves a problem raised in Comer [5] (see the remarks following Corollary 5 in [5]). This is a joint result of the present author with R.J.Thompson. The problem was probably motivated by the result in [5] that AP fails in PA_n. Concerning (i) below: by Corollary 4 of [5], AP fails for $RQPEA_n$ ($1 < n < \omega$); see also 2.3.8 in [30]. A modification of the argument in 2.2.15 of Pigozzi [30] and Corollary 5 in [5] yield that EP fails in $RQPEA_\alpha$ iff $1 < \alpha < \omega$.

THEOREM 4.

(i) ES *fails in* $RQPEA_\alpha$ *if* $\alpha > 1$.

(ii) PA_α *has the* strong EP *(for all* α*).*

Proof: One proves (ii) by showing that the conditions of Lemma 1.3 are satisfied by PA_α (note that $PA_\alpha \subseteq Df\,Bo_I$ with $I = (Sb\ \alpha) \cup {}^{\alpha}\alpha$).

We omit the proof of (i). ■

A $QPEA_\alpha$ $\mathfrak{A}$ is ***binary-generated*** if it is generated by elements x with the property $c_i x = x$ for all $1 < i < \alpha$. This concept was introduced in Monk [21], [22] and was investigated implicitly in Tarski-Givant [32] which motivates the following:

(Note that for $\alpha < \omega$, PEA_α is the same as $QPEA_\alpha$.)

PROPOSITION 5. ES *fails in* binary-generated $RPEA_3$ *and* PEA_3.

Proof: Let $R, E \subseteq U \times U$ be defined as in the proof of Theorem 9 below. Let $R^3 = R \times U$ and $E^3 = E \times U$. Let $\mathfrak{P}$ be the full set PEA_3 with unit $U \times U \times U$. Let $\mathfrak{A} \subseteq \mathfrak{P}$ be generated by R^3 and E^3, and let $\mathfrak{B} \subseteq \mathfrak{A}$ be generated by R^3. Then by 5.3.12 of [10] and by the proof of Theorem 9, $E^3 \notin B$. By the proof of Theorem 9, $Id_B : \mathfrak{B} \rightarrowtail \mathfrak{A}$ is an $RPEA_3$-epimorphism. This proves Proposition 5 for $RPEA$. For the PEA-case one uses the construction of [28] mentioned after the proof of Theorem 9. ■

Actually, we do not know if the construction described above proof can be used to prove that ES fails in binary-generated $RPEA_4$. In such a proof we would choose the unit of $\mathfrak{P}$ to be ${}^4U$ and would use $R^4 = R \times U \times U$, $E^4 = E \times U \times U$. (If E is not definable in $\langle U, R\rangle$ with 4 variables then the proof goes through for binary-generated $RPEA_4$.)

PROBLEM 6. *Does* ES *hold in* binary-generated $RPEA_\alpha$ and PEA_α *for* $4 < \alpha < \omega$ *? (For infinite* α*,* ES *easily holds by a result of Daigneault [7].) What is the answer for "n-ary-generated" instead of "binary-generated" (with* $1 < n < \alpha < \omega$*)?* ■

By Thm.5.12 of Johnson [12], ES holds in PEA_ω (and also in PA_ω). By Theorem 7 below, this does not extend to $RPEA_\omega$. As a consequence, the

SAP, while true in PEA_ω, fails in its subcategory $RPEA_\omega$. Therefore the "amalgamated products" of certain $RPEA_\omega$'s are necessarily *non*representable. This also implies the following: There is an epimorphism in $RPEA_\omega$ which is not an epimorphism in PEA_ω. It is an open problem whether this is also true for RA's and RRA's (relation algebras and representable relation algebras) and for CA_α's and RCA_α's (cylindric algebras and representable cylindric algebras). It is a logical consequence of Theorem 7(ii) together with [12] that Keisler's logic with infinitary predicates ([14]) has Beth's definability (and Craig's interpolation) property iff equality is omitted. (The proof-theoretic versions of Beth's and Craig's properties are true when equality is present however.)

A *full setPEA*$_\alpha$ is an $RPEA_\alpha$ consisting of all α-ary relations over some set U.

THEOREM 7.

(i) ES *fails in* RPA_n *and* $RQPA_\alpha$ *if* $n, \alpha > 1$ *and* $n < \omega$.

(ii) ES *and the* strong EP *fail for* $RPEA_\alpha$ *if* $\alpha > 1$, *but* EP *holds if* $\alpha \geq \omega$.

(iii) *For* $\alpha \geq \omega$, ES *fails in* $RPEA_\alpha$ *in the following very strong sense: Every nondiscrete*[6] *locally finite-dimensional* $RPEA_\alpha$ *is the domain of some epimorphism* $f : \mathfrak{A} \twoheadrightarrow \mathfrak{B}$ *which is not surjective. Moreover* (f *is a monomorphism and) every full setPEA*$_\alpha$ *is* $\{f\}$*-injective, i.e. every* $h : \mathfrak{A} \to \mathfrak{C}$ *with* $\mathfrak{C}$ *a full setPEA*$_\alpha$ *factors through* f ($h = k \circ f$ *for some* $k : \mathfrak{B} \to \mathfrak{C}$).

Outline of proof: (iii). Let $\mathfrak{A}$ be a nondiscrete, locally finite or dimension-complemented $RPEA_\alpha$. We will only treat the case when $\mathfrak{A}$ is subdirectly irreducible, but the same idea works in general. Then we may assume that $\mathfrak{A} \subseteq \mathfrak{B}$ for some full set algebra $\mathfrak{B}$ with a base, say, U. (So $B = Sb({}^\alpha U)$.) Let $X = \bigcap\{d_{ij} : i, j \in \alpha\}$. Then $X \in B \setminus A$ since $\Delta X = \alpha$. Let $\mathfrak{C} \subseteq \mathfrak{B}$ be generated by $A \cup \{X\}$. We claim that $Id_A : \mathfrak{A} \rightarrowtail \mathfrak{C}$ is the desired $RPEA_\alpha$-epimorphism w.r.t. which all the full set algebras are injective. Hint for the epi-part: For some τ, $s_\tau X = 1$ and $X \leq d_{ij}$ for all i, j. Id_A is an epi because there is at most one X with these properties, and these properties are preserved by homomorphisms.

(ii). For $\alpha \geq \omega$, failure of ES follows from (iii). For $1 < \alpha < \omega$, the construction quoted from [28] in the proof of Theorem 10 below can be used to disprove ES. Let $\alpha > 1$. Let $\mathfrak{A}$ be the full set PEA_α with a two-element base. Then it is impossible to embed two disjoint copies of $\mathfrak{A}$ into an $RPEA_\alpha \mathfrak{C}$ such that these copies would remain disjoint except on their minimal subalgebra. In particular, there are two atoms below $\bigcap\{d_{ij} : i, j \in \alpha\}$. These are not in the minimal subalgebra; therefore in $\mathfrak{C}$ there would be four atoms. At the same time, for $x \in At\mathfrak{A}$ we have

[6] A Boolean algebra with operators is discrete if all its extra Boolean operators are trivial, cf. [10].

$c_{(\alpha)}x = 1$ and $c^{\partial}_{(\alpha)}x = 0$. The characteristic of $\mathfrak{C}$ is 2 because $\mathfrak{C}$ inherits the property of $\mathfrak{A}$ that it cannot be represented over a unit element ${}^{\alpha}U$ with $|U| > 2$; but this is impossible.

We omit the proofs of (i) and the positive EP-part of (ii). ■

The strong EP-part of the $\alpha < \omega$ case of (ii) was already proved in [5]. We also note that $n < \omega$ is needed in Theorem 7(i) by [12] and a result of Daigneault and Monk.

* * *

The properties (AP etc.) we have been investigating for cylindric and related algebras were investigated for relation algebras (RA's) in McKenzie [20], p. 116, answering various problems raised in Jónsson [13] and Comer [5]. $QRA(\subseteq RA)$ is the class of relation algebras with a "quasi-pairing" element,[7] cf. Tarski-Givant [32],. and $WA(\supseteq RA)$ is the class of weakly associative RA's of Maddux [16]. WA's are the same as representable RA's (RRA's) relativized to reflexive and symmetric relations. ***Monadic-generated*** CA_α's, cf. [10], were investigated in Monk [21], [22] under the name "singulary CA's".

THEOREM 8. *QRA, WA, and monadic-generated CA_α's have the* SAP *and hence* ES.

Outline of proof: The idea of the proof in the case of WA is indicated in Problem 1.4 above.

For QRA and monadic-generated CA's the idea is the same. Namely, in both cases we find a functor F (in the category theoretic sense) connecting our algebras to Lf_ω of [10] such that F "preserves the properties under investigation". Therefore we outline the QRA-case only.

To any CA_α $\mathfrak{A}$ with $\alpha > 3$, the relation algebra $\mathfrak{Ra}(\mathfrak{A})$ associated with $\mathfrak{A}$ in the natural way was defined in §5.3 of [10] (e.g. the universe of $\mathfrak{Ra}(\mathfrak{A})$ is $\{x \in A : \Delta(x) \subseteq 2\}$, so it consists of the "binary" elements of $\mathfrak{A}$ — as it should). An Lf_ω $\mathfrak{A}$ is a QLf_ω iff $\mathfrak{Ra}(\mathfrak{A}) \in QRA$. Now, $\mathfrak{Ra} : QLf_\omega \rightarrowtail\!\!\!\!\rightarrow QRA$ is a functor with an inverse $F : QRA \longrightarrow QLf_\omega$. Further Daigneault's proof in [7] that Lf_ω has the SAP carries over to QLf_ω. Namely, if we have a diagram

$$\mathfrak{A}_2 \leftarrowtail \mathfrak{B} \rightarrowtail \mathfrak{A}_1$$

in QLf_ω, then the strong amalgam "computed in Lf_ω", say $\mathfrak{C}$, of this diagram will lie in QLf_ω. Now, QRA inherits the SAP from QLf_ω because $\mathfrak{Ra} \circ F = Id$ (and because $\mathfrak{Ra}$ and F are functors).

Concerning the monadic-generated case, only the $(\alpha < \omega)$-case is non-trivial, and then we should note that not all monadic-generated Lf_ω's are in the range of F. ■

[7] An RA is a QRA if it contains two functional elements p and q such that $q^{\smile} ; p = 1$, where x is functional if $x^{\smile} ; x \leq 1'$.

The result concerning WA was proved in Németi [28] where it was also proved that the variety ${}_{\omega}RRA$ defined below has the strong EP but has neither ES nor AP. As a contrast, we note that McKenzie [20] proved that EP fails in RRA. As mentioned in the remarks preceding Theorem 7, the RA counterpart of Theorem 7(ii) with [12] Thm.5.12 would say that there is an RRA-epimorphism which is not an RA-epimorphism. We do not know if this (or its CA-theoretic version) is true.

By a *symmetric RA* (*SyRA*) we understand an RA in which $x = x^{\smile}$ is valid. *Group RA*'s (*GRA*'s) are subalgebras of complex algebras of groups; cf. McKenzie [20]. For a cardinal κ, ${}_{\kappa}RRA$ is the variety generated by the full set RA $\mathcal{R}e(\kappa)$ whose universe is $\mathcal{P}(\kappa \times \kappa)$ cf. [10] Part II.

THEOREM 9. *(Németi [28]) There is an (injective) epimorphism* $f : \mathfrak{A} \rightarrowtail\!\!\!\!\twoheadrightarrow \mathfrak{B}$ *of* RRA *which is not surjective and satisfies (i)–(iv) below.*

(i) $\mathfrak{A}, \mathfrak{B} \in (GRA \cap SyRA)$.
(ii) $\mathfrak{A}, \mathfrak{B} \in {}_{\omega}RRA$.
(iii) $\mathfrak{B} \in {}_{45}RRA$.
(iv) $\mathfrak{B}$ *is embeddable into a non-representable* RA.

Outline of proof: Let R be the binary relation on a 45 element set U represented in Figure 1 below. A line like

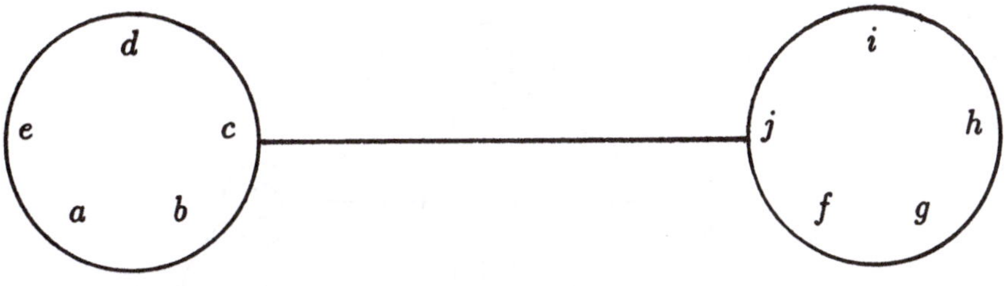

connecting two "bubbles" is meant to indicate that every element of each one of the bubbles is connected with every element of the other one. In other words, it indicates that $\{a,b,c,d,e\} \times \{f,g,h,i,j\} \subseteq R \cap R^{\smile}$. The way the elements in the same bubble are connected is standard, e.g. aRb but not aRd. If two bubbles are not connected the above way, then their elements are not related via R. So, in particular R is symmetric and irreflexive. Let $E \subseteq U \times U$ be that equivalence relation the equivalence classes of which are the bubbles of Figure 1. So in the diagram above, aEd but not aEh. If we take the quotient $\langle U, R\rangle / E$ of the structure in Figure 1 modulo E, then we get the 9-element structure (first used by Ulf Wostner) in Figure 2. But in Figure 2 the relation is reflexive because, by aRb, we get that a/E is related (or connected) to b/E but $a/E = b/E$.

Let $\mathfrak{A}$ and $\mathfrak{B}$ be the subalgebras of $\mathfrak{Re}(U)$ generated by $\{R\}$ and $\{R, E\}$, respectively. Let $f : \mathfrak{A} \rightarrowtail \mathfrak{B}$ be the identical embedding of $\mathfrak{A}$ into $\mathfrak{B}$. Now $A \neq B$ because of the following. $\mathfrak{A}$ has three atoms: R, $-R$, and 1' with $R;R = (-R);(-R) = 1$; so $\mathfrak{A}$ cannot contain E. So f is not surjective.

To see that $f : \mathfrak{A} \rightarrowtail \mathfrak{B}$ is an epi in RRA, observe the following.

(5) R is a symmetric graph, E is an equivalence relation, and $(R \cap E)$ is a disjoint union of "pentagons" (like a, b, c, d, e on the figure above). Further $E = (R \cap E) \cup [(R \cap E); (R \cap E)]$.

(6) E is a strong congruence relation of the structure $\langle U, R \cup E \rangle$ i.e. $(xEy \Longrightarrow [x(R \cup E)z \leftrightarrow y(R \cup E)z])$.

(5) and (6) are equational properties of E, R in $\mathfrak{B}$ since their first-order formulation does not require more than 3 variables. We note that "$(R \cap E)$ is a disjoint union of pentagons" can be expressed by the following equations: let $S = [(R \cap E); (R \cap E)] \setminus Id$. Then $R \cap E \cap S = \emptyset$, $(R \cap E); S = E \setminus Id$, $S; S = E \setminus S$. Hence (5),(6) are preserved under homomorphisms. Let $\mathfrak{B} \overset{h}{\underset{g}{\rightrightarrows}} \mathfrak{C} \in RRA$ be such that $h(R) = g(R)$. It is enough to prove that $h(E) = g(E)$. To this end, assume $\mathfrak{C} \subseteq \mathfrak{Re}(W)$ for some W. Then by the above consideration, (5) and (6) hold for $\langle W, h(R), h(E) \rangle$ and also for $\langle W, h(R), g(E) \rangle$. Now, a simple combinatorial lemma implies $h(E) = g(E)$. This lemma says that to any graph R^+ there is at most one relation E^+ satisfying (5),(6). This completes the proof that f is a nonsurjective epi in ${}_{45}RRA$, i.e. of (iii). It is easy to see that $\mathfrak{B} \in SyRA$. We omit the rest of the proof. ■

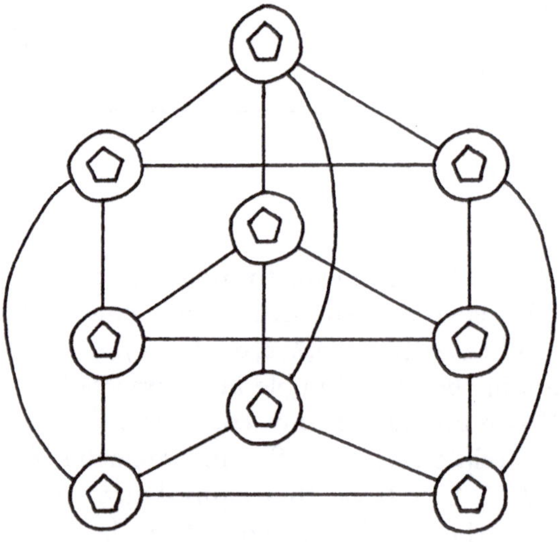

Figure 1

By the above theorem, for any class K with $RRA \supseteq K \supseteq GRA \cap SyRA \cap {}_{\omega}RRA \cap {}_{45}RRA$ we have that ES fails for K. For $SA(\supseteq RA)$ of Maddux [16], Németi [28] proved that for any K and $\kappa > 10$ with $SA \supseteq K \supseteq {}_{\kappa}RRA$, ES fails in K. For the class $CFG(\subset GRA)$ of complex algebras of finite groups McKenzie [20] proved that EP, AP fail in CFG. (The proof is

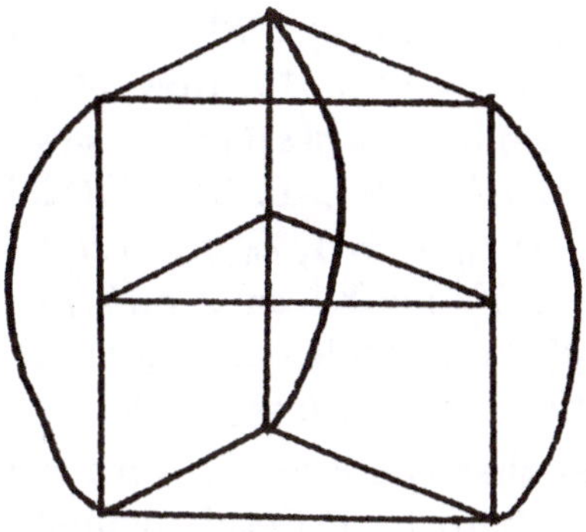

Figure 2

implicit in his proof of this for GRA.) Despite of this, ES holds for CFG. We conjecture that ES holds for complex algebras of groups in general. We note that there are several classes of groups with this property, e.g. ES does hold for (any class of) complex algebras of groups of prime order.

* * *

We now consider ***cylindric algebras*** (CA's). The AP and EP were thoroughly treated in Pigozzi [30], where earlier works of other authors, e.g. Comer and Daigneault, were systematically quoted. ES for CA's was investigated in [1] and [28] among others. By [1], ES fails in CA_n for $1 < n < \omega$, and by [28] together with [1], ES fails in RCA_α for $\alpha > 1$. The problem whether ES holds for CA_ω was left open in [1].

In view of the connections described in §4.3 of [10] between algebra and "pure logic", we see that Theorem 10 below implies that the ***proof-theoretic*** version of the Beth property also fails for the Henkin-Tarski infinitary logic in §4.3 of [10]. (The weak Beth case however remains open.)

THEOREM 10. *ES fails for* CA_α *with* $\alpha \geq \omega$.

Proof: Let $\alpha \geq \omega$. It was proved in [28] that there exists a nonsurjective RCA_α-epimorphism $Id_B : \mathfrak{B} \rightarrowtail \mathfrak{A}$ satisfying the following condition:

(7) $\mathfrak{A}$ is generated by three elements R, s, b with $\Delta s = \Delta b = 1$. Further $\mathfrak{A}$ is subdirectly irreducible, s is a 0-thin (in sense of [10] p. 60) element, and b is not. Further $R \leq s + b$, and letting $X = c_0(s \cap R)$ and $Y = c_0(b \cap R)$, we have $X \cap Y = 0 < X$, Y and $b \cap R = b \cap Y$. $\mathfrak{B}$ is generated by R and $(s + b)$. In addition $B \neq A$.

Here we prove that this Id_B is actually a CA_α epimorphism. Let $\mathfrak{A} \underset{g}{\overset{h}{\rightrightarrows}} \mathfrak{C}$ agree on B. For any $a \in A$ we write a^+ for $h(a)$. Assume $g \neq h$. We will derive a contradiction. Clearly, $g(s) \neq s^+$ (because $s + b \in B$ and $B \cup \{s\}$ generates $\mathfrak{A}$). Since in a CA it is impossible to have two different comparable 0-thin elements, we conclude $g(s) \cap s^+ = 0$ because $\mathfrak{A}$

is subdirectly irreducible. Then $g(s) < b^+$ because $g(s)$ is 0-thin but b^+ is not (and because $g(s)+g(b) = s^+ + b^+$). Now, $g(s) \cap R^+ \leq b^+ \cap R^+ \leq b^+ \cap Y^+$ (because $b \cap R = b \cap Y$) implies $g(s) \cap R^+ = g(s) \cap b^+ \cap Y^+ = g(s) \cap Y^+$. Let $b_0 = b^+ - g(s)$. Then $g(b) \geq b_0 > 0$. So $g(b) \cap R^+ \geq b_0 \cap R^+ \geq b_0 \cap b^+ \cap Y^+ = b_0 \cap Y^+$. Then, by $c_0(s) = 1$, $c_0(g(s \cap R)) = c_0(g(s) \cap Y^+) = Y^+$ and $c_0(g(b \cap R)) = c_0(b_0 \cap Y^+) = Y^+$ since $c_0(b_0) = 1$. But then $c_0 g(s \cap R)$ is not disjoint from $c_0 g(b \cap R)$ while our original X and Y *are* disjoint in $\mathfrak{A}$. This is a contradiction, proving $h = g$. ■

Theorem 7(i) implies that the logic L_n with n variables $(1 < n < \omega)$ and without equality fails to have Beth's definability property (answering a question of Henkin). The same for L_n with equality was proved in Andréka-Comer-Németi [1]; cf. also Theorem 4(i) above. By Theorem 5, this property of L_n continues to hold when we require the atomic formulas to be binary in the case of $n = 3$. The following is motivated by questions and suggestions of Henkin.

PROBLEM 11. *Does L_n (with or without equality) have the weak Beth property (in the sense of Barwise-Feferman [3] Def.II.7.3.1 p. 73)?* ■

This problem has a natural algebraic formulation in terms of cylindric algebras, cf. Theorem 7(iii). Theorem 7(iii) implies that even the *weak* Beth property fails for Keisler's logic [14] with equality added. As already implied, "RCA" abbreviates "representable CA". The algebraic form of Problem 11 is:

PROBLEM 12.

(i) *Is there a nonsurjective epimorphism f in RCA_α such that every full Cs_α is $\{f\}$-injective ? Cf. Theorem 7(iii) for explanation of $\{f\}$-injectivity.*

(ii) *The same as (i) but with RRA and full set RA (instead of RCA_α and full Cs_α, respectively).* ■

The following problem is relevant to Problem 12. By a category we always mean one given by a class of algebras and *all* homomorphisms between them.

PROBLEM 13.

(i) *Is there $K \supseteq Cs_\alpha$ with K a proper reflective subcategory of RCA_α? The same question with Cs_α^{reg} instead of Cs_α is also open.*

By a ***proper reflective subcategory*** we mean that K is reflective but not (regular epi)-reflective (the terminology is explained below). This also would mean the existence of a kind of a "universal" nonsurjective epi in RCA_α which is of the form of a K-reflection. This in turn would solve the RCA-part of Problem 12 above. C. Bergman's paper on saturated algebras may be relevant here.

(ii) *Is there $K = \mathbf{H}K \supseteq (Cs_\omega^{reg} \cap Lf_\omega)$ with K a proper reflective subcategory of RCA_ω ?*

(iii) *Is there a proper reflective subcategory K of RRA containing all simple RRA's ? Cf. Problem 12 above.* ■

Note on the *terminology* in Problem 13: The notion of a reflective subcategory can be found in any textbook on category theory. The concept of (regular epi)-reflective is defined e.g. in [11]. In our case "regular epi" is the same as "strong epi" which is the same as "surjective homomorphism".

PROBLEM 14. *Find a characterization of the weak Beth property that would be comparable to Thm.5.6.10 of* [10] *in its generality.* ■

For our concrete cases like CA_n's, L_n etc. the equivalence discussed in the remarks between Problems 11 and 12 does the job. But that is not as general as 5.6.10 of [10]. We conjecture that either (8) or (9) below might serve as such a general characterization.

Recall that for a logic $\mathcal{L}$ the associated class of algebras is $Alg(\mathcal{L})$. For a homomorphism f, we say that $Alg(\mathcal{L})$ *has enough* $\{f\}$*-injectives* iff every subdirectly irreducible member of $Alg(\mathcal{L})$ is embeddable into an $\{f\}$-injective one (cf. Theorem 7(iii)).

(8) For every epimorphism f of $Alg(\mathcal{L})$, if $Alg(\mathcal{L})$ has enough $\{f\}$-injectives, then f is surjective.

(9) There is no proper, reflective subcategory K of $Alg(\mathcal{L})$ such that $Alg(\mathcal{L}) \subseteq \mathbf{S}K$.

Now, our conjecture says that $\mathcal{L}$ has weak Beth iff (8) holds. The same applies for (9) (in place of (8)).

We note that (8) and (9) fail for $RPEA_\omega$ in place of $Alg(\mathcal{L})$. This is a consequence of Theorem 7(iii). For RCA_α and for RRA, (8) is equivalent with a positive answer for our Problem 12.

PROBLEM 15.

(i) *Give a simple, abstract (category-theoretic or universal-algebraic) characterization of full Cs_α's in RCA_α.*

(ii) *Give an abstract characterization of full set RA's in RRA.* ■

Only the finite ones are absolute retracts; see Andréka-Jónsson-Németi [2]. So absolute retracts cannot be used to give a characterization. Further, $\mathbf{SP}\{\mathfrak{Re}(U)\}$ is a variety iff U is finite. So Cs_α's and RA's with a finite base seem to behave quite differently from those without one, making a uniform characterization more difficult. Problem 15 can be found, implicitly, in Andréka-Jónsson-Németi [2].

Acknowledgments: This work has been supported by the Hungarian National Foundation for Scientific Research Grant No. 1810.

References

1. H.Andréka-S.D.Comer-I.Németi, *Epimorphisms in cylindric algebras*, Preprint (1982).
2. H.Andréka-B.Jónsson-I.Németi, *Relatively free relation algebras*, submitted to Proc. Conf. on Algebraic Logic and Universal Algebra in Computer Science, Ames (USA), 1988.
3. J.Barwise-S.Feferman (Eds.), *Model- Theoretic Logics*, Springer-Verlag (1985), xviii+893pp.
4. C.Bergman-R.McKenzie, *On the relationship of* AP, RS, *and* CEP *in congruence modular varieties. II.*, Preprint (1987).
5. S.D.Comer, *Classes without the amalgamation property*, Pacific J. Math. Vol **28** (1969), 309–318.
6. S.D.Comer, *Epimorphisms in discriminator varieties*, In: Lectures in Universal Algebra (Colloq. Math. Soc. J. Bolyai Vol **43**), Eds.: L.Szabó, Á.Szendrei, North-Holland (1986), 41–48.
7. A.Daigneault, *Freedom in polyadic algebras and two theorems of Beth and Craig*, Michigan J. Math. Vol **11** (1964), 129–135.
8. R.Goldblatt, *Varieties of Complex Algebras*, Preprint Victoria University, New Zealand (1988).
9. P. R.Halmos, *Algebraic Logic*, Chelsea Publishing Co., New York (1962), 271pp.
10. L.Henkin-J.D.Monk-A.Tarski, *Cylindric Algebras Part I and Part II*, North-Holland (1985).
11. H.Herrlich-G.E.Strecker, *Category Theory*, Allyn and Bacon Inc. (1973).
12. J.S.Johnson, *Amalgamation of polyadic algebras*, Transactions of the American Math. Society Vol **149** (June 1970), 627–652.
13. B.Jónsson, *Extensions of relational structures*, In: The theory of models. Eds.: J.W.Addison, L.Henkin, A.Tarski, North-Holland (1965), 146–157.
14. H.J.Keisler, *A complete first-order logic with infinitary predicates*, Fundamenta Mathematicae Vol **52** (1963),.
15. E.W.Kiss-L.Márki-P.Prőhle-W.Tholen, *Categorical algebraic properties. Compendium on amalgamation, congruence extension, epimorphisms, residual smallness, and injectivity*, Studia Sci. Math. Hungar. Vol **18** (1983), 79–141.
16. R.J.Maddux, *Some varieties containing relation algebras*, Trans. Amer. Math. Soc. Vol **272** (1982), 501–526.
17. R.J.Maddux, *Pair-dense relation algebras*, Submitted.
18. J.A.Makowsky, *Compactness, Embeddings and Definability*, in [3] (1985), 645–716.
19. L.L.Maksimova, *Craig's theorem in superintuitionistic logics and amalgamated varieties of pseudo-Boolean algebras*, Algebra i logika Vol **16** (1977), 643–681.

20. R.McKenzie, *The representation of relation algebras*, Doctoral Dissertation, University of Colorado, Boulder (1966), vi+128pp.
21. J.D.Monk, *Studies in cylindric algebra*, Doctoral Dissertation, University of California, Berkeley (1961), vi+83pp.
22. J.D.Monk, *Singulary cylindric and polyadic equality algebras*, Trans. Amer. Math. Soc. Vol **112** (1964), 185–205.
23. I.Németi, *Surjectiveness of epimorphisms is equivalent to Beth definability property in general algebraic logic*, Manuscript (1984).
24. I.Németi, *Cylindric-relativized set algebras have strong amalgamation*, The Journal of Symbolic Logic Vol **50,** No **3** (Sept. 1985), 689–700.
25. I.Németi, *Decidability of the equational theory of cylindric relativized set algebras*, Preprint, Math. Inst. Hungar. Acad. Sci. (Also available as a chapter of [26].) (1985).
26. I.Németi, *Free algebras and decidability in algebraic logic*, Dissertation for DSc with the Hungarian Academy of Sciences, Budapest (1985).
27. I.Németi, *Logic with three variables has Gödel's incompleteness property — thus free cylindric algebras are not atomic*, Math. Inst. Hungar. Acad. Sci., Preprint No 49/85 (1985).
28. I.Németi, *Epimorphisms and definability in relation-, polyadic-, and related algebras*, Invited lecture at the "Algebraic Logic in Comp. Sci." Conference, Ames, Iowa (USA) June 1988. (Also available as Seminar Notes for the Algebra Seminar in Ames, Fall 1987.) (1988).
29. H.Ono, *Interpolation and the Robinson property for logics not closed under the Boolean operations*, Algebra Universalis **23** (1986), 111–122.
30. D.Pigozzi, *Amalgamation, congruence-extension, and interpolation properties in algebras*, Algebra Universalis Vol **1** fasc.3 (1972), 269–349.
31. I.Sain, *Strong amalgamation and epimorphisms of cylindric algebras and Boolean algebras with operators*, Math. Inst. Hungar. Acad. Sci., Preprint No 17/82. Conditionally accepted by Studia Logica (1982).
32. A.Tarski-S.Givant, *A formalization of set theory without variables*, American Mathematical Society, Colloquium Publications Vol **41** (1987).
33. A.Wroński, *On a form of equational interpolation property, Foundations in logic and linguistics. Problems and solutions*, Selected contributions to the 7th International Congress, Plenum Press, London (1984).

The Resolution Rule: An Algebraic Perspective

Zbigniew Stachniak[1]

1 Introduction

Since their introduction in 1965, proof systems based on Robinson's resolution rule [8] (we shall call them *resolution proof systems*) have been extensively studied in the context of finding natural and efficient proof systems to support a wide spectrum of computational tasks. A number of important applications of such systems have been found in areas such as artificial intelligence, logic programming, database theory, specification, verification, and synthesis of sequential and distributed systems (cf. [1,3,4]). With a few notable exceptions, however, these systems are constructed and their properties studied in an *ad hoc* manner even though there are theoretical grounds for the development of general techniques.

The goal of this paper is to show that propositional resolution proof systems can be conveniently introduced and studied within an algebraic framework. We view a resolution proof system as a finite universal algebra (of formulas) augmented with an inference operator. We prove a number of fundamental methodological results. In Section 3 we introduce the key definition of a resolution counterpart of a propositional logic, and give a characterization of structural logics with resolution counterparts. In Section 4 we discuss the role of strongly finite logics in the methodological studies on resolution proof systems, and show that for these logics resolution counterparts can be effectively constructed. Finally, in Section 5, we tackle some of the problems concerning the efficiency of theorem proving methods based on the resolution principle. In this context we study the notion of refutational approximation of a propositional logic, and investigate closure operators on classes of resolution proof systems.

Only resolution proof systems for propositional logics are considered in this paper. We assume that the reader is familiar with automated theorem proving and algebraic terminology as covered in texts such as [2,6,7,8,12].

[1]Department of Computer Science York University, North York, Canada

This paper is in final form and no version of it will be submitted for publication elsewhere.

2 Logical Preliminaries

In this section we briefly review basic notions concerning propositional logics. Further details can be found in [5,11,13,14].

PROPOSITIONAL LOGICS: By a ***propositional language*** we understand an absolutely free algebra $\mathcal{L} = \langle L, f_0, ..., f_n \rangle$ having a countably infinite set $Var(L) = \{p_0, p_1, ...\}$ of free generators. We shall refer to the elements of $Var(L)$ and L as ***propositional variables*** and ***formulas***, respectively. The operations $f_0, ..., f_n$ are called ***logical connectives***. For every natural number $n \geq 0, \mathcal{L}^{(n)}$ denotes the subalgebra of $\mathcal{L}$ generated by $p_0, ..., p_n$, and $L^{(n)}$ denotes its base set. If $X \subseteq L$, then Var(X) denotes the set of all variables occurring in formulas of X. For every formula $\alpha(p_0, ..., p_n)$, $\alpha(p_0/\alpha_0, ..., p_n/\alpha_n)$ denotes the formula obtained from α by simultaneous replacement of every occurrence of every variable $p_i, 0 \leq i \leq n$, by the corresponding formula α_i.

Let $\mathcal{L}$ be a propositional language. A mapping $C : 2^L \to 2^L$ is said to be a ***consequence operation*** on $\mathcal{L}$ if for every $X, Y \subseteq L$ the following conditions hold:

(i) $X \subseteq C(X) = C(C(X))$;
(ii) $X \subseteq Y$ implies $C(X) \subseteq C(Y)$.

We say that a consequence operation C is:

> ***structural***, if for every $X \subseteq L$, and every endomorphism e of $\mathcal{L}$, $e(C(X)) \subseteq C(e(X))$ (henceforth we shall call endomorphisms of $\mathcal{L}$ ***substitutions***);
>
> ***logically compact***, if for every $X \subseteq L, C(X) = L$ implies $C(X_f) = L$, for some finite $X_f \subseteq X$;
>
> ***compact***, if for every $X \subseteq L, C(X) = \bigcup\{C(X_f) : X_f \subseteq X$ and X_f is finite$\}$.

A set X is called ***C-inconsistent*** (***C-consistent***) if $C(X) = L$ $(C(X) \neq L)$. Finally, by a ***propositional logic*** we understand a pair $\mathcal{P} = \langle \mathcal{L}, C \rangle$, where C is a consequence operation on $\mathcal{L}$. Through this paper we assume that all logics under consideration are not pathological, i.e. they have at least one consistent and at least one finite inconsistent set.

MATRIX SEMANTICS: (cf. [5,14]) If $\mathcal{A}$ is an algebra then by $|\mathcal{A}|$ we denote the base set of $\mathcal{A}$. A pair $\mathcal{M} = \langle \mathcal{A}, \mathcal{D} \rangle$ is called a ***logical matrix*** for a propositional language $\mathcal{L}$ if $\mathcal{A}$ is an algebra similar to $\mathcal{L}$ and $\mathcal{D}$ is a nonempty family of subsets of $|\mathcal{A}|$. Elements of $|\mathcal{A}|$ are called ***logical values***, and the elements of every set $D \in \mathcal{D}$ the ***designated logical values*** of $\mathcal{M}$. Homomorphisms of $\mathcal{L}$ into $\mathcal{A}$ are called ***logical valuations*** and the set of all such valuations is denoted by $Hom(\mathcal{L}, \mathcal{M})$.

With each matrix $\mathcal{M} = \langle \mathcal{A}, \mathcal{D} \rangle$ we associate the consequence operation $C_{\mathcal{M}}$ on $\mathcal{L}$ (the so-called *matrix consequence operation*) defined as follows: for every $X \subseteq L$ and every $\alpha \in L$,

> $\alpha \in C_{\mathcal{M}}(X)$ *iff for every* $h \in Hom(\mathcal{L}, \mathcal{M})$ *and every* $D \in \mathcal{D}, h(\alpha) \in D$ *whenever* $h(X) \subseteq D$.

We say that a logic $\langle \mathcal{L}, C \rangle$ is *strongly finite* (SF) if there is a finite matrix $\mathcal{M}$ for $\mathcal{L}$ such that $C = C_{\mathcal{M}}$.

PROOF TREES: Let $\mathcal{L}$ be a propositional language. A binary relation of the form $r(X, Y)$, where X is a sequence of formulas and $Y \subseteq L$, will be called a *multiple-conclusion rule of inference* on $\mathcal{L}$.

Given a set $\mathcal{R}$ of multiple-conclusion rules on $\mathcal{L}$ and a set $X \cup \{\alpha\} \subseteq L$ we shall write $X \overset{\mathcal{R}}{\Rrightarrow} \alpha$ iff there is a finite tree T (a *proof tree* of α *from* X) which satisfies the following properties:

(i) every node N of T is labeled with a set X_N of formulas and the label of the root of T is X;
(ii) for every leaf node $N, \alpha \in X_N$;
(iii) if $N_0, ..., N_n$ are the children of N, then there is a rule $r \in \mathcal{R}$ and a pair $\langle Y, \{\alpha_0, ..., \alpha_n\} \rangle \in r$ such that:
  (a) the elements of the sequence Y are in X_N;
  (b) $X_{N_i} = X_N \cup \{\alpha_i\}$, all $0 \leq i \leq n$.

We shall read the expression '$X \overset{\mathcal{R}}{\Rrightarrow} \alpha$' as '$\alpha$ is derivable from X on the basis of $\mathcal{R}$'.

3 Resolution Proof Systems

Resolution proof systems are refutational deductive systems which means that during the deductive process we attempt to show that a certain set of formulas is inconsistent. If a logic $\mathcal{P} = \langle \mathcal{L}, C \rangle$ is associated with a resolution proof system Rs on $\mathcal{L}$, then to show that a formula α is a consequence of a finite set X in $\mathcal{P}$ we first convert X and α to a set X_α of formulas so that:

$$\alpha \in C(X) \textit{ iff } C(X_\alpha) = L.$$

Then, we use the inference rules of Rs to determine the consistency status of X_α. Let us note that if $\mathcal{P}$ is decidable, then X_α can always be effectively selected. We call X_α the *deduced set*. Hence, during the deductive process we apply the inference rules of Rs to formulas of the deduced set. This process terminates successfully if the deduced set contains a preselected inconsistent set of formulas.

Let $\mathcal{L}$ be a propositional language. By a *resolution proof system* on $\mathcal{L}$ we understand a deductive system of the form $Rs = \langle \mathcal{A}, \mathcal{F} \rangle$, where $\mathcal{A}$ is a finite algebra of formulas similar to $\mathcal{L}$ and $\mathcal{F}$ is a family of subsets of the base set Ver of $\mathcal{A}$.[2] Every resolution proof system Rs on $\mathcal{L}$ determines three types of inference rules in terms of which the deductive process in Rs is carried out. These rules are: the resolution rule, the transformation rules, and the $\Box$-rules. We shortly discuss these rules below.

THE RESOLUTION RULE: Let $v_0, ..., v_1$ be a fixed enumeration without repetitions of Ver. The *resolution rule* is the set of all sequents of the form:

$$Res_{Ver}: \quad \frac{\alpha_0(p), ..., \alpha_n(p)}{\alpha_0(p/v_0), ..., \alpha_n(p/v_n)},$$

where p is a variable which does not occur in formulas of Ver. Intuitively, if a set $X = \{\alpha_0(p), ..., \alpha_n(p)\}$ of formulas is consistent, then so is $X \cup \{\alpha_i(p/v_i)\}$, for some $0 \leq i \leq n$.

THE TRANSFORMATION RULES: A *transformation rule* is an expression of the form $f(v_1, ..., v_t) \Rightarrow v$, where f is a t-ary connective, $v_1, ..., v_t, v \in Ver$, and $f(v_1, ..., v_t) = v$ holds in $\mathcal{A}$. The role of transformation rules during the deductive process is to simplify formulas by replacing occurrences of $f(v_1, ..., v_t)$ with inferentially equivalent verifier v. We read the expression '$f(v_1, ..., v_t) \Rightarrow v$' as '$f(v_1, ..., v_t)$ reduces to v'.

THE $\Box$-RULES: A $\Box$-*rule* is an expression of the form $Y \Rightarrow \Box$, where $Y \in \mathcal{F}$ and $\Box$ is a designated constant symbol (not in L) denoting falsehood. The role of $\Box$-rules during the deductive process is to designate the elements of $\mathcal{F}$ as inconsistent.

Let $\mathcal{P} = \langle \mathcal{L}, C \rangle$ be a propositional logic and let Rs be a resolution proof system on $\mathcal{L}$. We say that a finite set X has a *refutation* in Rs (in symbols, $X \overset{Rs}{\Rightarrow} \Box$) if $\Box$ is derivable from X on the basis of the inference rules of Rs, i.e. if there is a proof tree of $\Box$ from X. We shall call Rs a *resolution counterpart* of $\mathcal{P}$ if the following three conditions are satisfied (cf. [10]):

(r1) for every finite $X \subseteq L, C(X) = L$ *iff* $X \overset{Rs}{\Rightarrow} \Box$, provided that formulas of X and Ver do not share variables;

(r2) if $w_0 \Rightarrow w_1$ is a transformation rule, then for every $\alpha(p) \in L$, $C(\alpha(p/w_0)) = C(\alpha(p/w_1))$;

(r3) for every $V \subseteq Ver, C(V) = L$ *iff* for some $U \in \mathcal{F}, U \subseteq V$.

The condition (r1), called *refutational completeness*, expresses the direct correspondence between finite inconsistent and refutable sets. We shall call

[2] In what follows, we shall identify the connectives of $\mathcal{L}$ with the respective operations of $\mathcal{A}$.

every set X satisfying $Var(X) \cap Var(Ver) = \emptyset$ *clean* in Rs. The restriction of resolution to clean sets only is merely technical; it enables us to give simple proofs of soundness of the resolution rule (cf. [11]). Moreover, in many cases we can associate simpler resolution proof systems with a given propositional logic. Let us note that the restriction in question only advises us to write formulas using variables that do not occur in verifiers or, equivalently, to treat variables occurring in verifiers as constants. The meaning of (r2) is obvious. The condition (r3) says that the closure $\mathcal{F}^*$ of $\mathcal{F}$ under supersets consists of all inconsistent sets of verifiers. For reasons of simplicity, in what follows we shall assume that $\mathcal{F} = \mathcal{F}^*$. Hence, (r3) can be rewritten as:

$$\text{for every } V \subseteq Ver, C(V) = L \text{ iff } V \in \mathcal{F}.$$

Before we present the main result of this section, the characterization of structural logics with resolution counterparts, we shall briefly discuss the notion of the matrix induced by a resolution proof system. Let $Rs = \langle \mathcal{A}, \mathcal{F} \rangle$ be a resolution proof system on $\mathcal{L}$. A set $d \subseteq |\mathcal{A}|$ is said to be *maximal consistent* in Rs if $d \notin \mathcal{F}$ and for every $d' \subseteq |\mathcal{A}|$, if $d' \notin \mathcal{F}$ and $d \subseteq d'$, then $d = d'$. By $\mathcal{D}$ let us denote the family of all maximal Rs-consistent sets. The matrix $M_{Rs} = \langle \mathcal{A}, \mathcal{D} \rangle$ is called the *matrix induced by Rs* and some of its properties are listed in the following lemma.

LEMMA 3.1: [10] Let Rs be a resolution counterpart of a structural logic $\langle \mathcal{L}, C \rangle$. Then:
(i) for every finite $X \subseteq L, C(X) = L$ *iff* $C_{M_{Rs}}(X) = L$.
(ii) Rs is a resolution counterpart of $\langle \mathcal{L}, C_{M_{Rs}} \rangle$.

EXAMPLE A: Let $\mathcal{P}_2 = \langle \mathcal{L}_2, C_2 \rangle$ be the classical propositional logic with the connectives $\neg$ (negation), $\vee$ (disjunction), and $\wedge$ (conjunction). Let 1 and 0 stand for $p \vee \neg p$ and $p \wedge \neg p$, respectively. Moreover, let $\mathcal{F}_2 = \{\{0\}, \{0,1\}\}$ and let $\mathcal{A}_2 = \langle \{0,1\}, \neg, \vee, \wedge \rangle$ be the two-element Boolean algebra (of truth-values) with usual definitions of the operations $\neg, \vee$, and $\wedge$. The resolution rule of Rs_2 has the form:

$$Res_{\{0,1\}} : \frac{\alpha_0(p), \alpha_1(p)}{\alpha_0(p/0), \alpha_1(p/1)}.$$

$Rs_2 = \langle \mathcal{A}_2, \mathcal{F}_2 \rangle$ is a resolution counterpart of $\mathcal{P}_2$ and we can justify this fact as follows. Let M be the matrix induced by Rs_2. Clearly, M is the two-element matrix adequate for $\mathcal{P}_2$ and hence, $C_2 = C_M$. By Lemma 3.1(ii), Rs_2 is a counterpart of $\mathcal{P}_2$. □

THEOREM 3.2: (O'HEARN, STACHNIAK) A structural logic $\langle \mathcal{L}, C \rangle$ has a resolution counterpart *iff* there is a natural number k and a subalgebra $\mathcal{A}$ of $\mathcal{L}^{(k)}$ such that:

(i) the quotient set $|\mathcal{A}|/\Theta_C$ is finite;

(ii) for every finite $X \subseteq L$, $C(X) = L$ *iff* for every $e \in Hom(\mathcal{L}, \mathcal{A})$, $C(e(X)) = L$,

where Θ_C is the congruence of $\mathcal{L}$ defined in the following way: $\alpha\Theta_C\beta$ *iff* $C(\gamma(p/\alpha)) = C(\gamma(p/\beta))$, all $\gamma \in L$.

PROOF: Let $\mathcal{P} = \langle \mathcal{L}, C \rangle$ be a structural propositional logic, and let $f_1, ..., f_l$ be the list of logical connectives of $\mathcal{P}$. We assume that for every $1 \leq i \leq l$, f_i is m_i-ary.

First, let us suppose that $Rs = \langle \mathcal{V}, \mathcal{F} \rangle$ is a resolution counterpart of $\mathcal{P}$. Let $k = card(Var(|\mathcal{V}|))$, and let $A = \{a \in L^{(k)} : a\Theta_C b$, for some $b \in |\mathcal{V}|\}$. Moreover, for every $1 \leq i \leq l$, let f_i^A be the function defined as follows: $f_i^A(a_1, ..., a_{m_i}) = f_i(a_1, ..., a_{m_i})$, for every $a_1, ..., a_{m_i} \in A$. We claim that every f_i^A is an operation on A. For, let $a_1, ..., a_{m_i} \in A$, and let $b_1, ..., b_{m_i} \in |\mathcal{V}|$ be such that for every $1 \leq j \leq m_i, b_j\Theta_C a_j$. Since Rs is a resolution counterpart of $\mathcal{P}$, there exists $b \in |\mathcal{V}|$ such that $f_i(b_1, ..., b_{m_i}) \Rightarrow b$ is a transformation and hence, $f_i(b_1, ..., b_{m_i})\Theta_C b$. This shows that $f_i^A(a_1, ..., a_{m_i}) = f_i(a_1, ..., a_{m_i})\Theta_C b$, and concludes the proof of the claim. Now, (i) follows immediately from the fact that $card(A/\Theta_C) \leq card(|\mathcal{V}|) < \omega$.

Let X be a finite C-inconsistent set, let e be a substitution, and let C_{Rs} denote the consequence operation defined by the matrix $M_{Rs} = \langle \mathcal{V}, \mathcal{D} \rangle$ induced by Rs. By Lemma 3.1(i), $C_{Rs}(X) = L$. Since C_{Rs} is structural, $C_{Rs}(e(X)) = L$ which, again by Lemma 3.1(i), shows that $C(e(X)) = L$.

To conclude the proof of (ii), let us assume that X is finite and C-consistent. Then there exist $h \in Hom(\mathcal{L}, M_{Rs}), d \in \mathcal{D}$, and $\alpha \in L$ such that

(a) $h(X) \subseteq d$ and $h(\alpha) \notin d$.

Let e_h be a substitution such that for every variable p, $e_h(p) = h(p)$. Clearly, $e_h \in Hom(\mathcal{L}, \mathcal{A})$. We claim that

(b) $C(e_h(X)) \subseteq C(h(X))$.

To prove the claim let us note first that for every $\beta \in L, e_h(\beta)\Theta_C h(\beta)$. Clearly, this fact is true if β is a propositional variable. Let us suppose that $\beta = f_i(a_1, ..., a_{m_i})$ and that the result is true for every $a_j, 1 \leq j \leq m_i$. Then

$$e_h(f_i(a_1, ..., a_{m_i})) = \\ f_i(e_h(a_1), ..., e_h(a_{m_i}))\Theta_C f_i(h(a_1), ..., h(a_{m_i}))\Theta_C h(f_i(a_1, ..., a_{m_i})).$$

Thus $e_h(\beta)\Theta_C h(\beta)$. From the definition of Θ_C it follows that $e_h(X) \subseteq C(h(X))$ which implies (b).

Next, let us observe that

(c) $C(h(X)) \cap |\mathcal{V}| \subseteq d$.

The truth of (c) follows from the fact that $h(X) \subseteq d$ and that $C(d) \cap |\mathcal{V}| = d$. From (b) and (c) we conclude that $C(e_h(X)) \cap |\mathcal{V}| \subseteq d$. If $C(e_h(X))$ were C-inconsistent, then we would have $C(e_h(X)) \cap |\mathcal{V}| = |\mathcal{V}| \not\subseteq d$, which is impossible. Hence, $C(e_h(X)) \neq L$, which completes our proof of the first half of this theorem.

For the other half, let us assume that for some k and for some subalgebra $\mathcal{A}$ of $\mathcal{L}^{(k)}$, the conditions (i) and (ii) are satisfied. Let $Rs = \langle \mathcal{V}, \mathcal{F} \rangle$ be a resolution proof system on $\mathcal{L}$ in which $|\mathcal{V}|$ consists of representatives of all classes of $|\mathcal{A}|/\Theta_C$, and for every connective f_i the corresponding operation f_i^V on $|\mathcal{V}|$ is defined by:

$$f_i^V(a_1, ..., a_{m_i}) = a \textit{ iff } f_i(a_1, ..., a_{m_i}) \Theta_C a,$$

for all $a_1, ..., a_{m_i}, a \in |\mathcal{V}|$. The family $\mathcal{F}$ consists of all C-inconsistent subsets of $|\mathcal{V}|$.

Let C_{Rs} be the consequence operation defined by the matrix $M_{Rs} = \langle \mathcal{V}, \mathcal{D} \rangle$ induced by Rs. Let $X \subseteq L$ be a finite set clean in Rs. We claim that

(d) $C_{Rs}(X) = L$ *iff* $C(X) = L$.

Let us consider two matrices: $\mathcal{M} = \langle \mathcal{A}, \{C(d) \cap |\mathcal{A}| : d \in \mathcal{D}\} \rangle$ and $\mathcal{M}^* = \langle \mathcal{A}, \{C(Y) \cap |\mathcal{A}| : Y \subseteq L\} \rangle$. Clearly, $C_{\mathcal{M}^*} \leq C_{\mathcal{M}}$, and since $\mathcal{M}^*$ is a submatrix of the matrix $\langle \mathcal{L}, \{C(X) : X \subseteq L\} \rangle$ adequate for $\mathcal{P}$, we have also $C \leq C_{\mathcal{M}^*}$. Let us note that the matrices M_{Rs} and $\mathcal{M}/\Theta_C$ are isomorphic, and therefore

$$C \leq C_{\mathcal{M}^*} \leq C_{\mathcal{M}} = C_{\mathcal{M}/\Theta_C} = C_{Rs}.$$

This shows that $C(X) = L$ implies $C_{Rs}(X) = L$.

Conversely, suppose that $C(X) \neq L$. By (ii), there is a substitution e such that $C(e(X)) \neq L$. Let $h_e \in Hom(\mathcal{L}, M_{Rs})$ be defined as follows: for every variable p, $h_e(p) = v$ *iff* $e(p) \Theta_C v$. It is an easy exercise to show that for every $\alpha \in L$:

(e) $e(\alpha) \Theta_C h_e(\alpha)$.

Since $C(C(e(X)) \cap |\mathcal{V}|) \subseteq C(e(X)) \neq L$, there is $d \in \mathcal{D}$ such that $C(e(X)) \cap |\mathcal{V}| \subseteq d$. From the definition of Θ_C and from (e) it follows that $h_e(X) \subseteq C(e(X)) \cap |\mathcal{V}| \subseteq d$, which demonstrates that $C_{Rs}(X) \neq L$. This finishes the proof of (d).

Now, let us note that for every $h \in Hom(\mathcal{L}, M_{Rs})$ and every $d \in \mathcal{D}$,

(f) $h(L) \not\subseteq d$.

Namely, if for some homomorphism h and some maximal consistent set $d, h(L) \subseteq d$, then, by (ii), $C(h(L)) = L$, which contradicts the assumption that $C(d) \neq L$. Following the proof of Theorem 4.6 in [10] we can show that (f) implies

$$C_{Rs}(X) = L \textit{ iff } X \overset{Rs}{\Rightarrow} \square,$$

which combined with (d) gives (r1).

(r2) can be demonstrated as follows. If $f_i(a_1, ..., a_{m_i}) \Rightarrow a$ is a transformation, then $f(a_1, ..., a_{m_i})\Theta_C a$ which, by the definition of Θ_C, translates to $C(\alpha(p/f(a_1, ..., a_k))) = C(\alpha(p/a))$, for every $\alpha \in L$.

Finally, to demonstrate (r3) it suffices to show that $\mathcal{F} \neq \emptyset$. Let $e \in Hom(\mathcal{L}, \mathcal{A})$ be an arbitrary homomorphism. Clearly, $e(L) \subseteq |\mathcal{A}|$, and by (ii), $C(|\mathcal{A}|) = L$. Since $|\mathcal{A}| \subseteq C(|\mathcal{V}|), C(|\mathcal{A}|) \subseteq C(|\mathcal{V}|)$, and hence $|\mathcal{V}|$ is C-inconsistent. □

EXAMPLE B: Let $\mathcal{L}_3$ be the propositional language obtained from $\mathcal{L}_2$ of Example A by the addition of the implication connective $\rightarrow$, and let $M_3 = \langle\{0, \frac{1}{2}, 1\}, \rightarrow, \vee, \wedge, \neg, \{1\}\rangle$ be the three-element Lukasiewicz matrix for $\mathcal{L}_3$. (cf. [13]). By $\mathcal{P}_3$ let us denote the three-valued Lukasiewicz logic, i.e. the logic determined by M_3, and let C_3 denote its consequence operation. Finally, let us define the congruence Θ of $\mathcal{L}_3^{(0)}$ in the following way: for every $\alpha, \beta \in L_3^{(0)}$,

$$\alpha\Theta\beta \text{ iff for every } h \in Hom(\mathcal{L}_3^{(0)}, M_3), h(\alpha) = h(\beta).$$

From Lemma 1 in [13] it follows that Θ and $\Theta_{C_{M_3}}$ restricted to $\mathcal{L}_3$ are identical and that $L_3^{(0)}/\Theta$ consists of twelve congruence classes. This shows that Theorem 3.2(i) is satisfied for $k = 0$ and $\mathcal{A} = \mathcal{L}_3^{(0)}$. Now, let $X \subseteq L$ be finite. If $C_3(X) = L_3$, then by structurality of C_3, $C_3(e(X)) = L_3$, for every $e \in Hom(\mathcal{L}_3, \mathcal{L}_3^{(0)})$. Conversely, if $C_3(X) \neq L_3$, then there is a valuation h in M_3 such that $h(X) \subseteq \{1\}$. Let e_h be the substitution defined as follows: for every $q \in Var(X)$,

$$e_h(q) = \begin{cases} p_0 \rightarrow p_0 & \text{if } h(q) = 1, \\ \neg(p_0 \rightarrow p_0) & \text{if } h(q) = 0, \\ p_0 & \text{if } h(q) = \frac{1}{2}. \end{cases}$$

Let h' be any valuation in M_3 such that $h'(p_0) = \frac{1}{2}$. Since $h'(e_h(X)) = h(X)$, $C_3(e_h(X)) \neq L_3$. This shows that Theorem 3.2(ii) is satisfied, and hence that $\mathcal{P}_3$ has a resolution counterpart. □

If C is a consequence operation on $\mathcal{L}$, then let I_C denote the characteristic consequence operation on $\mathcal{L}$ defined as follows: for every set $X \subseteq L, I_C(X) = X$, if $C(X) \neq L$, and $I_C(X) = L$, otherwise (cf. [9]).

COROLLARY 3.3: Let C_0, C_1 be two consequence operations on $\mathcal{L}$ such that $C_0 \leq C_1$ and $I_{C_0} = I_{C_1}$. Moreover, suppose that there exists a resolution counterpart of C_0. Then:

(i) C_1 has a resolution counterpart.

(ii) The cardinality of a minimal resolution counterpart of C_0 is greater than or equal to the cardinality of a minimal resolution counterpart of C_1.

4 Resolution Counterparts of SF Logics

Lemma 3.1(ii) states that every resolution counterpart of a structural logic constitutes also a resolution counterpart of a certain *SF* logic, and hence expresses the importance of *SF* logics in the methodological studies on resolution proof systems. In fact, in [11] it is shown that resolution counterparts of *SF* logics can be effectively constructed from the semantic descriptions of these logics. What follows is another proof of this fact based on the criterion presented in Theorem 3.2.

THEOREM 4: There is an effective procedure which for every *SF* logic $\mathcal{P}$ constructs a resolution counterpart of $\mathcal{P}$.

PROOF: Let $\mathcal{P} = \langle \mathcal{L}, C \rangle$ be a logic defined by a finite matrix of cardinality k. By Wójcicki's criterion of strong finiteness, for every $X \cup \{\alpha\} \subseteq L$, the following two conditions are satisfied:

(a) $\alpha \in C(X)$ *iff* for every $e \in Hom(\mathcal{L}, \mathcal{L}^{(k)}), e(\alpha) \in C(e(X))$;

(b) the quotient set $L^{(k)}/\Theta_C$ is finite

(cf. [13]). From (b) we immediately obtain the condition (i) of Theorem 3.2, whereas (a) and the existence of a finite C-inconsistent set give us Theorem 3.2(ii). Therefore, there exists a resolution counterpart $Rs = \langle \mathcal{V}, \mathcal{F} \rangle$ of $\mathcal{P}$. Moreover, from the proof of Theorem 3.2 we can assume that $\mathcal{V}$ and $\mathcal{L}^{(k)}/\Theta_C$ are isomorphic and that $\mathcal{F}$ consists of all C-inconsistent subsets of $|\mathcal{V}|$.

To show that Rs can be constructed effectively, it suffices to provide a construction of $\mathcal{V}$. To this end, let us arrange all formulas of $L^{(k)}$ in an arbitrary manner into a sequence

(c) $\alpha_0, \alpha_1, ..., \alpha_i, ...$

Let us select a finite subsequence $\beta_0, ..., \beta_r$ of (c) such that $L^{(k)}/\Theta_C = \{[\beta_0]_{\Theta_C}, ..., [\beta_r]_{\Theta_C}\}$. A subsequence $\beta_0, ..., \beta_r$ of (c) possesses the required property if the following two conditions are satisfied:

(d) for every variable $p_i, 0 \leq i \leq k$, there is a formula β_j in the sequence such that $p_i \Theta_C \beta_j$;

(e) for every m-ary connective f and for every $\beta_{i_1}, ..., \beta_{i_m}, 0 \leq i_1, ..., i_m \leq r$, there is a formula β_j in the sequence such that $f(\beta_{i_1}, ..., \beta_{i_m}) \Theta_C \beta_j$.

In light of Theorem 5.13 in [10], we can effectively determine whether two formulas $\alpha, \beta \in L^{(k)}$ are Θ_C-congruent or not, and hence, a subsequence in question satisfying both (d) and (e) can be found in a finite number of steps.

Finally, the operations of $\mathcal{V}$ are defined in the usual manner. Namely, for every m-ary connective f the corresponding operation $f^{\mathcal{V}}$ on $|\mathcal{V}|$ is defined by

$$f^V(a_1, ..., a_m) = a \textit{ iff } f(a_1, ..., a_m)\Theta_C a, \text{ all } a_1, ..., a_m, a \in |\mathcal{V}|. \qquad \square$$

EXAMPLE C: Let $\mathcal{P}_3$ and Θ be as in Example B. Let $Rs_3 = \langle \mathcal{A}_3, \mathcal{F}_3 \rangle$ be the resolution proof system on $\mathcal{L}_3$ defined in the following way. The set Ver of verifiers of Rs_3 consists of representatives of all congruence classes of $\mathcal{L}_3/\Theta$. The operations of $\mathcal{A}_3$ are defined as follows. For every $v_0, v_1, v_3 \in Ver, \neg v_0 = v_1$ if the congruence classes of $\neg v_0$ and v_1 are identical; $v_0 \vee v_1 = v_3$ if the congruence classes of $v_0 \vee v_1$ and v_3 are identical, etc. Finally, $\mathcal{F}_3$ consists of all C_3-inconsistent subsets of Ver. From the proof of Theorem 4 it follows that Rs_3 is a counterpart of $\mathcal{P}_3$. $\square$

5 Refutational Approximation of Propositional Logics

Let $\mathcal{P} = \langle \mathcal{L}, C \rangle$ be a propositional logic. In this section we study resolution proof systems on $\mathcal{L}$ that are refutationally stronger than C, i.e. we concentrate on the class

$$\mathcal{R}(C) = \{Rs : \textit{for every finite clean } X \textit{ in } Rs, \textit{ if } C(X) = L, \textit{ then } X \overset{Rs}{\Rightarrow} \square\}.$$

We assume that for every $Rs \in \mathcal{R}(C)$, the matrix M_{Rs} induced by Rs is proper, i.e. L is not satisfiable in M_{Rs}. Hence, every $Rs \in \mathcal{R}(C)$ is refutationally complete with respect to $C_{M_{Rs}}$ [3]. For instance, if Rs is a counterpart of a structural logic, then M_{Rs} is proper. There are at least two reasons for studying the class $\mathcal{R}(C)$. First, this class contains an array of resolution proof systems which can be used as refutational approximations of C. Second, $\mathcal{R}(C)$ provides us with a useful tool for expressing and exploring refutational properties of the logic.

5.1 REFUTATIONAL APPROXIMATION

The efficiency of resolution procedures depends, among others, on the size of a given resolution proof system, i.e. on the number of verifiers. Even

[3]The proof of this fact can follow the proof of Theorem 4.6 in [10].

minimal resolution counterparts of some propositional logics may prove to be too large to be efficiently implemented. One possible solution to the problem of the efficiency of resolution based theorem proving procedures is to select a finite class $\mathcal{K} \subseteq \mathcal{R}(C)$ of small systems which are refutationally as 'close' to C as possible. Systems in $\mathcal{K}$ can be then executed in parallel to determine the consistency status of deduced sets. The idea of refutational approximation of a propositional logic is similar to the application of logical matrices in automated theorem proving known as the node-pruning method (cf. [12]).

Let us introduce the key notion of this section. We say that a finite class $\mathcal{K} \subseteq \mathcal{R}(C)$ ***refutationally approximates*** C if for every finite set $X \subseteq L$ clean in all the systems of $\mathcal{K}$,

$$C(X) = L \text{ iff for every } Rs \in \mathcal{K}, X \overset{Rs}{\Rightarrow} \square.$$

We shall write '$RA(C)$' if C can be refutationally approximated by a finite set of resolution proof systems from $\mathcal{R}(C)$. Clearly, every non-empty class $\mathcal{K} \subseteq \mathcal{R}(C)$ approximates at least one logic, for instance the logic $\langle \mathcal{L}, C_{\mathcal{K}} \rangle$, where $C_{\mathcal{K}}$ is defined as follows: $C_{\mathcal{K}}(X) = L$, if there is a finite $X_f \subseteq X$ clean in all the systems of $\mathcal{K}$ and such that $X_f \overset{Rs}{\Rightarrow} \square$, otherwise $C_{\mathcal{K}}(X) = X$.

THEOREM 5.1: Let C be a structural consequence operation. Then the following conditions are equivalent:

(i) there is $Rs \in \mathcal{R}(C)$ such that $\{Rs\}$ refutationally approximates C;

(ii) there is a finite set $\mathcal{K} \subseteq \mathcal{R}(C)$ which refutationally approximates C.

PROOF: Since (i)⇒(ii) is trivial, let us suppose that there exists a finite set $\{Rs_i : i \leq n\} \subseteq \mathcal{R}(C)$ which refutationally approximates C. For every $i \leq n$, let M_i be the matrix induced by Rs_i. From Theorem 1 in [14] it follows that the consequence operation C_M determined by the direct product M of the matrices M_i satisfies the following condition: for every $X \subseteq L, C_M(X) = L$ iff for every $i \leq n, C_{M_i}(X) = L$. Since C_M is SF and logically compact, Theorem 4 guarantees the existence of a resolution counterpart $Rs = \langle \mathcal{A}, \mathcal{F} \rangle$ of $\langle \mathcal{L}, C_M \rangle$. To show that $\{Rs\}$ refutationally approximates C let us first note that:

(a) if C' is a structural consequence operation, X is a finite set of formulas, and e is an automorphism of $\mathcal{L}$, then $C'(X) = L$ *iff* $C'(e(X)) = L$.

Next, let X be finite and clean in Rs. Moreover, let us select an automorphism e of $\mathcal{L}$ such that the verifiers of every system in $\mathcal{K}$ as well as the verifiers of Rs do not share variables with formulas of $e(X)$. If $C(X) = L$, then by (a), $C(e(X)) = L$. This means that $C_{M_i}(e(X)) = L$ and hence that $C_M(e(X)) = L$. By (a) and by Lemma 3.1(ii) we may conclude that

$X \overset{Rs}{\Rightarrow} \square$. By a similar argument we can demonstrate that $C(X) \neq L$ implies $X \overset{Rs}{\not\Rightarrow} \square$. To conclude, $\{Rs\}$ refutationally approximates C. □

Although we are able to reduce finite refutational approximation classes to one element sets, this however can be done at the risk of increasing the size of resolution proof systems.

Refutational approximation can also be expressed in terms of properties of logical matrices. We say that a class $\mathcal{W}$ of matrices for $\mathcal{L}$ refutationally approximates C if for every finite set $X \subseteq L$,

$$C(X) = L \quad \textit{iff} \quad \bigcap\{C_M(X) : M \in \mathcal{W}\} = L.$$

We shall write '$RAM(C)$', if there is a finite class $\mathcal{W}$ of finite matrices which refutationally approximates C.

THEOREM 5.2: Let $\mathcal{P} = \langle \mathcal{L}, C \rangle$ be a structural logic. Then $RA(C)$ *iff* $RAM(C)$.

PROOF: This theorem immediately follows from Lemma 3.1 and Theorem 3.2. □

THEOREM 5.3: Suppose that C_0 and C_1 are two structural consequence operations on $\mathcal{L}$ such that $RAM(C_0)$ and $RAM(C_1)$ hold. Then:

(i) $\mathcal{R}(C_0) = \mathcal{R}(C_1)$ *iff* $I_{C_0}(X) = I_{C_1}(X)$, for every finite $X \subseteq L$.

(ii) If both C_0 and C_1 are logically compact, then $\mathcal{R}(C_0) = \mathcal{R}(C_1)$ *iff* $I_{C_0} = I_{C_1}$.

PROOF: (i) Suppose that for some finite $X \subseteq L, I_{C_0}(X) \neq X$, while $I_{C_1}(X) = L$. Let $Rs \in \mathcal{R}(C_0)$ be a resolution proof system such that $\{Rs\}$ refutationally approximates C_0. The existence of Rs is guaranteed by Theorems 5.1 and 5.2, and by the fact that $RAM(C_0)$ holds. We can assume that X is clean in Rs (otherwise we can rename variables occurring in verifiers of Rs). Clearly, $X \overset{Rs}{\not\Rightarrow} \square$ and $Rs \notin \mathcal{R}(C_1)$. The other part of (i) is obvious.

(ii) immediately follows from (i). □

5.2 Closure Operators on $\mathcal{R}(C)$

The search for refutation approximations of propositional logics by finite classes of 'small' resolution systems brings us to the study of algebraic operators on $\mathcal{R}(C)$ which reduce the size of resolution systems. This section is devoted to the study of some of these operators. We start with the following definitions.

Let $Rs_i = \langle \mathcal{A}_i, \mathcal{F}_i \rangle, i = 0, 1$, be two resolution proof systems on $\mathcal{L}$. A homomorphism h of $\mathcal{A}_0$ into $\mathcal{A}_1$ is said to be an *r-homomorphism* of Rs_0

into Rs_1 if $\mathcal{F}_1 = \{h(f) : f \in \mathcal{F}_0\}$. If in addition $a \in f$ iff $h(a) \in h(f)$, for every $a \in |\mathcal{A}_0|$ and every $f \in \mathcal{F}_0$, then h is called a ***strong homomorphism***. One-to-one and onto r-homomorphisms are called ***isomorphisms***. Rs_1 is said to be a ***subsystem*** of Rs_0 if $\mathcal{A}_1$ is a subalgebra of $\mathcal{A}_0$ and $\mathcal{F}_1 = \{f \cap |\mathcal{A}_1| : f \in \mathcal{F}_0\}$.

Finally, let Θ be a congruence of $\mathcal{A}_0$ consistent with $\mathcal{F}_0$, i.e. such that for every $f \in \mathcal{F}_0$ and every $a, b \in |\mathcal{A}_0|, a\Theta b$ implies $(a \in f \Leftrightarrow b \in f)$. Let $Rs_\Theta = \langle \mathcal{A}_\Theta, \mathcal{F}_\Theta \rangle$ be a resolution proof system on $\mathcal{L}$ such that $|\mathcal{A}_\Theta|$ consists of representatives of equivalence classes of $|\mathcal{A}_0|/\Theta, \mathcal{F}_\Theta = \{f \cap |\mathcal{A}_\Theta| : f \in \mathcal{F}_0\}$, and for every operation G of $\mathcal{A}_0$, the corresponding operation G_Θ of $\mathcal{A}_\Theta$ is defined in the following way:

$$G_\Theta(a_0, ..., a_k) = a \text{ iff } G(a_0, ..., a_k)\Theta a.$$

We call any resolution system Rs_Θ obtained in that way a *quotient* of Rs_0 modulo Θ.

Finally, we define the operators **H**, **S**, **Q** on classes of resolution proof systems on $\mathcal{L}$ in the following way:

H($\mathcal{K}$): r-homomorphic images of proof systems of $\mathcal{K}$;

S($\mathcal{K}$): substructures of systems of $\mathcal{K}$;

Q($\mathcal{K}$): quotients of systems of $\mathcal{K}$.

THEOREM 5.4: Let h be an r-homomorphism of Rs_0 onto Rs_1 and let C_0, C_1 be the consequence operations defined by the matrices induced by Rs_0 and Rs_1, respectively. Then, $I_{C_0} \leq I_{C_1}$.

PROOF: Let $M_i = \langle \mathcal{A}_i, \mathcal{D}_i \rangle, i = 0, 1$, be the matrices induced by Rs_0 and Rs_1, respectively. Suppose that for some $X \subseteq L, C_1(X) \neq L$. Then there is $h_1 \in Hom(\mathcal{L}, M_1)$ and $d_1 \in \mathcal{D}_1$, such that $h_1(X) \subseteq d_1$. Let $h_0 \in Hom(\mathcal{L}, M_0)$ be a homomorphism such that for every variable p, $h(h_0(p)) = h_1(p)$. One can easily show that for every $\alpha \in L$,

(1) $h(h_0(\alpha)) = h_1(\alpha)$.

Since $h_1(X) \subseteq d_1$, we have $h^{-1}(h_1(X)) \subseteq h^{-1}(d_1)$ and, by (1), $h_0(X) \subseteq h^{-1}(d_1)$. If $h^{-1}(d_1)$ were in $\mathcal{F}_0$, then $h(h^{-1}(d_1)) = d_1 \in \mathcal{F}_1$, which is impossible. Hence, there is a maximal consistent set d_0 in Rs_0 containing $h_0(X)$. Since L is not satisfiable in $M_0, C_0(X) \neq L$. □

COROLLARY 5.5: $\mathbf{H}(\mathcal{R}(C)) \subseteq \mathcal{R}(C)$.

THEOREM 5.6: Let Rs_1 be a substructure of Rs_0, and let C_0 and C_1 be the consequence operations defined by the matrices induced by Rs_0 and Rs_1, respectively. Then, $I_{C_0} \leq I_{C_1}$.

PROOF: Let $Rs_i = \langle \mathcal{A}_i, \mathcal{F}_i \rangle$ and $C_i, i = 0,1$, be as required, and let $A = |\mathcal{A}_0| - |\mathcal{A}_1|$. If $\mathcal{D}_1 = \{\emptyset\}$, then $I_{C_1}(X) \neq L$ iff $X = \emptyset$, and the theorem holds.

Let us suppose that $\mathcal{D}_1 \neq \{\emptyset\}$. We claim that for every $d_1 \in \mathcal{D}_1$,

(1) $d_1 = d_0 \cap |\mathcal{A}_1|$,

for some $d_0 \in \mathcal{D}_0$. Let $d_1 \in \mathcal{D}_1$. Let us note that $d_1 \cup A \notin \mathcal{F}_0$ (otherwise $d_1 = (d_1 \cup A) \cap |\mathcal{A}_1| \in \mathcal{F}_1$). Hence, there exists $d_0 \in \mathcal{D}_0$ such that $d_1 \cup A \subseteq d_0$. We show that

(2) $d_0 \cap |\mathcal{A}_1| \in \mathcal{D}_1$.

Suppose (2) is false. We have two cases.

CASE 1: Suppose that there is $d' \in \mathcal{D}_1$ such that $d_0 \cap |\mathcal{A}_1|$ is properly included in d'. Then d' also properly includes d_1, which is impossible, since d_1 is maximal consistent.

CASE 2: Suppose that $d_0 \cap |\mathcal{A}_1| \in \mathcal{F}_1$. Then for some $f \in \mathcal{F}_0, d_0 \cap |\mathcal{A}_1| = f \cap |\mathcal{A}_1|$. Since $A \subseteq d_0, f \subseteq d_0$, which is impossible.

Therefore (2) is true. To show (1) let us note only that $d_1 \subseteq d_0 \cap |\mathcal{A}_1|$ and that d_1 is maximal consistent in Rs_1.

Using (1) the theorem can be demonstrated as follows. Let $X \cup \{\alpha\} \subseteq L$ be such that $\alpha \notin C_1(X)$. Then for some $h \in Hom(\mathcal{L}, M_1)$ and some $d_1 \in \mathcal{D}_1, h(X) \subseteq d_1$ while $h(\alpha) \notin d_1$. By (1), there is $d_0 \in \mathcal{D}_0$, such that $h(X) \subseteq d_0$ and $h(\alpha) \notin d_0$. This shows that $\alpha \notin C_0(X)$. □

COROLLARY 5.7: $\mathbf{S}(\mathcal{R}(C)) \subseteq \mathcal{R}(C)$.

Before we state results concerning congruence relations let us introduce the following notation. If $Rs_\Theta = \langle \mathcal{A}_\Theta, \mathcal{F}_\Theta \rangle$ is a quotient of $Rs = \langle \mathcal{A}, \mathcal{F} \rangle$ modulo Θ and $d \subseteq |\mathcal{A}|$, then let $r(d) = \{a \in |\mathcal{A}_\Theta| : \text{for some } b \in d, a\Theta b\}$. In other words, $r(d)$ denotes the set of representatives of congruence classes in d/Θ.

LEMMA 5.8: Let Θ be a congruence of a resolution proof system $Rs = \langle \mathcal{A}, \mathcal{F} \rangle$, and let $M = \langle \mathcal{A}, \mathcal{D} \rangle$ be the matrix induced by Rs. Then for every $d_\Theta \in \mathcal{D}_\Theta$, there exists $d \in \mathcal{D}$ such that $d_\Theta = r(d)$ and for every $b \in d, [b]_\Theta \subseteq d$.

PROOF: Let $d_\Theta \in \mathcal{D}_\Theta$, and let $d = \{a \in |\mathcal{A}| : \text{for some } b \in d_\Theta, a\Theta b\}$. Clearly, $d_\Theta = r(d) \cap |\mathcal{A}_\Theta|$. We claim that $d \in \mathcal{D}$. Let us note that $d \notin \mathcal{F}$ (otherwise d_Θ would be in $\mathcal{F}_\Theta$). Suppose now that for some $b \notin d, d \cup \{b\} \notin \mathcal{F}$. Then $r(d \cup \{b\}) = d_\Theta \cup r(\{b\}) \in \mathcal{F}_\Theta$. This means, however, that $d \cup \{b\} \in \mathcal{F}$ – contradiction. To conclude, $d \in \mathcal{D}$.

Finally, using a similar argument we can demonstrate that for every $b \in d, [b]_\Theta \subseteq d$.

□

THEOREM 5.9: $\mathbf{Q}(\mathcal{R}(C)) \subseteq \mathcal{R}(C)$.

PROOF: Let Θ be a congruence of a resolution system $Rs \in \mathcal{R}(C)$. Let M and M_Θ denote the matrices induced by Rs and Rs_Θ, respectively, for some quotient Rs_Θ of Rs. We claim that for every $X \subseteq L$,

(1) $C_M(X) = L$ implies $C_{M_\Theta}(X) = L$.

Let $\alpha \notin C_{M_\Theta}(X)$, i.e. let $h \in Hom(\mathcal{L}, M_\Theta)$, and let $d_\Theta \in \mathcal{D}_\Theta$ be such that $h(X) \subseteq d_\Theta$ while $h(\alpha) \notin d_\Theta$. Let $h' \in Hom(\mathcal{L}, M)$ be defined as follows: for every $p \in Var(L), h'(p) = h(p)$. Then for every $\alpha \in L$,

(2) $h'(\alpha)\Theta h(\alpha)$.

By Lemma 5.8, $d_\Theta = r(d)$, for some $d \in \mathcal{D}$. Since $d_\Theta \subseteq d, h'(X) \subseteq d$ and since $h(\alpha) \notin d$, by (2), $h'(\alpha) \notin d$. This proves (1) from which the theorem follows immediately. □

As it has been mentioned at the beginning of this section the study of closure operators on $\mathcal{R}(C)$ is motivated by the search for refutation approximations of propositional logics consisting of small resolution proof systems. All three operators discussed in this section can be used in such search since they may substantially reduce the size of a resolution proof system. They can also be used in the study of resolution counterparts of structural logics. For instance, if Rs is a minimal counterpart of a structural logic $\langle \mathcal{L}, C_0 \rangle$ which is refutationally maximal, that is, for no structural consequence operation C_1, I_{C_0} is strictly smaller than I_{C_1}, then Rs has no proper subsystems and every strong homomorphic image of Rs is r-isomorphic with Rs. Since the construction of a minimal resolution counterpart can be computationally expensive (cf. [10]), the study of **H**, **S**, **Q**, and other closure operators may exhibit the structure and properties of minimal resolution counterparts of structural logics.

Acknowledgements: This research was supported by NSERC under grant A9136.

6 References

[1] M. ABADI AND Z. MANNA: A Timely Resolution, *Stanford University TR* STAN-CS-86-1106 (1986).

[2] G. GRÄTZER: *Universal Algebra*, D. Van Nostrand (1968).

[3] L. HENSCHEN AND A. NAQVI: Representing Infinite Sequences of Resolvents in Recursive First-Order Horn Databases, *LNCS* 138 (1982), 342-359.

[4] K. KONOLIGE: Resolution and Quantified Epistemic Logics, *LNCS* 230 (1986), 199-208.

[5] J. ŁOŚ AND R. SUSZKO: Remarks on Sentential Logics, *Indagationes Mathematicae* 20 (1958), 177-183.

[6] Z. MANNA AND R. WALDINGER: Special Relations in Automated Deduction, *JACM* 33 (1986), 1-59.

[7] N. MURRAY: Completely Non-Clausal Theorem Proving, *Artificial Intelligence* 18 (1982), 67-85.

[8] J.A. ROBINSON: A Machine-Oriented Logic Based on the Resolution Principle, *JACM* 12 (1965), 23-41.

[9] Z. STACHNIAK: Some Notes on Characteristic Consequence Operations, *Bulletin of The Section of Logic, PAN* 7 (1978), 159,168.

[10] Z. STACHNIAK: Minimization of Resolution Proof Systems (to appear); see also *York University TR* CS-88-07 (1988).

[11] Z. STACHNIAK AND P. O'HEARN: Resolution in the Domain of Strongly Finite Logics, to appear in *Fundamenta Informaticae* (1989); see also *York University Technical Report* #CS-87-14 (1987).

[12] P. THISTLEWAITE, M. MCROBBIE AND R. MEYER: *Automated Theorem-Proving in Non-Classical Logics*, Pitman (1988).

[13] R. WÓJCICKI: Strongly Finite Sentential Calculi, in *Selected Papers on Lukasiewicz Sentential Calculi* (ed. R. Wójcicki) (1977), 53-77.

[14] J. ZYGMUNT: A Note on Direct Products and Ultraproducts of Logical Matrices, *Studia Logica* 33 (1974), 349-357.

Incremental Models of Updating Data Bases

Marek A. Suchenek[1]

ABSTRACT

This paper introduces a generalization of weak model-theoretic forcing of [Rob71] and [Kei73]. This generalized forcing preserves classic properties of weak model-theoretic forcing, e.g. Generic Set Theorem (Thm. 3.23), Generic Model Theorem (Thm. 3.24, Thm. 4.8), and Henrard's Theorem (Thm. 5.8). It is applied in this paper to investigate a deductive model for updating a deductive data base with incomplete information, whose possible variations are restricted to certain finite sets of atomic or negated atomic first-order sentences. Moreover, the paper introduces the notion of pragmatic truth pertinent to those models, and characterizes it in terms of generalized forcing (Thm. 5.11).

In conclusion, the paper offers (Thm. 8.4) two semantic and two syntactic characterizations of the $\forall$-fragment of minimal entailment.

1 Introduction

Non-standard treatment of negation has a short but tempestuous history in Artificial Intelligence. A wide spectrum of solutions, from *negation as failure* to *closed world assumption* to *circumscription* to *non-monotonic logics*, has been proposed as a remedy for lack of complete information in the process of machine reasoning. Up to now, none of them seems fully satisfactory.

In this paper we present a different approach which is based on a syntactic construction of prioritized, conservative expansion. Inspired by ideas of Henrard [Hen73], we employ conservative expansions to introduce a family of weak generalized model-theoretic forcings. These forcings show promise as useful tools for formulation and analysis of reasoning paradigms pertinent to non-standard negation.

In order to describe their possible applications to the theory of *update* in deductive data bases, we consider a forcing-based incremental model of a data base introduced in [Suc84]. In this model the data base consists of a fixed part Σ, represented by a set of universally quantified sentences in a first-order language L, and of a variable part p (called increment in this paper), represented by a set of atomic and/or negated atomic sentences of L. The range of variableness of p, fixed for every particular model and

[1]Dept. of Computer Science, Wichita State University, Wichita KA 67208

for every Σ, defines the class of updates which may virtually happen in the lifetime of the data base. The fact that not all finite sets of atomic and negated atomic sentences are allowed as increments motivates us to introducing the notion of ***pragmatic truth***.

Pragmatic truth is the key concept in our model of updating data base. It appears, at least within this model, to be more adequate and useful than the classic notion of truth. Therefore, we construct a family of non-monotonic logics for reasoning from data bases of the form $\Sigma \cup p$, complete with respect to pragmatic truth within the scope of universal sentences. As the reader will see, ***not*** in these logics means ***it cannot happen that***.

In the last part of the paper we focus on two particular cases of such general models. Firstly, we allow as increments finite sets of atomic or negated atomic sentences of L. The weak model-theoretic forcing of [Rob71], a special instance of our general one, forms an adequate basis for this case. This case was a subject of initial study in [Suc84] and [Suc89c]. Secondly, we restrict increments to negative sentences, obtaining a proper generalization of the closed world data bases of [Rei78] and [Min82]. In this second case, negative forcing, a construction formally beyond the general scheme of [Kei73], proves its usefulness. In particular, we demonstrate a direct relationship between pragmatic truth, closed world assumption, and another well known source of non-monotonism and non-standard behavior of negation: the minimal model semantics. These results, which we prove by applying our negative forcing technique, are refinements of those known in the literature, e.g. [She88].

The paper is organized as follows. Sections 2 and 3 provide technical results used in the rest of the paper. They introduce concepts of generalized forcing and pragmatic truth, and investigate their abstract properties. Section 4 concerns a special case of generalized forcing, whose instances are used in the rest of the paper. Sections 5, 6, and 7 focus on models of updating deductive data base. The role of pragmatic truth in these models and its relationships with generalized forcing are exposed. A strong and consistent version cwa_S of the closed world assumption is characterized there. Section 8 investigates properties of minimal model semantics. Its main result (Thm. 8.4) states that for $\forall$-sentences, minimal entailment is equivalent to positive pragmatic entailment, to negative forcibility, and to entailment under cwa_S.

2 Prioritized Conservative Expansions

We consider a first-order countable language L. We usually follow the notations of [Bar78], Chaps. A1, A2, A4, and B1. Because we have to use logical connectives and quantifiers both in L and in the language of this article, outside L we occasionally use $\bigwedge$, $\bigvee$, and $\sim$ in the sense of bounded quantifiers (*for all* and *there exists*, respectively) and negation. E.g.,

$$\bigwedge_{c \in C} \sim \Sigma \vdash (\forall x)(\varphi(x,c))$$

means: "for all $c \in C$, Σ does not prove $(\forall x)(\varphi(x,c))$". For the purpose of saving paper we will use also $(\bigwedge_{c \in C}) \sim (\Sigma \vdash (\forall x)(\varphi(x,c)))$ with the same meaning. Moreover, we apply boldface type to indicate the least binding logical connective in formulas from outside L.

Through this paper Σ denotes a **consistent** theory in L. By Σ_Γ we denote the set $Cn(\Sigma) \cap \Gamma$. In particular, Σ_L means the closure of Σ under first-order provability relation $\vdash$ in L.

We start by introducing the key technical concept used in the paper. For the reader familiar with model theory it may be helpful to know that this concept generalizes the notion of *companion operator* (see [Bar78], Chap. A4, Sec.3).

Let η be a (transfinite) enumeration of certain sentences of L, i.e., $\eta : \kappa \to L$, where κ is a countable ordinal number.

Definition 2.1 *of η-prioritized Γ-conservative expansion $\Sigma^{\Gamma,\eta}$ of Σ* (by induction on ordinal numbers from $Dm(\eta)$).

$$\Sigma_0^{\Gamma,\eta} = \Sigma$$

$$\Sigma_{\alpha+1}^{\Gamma,\eta} = \begin{cases} \Sigma_\alpha^{\Gamma,\eta} \cup \{\eta_\alpha\} & \text{if } (\Sigma_\alpha^{\Gamma,\eta} \cup \{\eta_\alpha\})_\Gamma = \Sigma_\Gamma \\ \Sigma_\alpha^{\Gamma,\eta} & \text{otherwise} \end{cases}$$

$$\Sigma_\lambda^{\Gamma,\eta} = \bigcup \{\Sigma_\alpha^{\Gamma,\eta} \mid \alpha \in \lambda\}, \text{ where } \lambda \text{ is a positive limit ordinal}$$

$$\Sigma^{\Gamma,\eta} = \bigcup \{\Sigma_\alpha^{\Gamma,\eta} \mid \alpha \in Dm(\eta)\}.$$

□

Let us state some basic properties of $\Sigma^{\Gamma,\eta}$.

Theorem 2.2 (i) If *false* $\in \Gamma$ **then** $\Sigma^{\Gamma,\eta}$ is consistent.

(ii) If $\Lambda = \{\varphi \wedge \psi \mid \varphi, \psi \in \Gamma\}$ **then** $\Sigma^{\Gamma,\eta} = \Sigma^{\Lambda,\eta}$.

(iii) $\Sigma^{L,\eta} = \Sigma_{Rg(\eta)}$.

(iv) $(\Sigma^{\Gamma,\eta})_\Gamma = \Sigma_\Gamma$.

(v) For each $\varphi \in Rg(\eta)$, $\varphi \in \Sigma^{\Gamma,\eta}$ **iff** $(\Sigma^{\Gamma,\eta} \cup \{\varphi\})_\Gamma = \Sigma_\Gamma$.

(vi) If $Rg(\eta) = L$ **then** $\Sigma^{\Gamma,\eta}$ is closed under $\vdash$.

(vii) If $Rg(\eta) = L$ **then** $\Sigma^{\{false\},\eta}$ forms a complete theory.

(viii) $\Sigma \subseteq \Sigma^{\Gamma,\eta}$.

(ix) If η^{-1} and ζ^{-1} coincide on $L \setminus \Gamma$ **then** $\Sigma^{\Gamma,\eta} = \Sigma^{\Gamma,\zeta}$.

(x) If $Rg(\xi) \subseteq Rg(\eta)$ **then** $(\Sigma^{\Gamma,\eta})^{\Gamma,\xi} = \Sigma^{\Gamma,\eta}$.

(xi) If $\Sigma \subseteq \Theta \subseteq \Pi$ and $\Sigma^{\Gamma,\eta} = \Pi^{\Gamma,\eta}$ **then** $\Sigma^{\Gamma,\eta} = \Theta^{\Gamma,\eta}$.

(xii) If $Rg(\xi) \subseteq Rg(\eta)$ **then** $\Sigma^{\Gamma,\eta} \subseteq \Sigma^{\Gamma,\xi}$ implies $\Sigma^{\Gamma,\eta} = \Sigma^{\Gamma,\xi}$.

Proof. (i) Suppose $\Sigma^{\Gamma,\eta}$ is inconsistent. Because $\Sigma_0^{\Gamma,\eta}$ is consistent, by compactness theorem (cf e.g. [Bar78] Chap. A1, thm. 2.4) there is $\alpha \in Dm(\eta)$ such that $\Sigma_{\alpha+1}^{\Gamma,\eta}$ is inconsistent, but $\Sigma_\alpha^{\Gamma,\eta}$ is consistent. Hence *false* $\in (\Sigma_\alpha^{\Gamma,\eta} \cup \{\eta_\alpha\})_L$. Since *false* $\notin \Sigma_\Gamma$ then *false* $\notin \Gamma$.

(ii) For each $\Pi \subseteq L$, $\Pi_\Gamma \subseteq \Sigma_\Gamma$ iff $\Pi_\Lambda \subseteq \Sigma_\Lambda$. Then apply induction.

(iii) For each $\Pi \subseteq Rg(\eta), \Pi_L \subseteq \Sigma_L$ iff for each $\varphi \in \Pi$, $\Sigma \vdash \varphi$. Then apply induction.

(iv) is obvious.

(v) Let $\varphi = \eta_\alpha$. We have $\varphi \in \Sigma^{\Gamma,\eta}$ iff $(\Sigma_\alpha^{\Gamma,\eta} \cup \{\varphi\})_\Gamma \subseteq \Sigma_\Gamma$, which holds iff $(\Sigma^{\Gamma,\eta} \cup \{\varphi\})_\Gamma \subseteq \Sigma_\Gamma$.

(vi) If $\Sigma^{\Gamma,\eta} \vdash \eta_\alpha$ then by (iv) $(\Sigma^{\Gamma,\eta} \cup \{\eta_\alpha\})_\Gamma = \Sigma_\Gamma$, hence $(\Sigma_\alpha^{\Gamma,\eta} \cup \{\eta_\alpha\})_\Gamma \subseteq \Sigma_\Gamma$.

(vii) $(\Sigma_\alpha^{\{false\},\eta} \cup \{\eta_\alpha\})_{\{false\}} \subseteq \Sigma_{\{false\}}$ means that $\Sigma_\alpha^{\{false\},\eta} \cup \{\eta_\alpha\}$ is consistent. Therefore for each φ, either $\varphi \in \Sigma^{\Gamma,\eta}$ or $\neg\varphi \in \Sigma^{\Gamma,\eta}$.

(viii) is obvious.

(ix) By (iv), elements of Γ are in $\Sigma^{\Gamma,\eta}$ iff they are provable from Σ, so, if one moves them to the beginning of η, resulting enumeration ξ satisfies $\Sigma^{\Gamma,\eta} = \Sigma^{\Gamma,\xi}$. The same argument shows that $\Sigma^{\Gamma,\zeta} = \Sigma^{\Gamma,\xi}$.

(x) By (viii) we have $\Sigma^{\Gamma,\eta} \subseteq (\Sigma^{\Gamma,\eta})^{\Gamma,\xi}$. Let $\varphi \in (\Sigma^{\Gamma,\eta})^{\Gamma,\xi}$. Of course, $\varphi \in Rg(\xi)$, and therefore $\varphi \in R(\eta)$. By (v) we have $((\Sigma^{\Gamma,\eta})^{\Gamma,\xi} \cup \{\varphi\})_\Gamma = (\Sigma^{\Gamma,\eta})_\Gamma$, and by (iv), $((\Sigma^{\Gamma,\eta})^{\Gamma,\xi} \cup \{\varphi\})_\Gamma = \Sigma_\Gamma$. Hence $\Sigma^{\Gamma,\eta} \subseteq (\Sigma^{\Gamma,\eta})^{\Gamma,\xi}$ implies $(\Sigma^{\Gamma,\eta} \cup \{\varphi\})_\Gamma \subseteq \Sigma_\Gamma$. Since, by (iv), $\Sigma_\Gamma \subseteq (\Sigma^{\Gamma,\eta} \cup \{\varphi\})_\Gamma$, therefore $(\Sigma^{\Gamma,\eta} \cup \{\varphi\})_\Gamma = \Sigma_\Gamma$. Using (v) again we obtain that $\varphi \in \Sigma^{\Gamma,\eta}$.

(xi) By (iv) we have $\Sigma_\Gamma = \Pi_\Gamma$. On the other hand, $\Sigma_\Gamma \subseteq \Theta_\Gamma \subseteq \Pi_\Gamma$, therefore $\Theta_\Gamma = \Sigma_\Gamma$. Assume for some $\alpha \in Dm(\eta)$ that $\Sigma_\alpha^{\Gamma,\eta} \subseteq \Theta_\alpha^{\Gamma,\eta} \subseteq \Pi_\alpha^{\Gamma,\eta}$. If $\eta_\alpha \in \Sigma^{\Gamma,\eta}$ then $\eta_\alpha \in \Pi^{\Gamma,\eta}$; hence $(\Pi^{\Gamma,\eta} \cup \{\eta_\alpha\})_\Gamma = \Pi_\Gamma$, and therefore $(\Theta_\alpha^{\Gamma,\eta} \cup \{\eta_\alpha\})_\Gamma = \Theta_\Gamma$, which, by (v), implies $\eta_\alpha \in \Theta^{\Gamma,\eta}$. If $\eta_\alpha \in \Theta^{\Gamma,\eta}$ then $(\Sigma_\alpha^{\Gamma,\eta} \cup \{\eta_\alpha\})_\Gamma = \Sigma_\Gamma$, hence, by (v), $\eta_\alpha \in \Sigma^{\Gamma,\eta}$, and $\eta_\alpha \in \Pi^{\Gamma,\eta}$. Thus $\Sigma_{\alpha+1}^{\Gamma,\eta} \subseteq \Theta_{\alpha+1}^{\Gamma,\eta} \subseteq \Pi_{\alpha+1}^{\Gamma,\eta}$. Then apply induction.

(xii) We have $\Sigma \subseteq \Sigma^{\Gamma,\eta} \subseteq \Sigma^{\Gamma,\xi}$. By (x), $\Sigma^{\Gamma,\xi} = (\Sigma^{\Gamma,\xi})^{\Gamma,\xi}$. Therefore by (xi), $\Sigma^{\Gamma,\xi} = (\Sigma^{\Gamma,\eta})^{\Gamma,\xi}$. Using (x) again, we obtain $\Sigma^{\Gamma,\xi} = \Sigma^{\Gamma,\eta}$. □

One may observe that the operation $^{\Gamma,\eta}$ is a generalization of *normal default* of [Rei80]. For example, Reiter's rule

$$\frac{\psi : M\varphi}{\varphi}$$

may be equivalently expressed by putting $\eta = \langle \varphi \wedge \psi \rangle$ and $\Gamma = \{\psi, false\}$. However, the operation $^{\Gamma,\eta}$ is more powerful in its general form.

3 Generalized Forcing and Pragmatic Truth

We focus our attention on cases of $\Gamma \subseteq \forall$ which are generated by some classes of atomic and negated atomic sentences. We use, occasionally, two restrictive assumptions on Σ: an equivalence of Σ to a set of universal sentences, and the existence of an infinite set C of constants of L, not appearing in Σ. We make them easily visible by using symbols $\boxed{\forall}$ and $\boxed{C}$, respectively. Moreover, we require $Rg(\eta) = L$ from this point on.

We say that enumeration η of sentences of L is *consistent with Kleene-Mostowski hierarchy* iff for every $n \in \omega$ exists an ordinal number $\alpha_n \in Dm(\eta)$ such that $Rg(\eta \restriction \alpha_n) = \forall_{n-1}$, where $\forall_{-1}$ denotes the set of all quantifier-free sentences of L.

Let Δ be a class of certain finite sets of atomic and/or negated atomic sentences of L, containing the empty set 0, closed under $\cup$ and substitutions of constants from C, and such that for every atomic sentence φ of L, either $\{\varphi\}$ or $\{\neg\varphi\}$ is in Δ. Let $\Gamma(\Delta)$ be the least class containing $\bigvee n(\Delta) = \{\bigvee np \mid p \in \Delta\}$ (where $\bigvee np$ means the disjunction of complements[2] of elements of p) closed under $\vee$ and under introducing of quantifier $\forall$ (i.e. if $\varphi(c) \in \Gamma(\Delta)$ then $\forall x\varphi(x) \in \Gamma(\Delta)$). E.g. because $0 \in \Delta$, $\bigvee n0$ (the *false* formula) $\in \Gamma(\Delta)$. Also, for every atomic sentence φ, either φ or $\neg\varphi$ is in $\Gamma(\Delta)$. Let Σ be a consistent theory, such that infinitely many constants of C does not appear in Σ, and let $Cond_\Delta(\Sigma)$ denote the class of all these elements of Δ which are consistent with Σ. We start from a simple observation.

Lemma 3.1 $Cond_\Delta(\Sigma) = Cond_\Delta(\Sigma_{\Gamma(\Delta)}) = Cond_\Delta((\Sigma_{\Gamma(\Delta)})^{\Gamma(\Delta),\eta})$

Proof. For each $\Pi \subseteq L$, $\Pi \cup p$ is inconsistent iff $\Pi \vdash \bigvee np$. Since $\bigvee np \in \Gamma(\Delta)$ then $\Sigma \cup p$ is consistent iff $\Sigma_{\Gamma(\Delta)} \cup p$ is consistent. This proves the first equality. The second follows from the first (by putting $(\Sigma_{\Gamma(\Delta)})^{\Gamma(\Delta),\eta}$ instead of Σ) and from Theorem 2.2 (iv). □

The main technical object of our study is defined as follows.

Definition 3.2 *of generalized forcing.*

For every $\Sigma \subseteq L$ and enumeration η of L, the relation Δ^{η}_{Σ} of generalized forcing between elements p of $Cond_\Delta(\Sigma)$ and sentences φ of L is defined by:

$$p\Delta^{\eta}_{\Sigma}\varphi \text{ iff } (\Sigma_{\Gamma(\Delta)})^{\Gamma(\Delta),\eta} \cup p \vdash \varphi. \qquad \square$$

The relation Δ^{η}_{Σ} is defined in terms of $(\Sigma_{\Gamma(\Delta)})^{\Gamma(\Delta),\eta}$. Also the converse is true.

Lemma 3.3 $(\Sigma_{\Gamma(\Delta)})^{\Gamma(\Delta),\eta} = \{\varphi \in L \mid 0\Delta^{\eta}_{\Sigma}\varphi\}$.

Proof. Immediate from Definition 3.2. □

Before giving a syntactical characterization of Δ^{η}_{Σ} we investigate the notion of pragmatic $\Gamma(\Delta)$-truth.

---

[2] if φ is of the form $\neg\psi$ then the complement of φ is ψ; otherwise it is $\neg\varphi$

Definition 3.4 *of pragmatic* $\Gamma(\Delta)$*-truth and pragmatic* $\Gamma(\Delta)$*-entailment.*

A sentence $\varphi \in L$ is pragmatically $\Gamma(\Delta)$-true in model $\mathcal{M}$ for theory Σ, which is denoted by $\mathcal{M} \Delta\!\!\models \varphi$, **iff** for each $p \in \Delta$ with $\mathcal{M} \models p$, $\Sigma \cup p \cup \{\varphi\}$ is consistent. We write $\Sigma \Delta\!\!\models \varphi$ **iff** φ is pragmatically $\Gamma(\Delta)$-true in every model for Σ. We call the relation $\Delta\!\!\models$ a ***pragmatic*** $\Gamma(\Delta)$***-entailment.*** □

If Δ is interpreted as a collection of all possible outcomes of a finite direct experiment, which seems indeed a reasonable interpretation, then pragmatic $\Gamma(\Delta)$-truthfulness of a sentence is equivalent to its experimental irrefutability. In the sequel we will use terms ***pragmatic truth*** (***pragmatic entailment***), or $\Gamma(\Delta)$***-truth*** ($\Gamma(\Delta)$***-entailment***), interchangeably with ***pragmatic*** $\Gamma(\Delta)$***-truth*** (***pragmatic*** $\Gamma(\Delta)$***-entailment***). If Σ is known in the context then we call a model $\mathcal{M}$ for Σ a pragmatic model of φ iff $\mathcal{M} \Delta\!\!\models \varphi$. The following facts help us to characterize the $\Gamma(\Delta)$-entailment.

Lemma 3.5 $\Sigma \vdash \varphi$ **implies** both $\Sigma \Delta\!\!\models \varphi$ and $\Sigma \Delta\!\!\not\models \neg\varphi$.

Proof. Straightforward. □

Lemma 3.6 For each sentence $\varphi \in L$ and every $p \in Cond_\Delta(\Sigma)$, $\Sigma \cup p \Delta\!\!\models \varphi$ iff $\Sigma \Delta\!\!\models \bigwedge p \supset \varphi$.

Proof. $\bigwedge p \supset \varphi$ is $\Gamma(\Delta)$-false in model $\mathcal{M}$ for Σ, iff for some $q \in \Delta$ with $\mathcal{M} \models q$, $\Sigma \cup q \vdash \neg(\bigwedge p \supset \varphi)$ iff for some $q \in \Delta$ with $\mathcal{M} \models q$, $\Sigma \cup q \vdash \bigwedge p \wedge \neg\varphi$ iff φ is $\Gamma(\Delta)$-false in model $\mathcal{M}$ for $\Sigma \cup p$. □

The following two results provide a neat characterization of pragmatic entailment.

Theorem 3.7 $\boxed{C}$ For each sentence $\varphi \in L$,

$$\Sigma \Delta\!\!\models \varphi \text{ iff } (\Sigma \cup \{\varphi\})_{\Gamma(\Delta)} = \Sigma_{\Gamma(\Delta)}.$$

Proof. Implication to the right.

Suppose conversely. Let $p(\vec{c}) \in \Delta$ be such that for $\psi = (\forall \vec{x})(\bigvee n p(\vec{x}))$, $\psi \in (\Sigma \cup \{\varphi\})_{\Gamma(\Delta)} \setminus \Sigma_{\Gamma(\Delta)}$. Hence there exists canonic[3] $\mathcal{M} \models \Sigma_{\Gamma(\Delta)} \cup \{\neg\psi\}$, i.e. $\mathcal{M} \models \Sigma_{\Gamma(\Delta)} \cup \{(\exists \vec{x})(\bigwedge p(\vec{x}))\}$, and a combination $\vec{m}$ of its elements such that $\mathcal{M} \models p(\vec{x})[\vec{m}]$, and hence $\mathcal{M} \models \Sigma \cup p(\vec{x})[\vec{m}]$. Since Δ is closed under substitutions of constants, and $\mathcal{M}$ is a canonic structure, we have $p(\vec{m}) \in \Delta$ and $\mathcal{M} \models \Sigma \cup p(\vec{m})$. On the other hand, $p(\vec{m}) \vdash \neg\psi$ and then ψ is not $\Gamma(\Delta)$-true in $\mathcal{M}$. Because $\Sigma \cup \{\varphi\} \vdash \psi$ therefore $\Sigma \cup p(\vec{m}) \vdash \neg\varphi$, which means φ is not $\Gamma(\Delta)$-true in $\mathcal{M}$, either.

Implication to the left.

---

[3]by a canonic structure for L we mean any first-order structure $\mathcal{N}$ such that the interpretation of C in $\mathcal{N}$ is *on* $\mathcal{N}$

Let φ be a $\Gamma(\Delta)$-false sentence in a canonic $\mathcal{M} \models \Sigma$. Let $p \in \Delta$ be such that $\mathcal{M} \models p$ and $\Sigma \cup p \vdash \neg\varphi$. We have $\Sigma \cup \{\varphi\} \vdash \vee np$. $\vee np \in \Gamma(\Delta)$, and moreover $\Sigma \not\vdash \vee np$ since $\mathcal{M} \models \Sigma \cup p$. Hence $\vee np \in (\Sigma \cup \{\varphi\})_{\Gamma(\Delta)} \setminus \Sigma_{\Gamma(\Delta)}$. □

Theorem 3.8 For each sentence $\varphi \in L$,

$$\Sigma \overset{\Delta}{\models} \varphi \text{ iff } \bigwedge_{p \in Cond_\Delta(\Sigma)} \sim (\Sigma \cup p \vdash \neg\varphi).$$

Proof. Implication to the right. If $\Sigma \cup p \vdash \neg\varphi$ then, by Lemma 3.5, φ is $\Gamma(\Delta)$-false in models of $\Sigma \cup p$.

Implication to the left follows immediately from Definition 3.4. □

They prove a technical result we will need later in this paper.

Lemma 3.9 $\boxed{C}$ For each sentence $\varphi \in L$,

$$(\Sigma \cup \{\varphi\})_{\Gamma(\Delta)} = \Sigma_{\Gamma(\Delta)} \text{ iff } \bigwedge_{p \in Cond_\Delta(\Sigma)} \sim (\Sigma \cup p \vdash \varphi).$$

Proof. Follows from theorems 3.7 and 3.8. □

The following facts relate generalized forcing and $\Gamma(\Delta)$-entailment.

Lemma 3.10 If $\Sigma \subseteq (\Sigma_{\Gamma(\Delta)})^{\Gamma(\Delta),\eta}$ and $0\Delta^{\eta}_{\Sigma}\varphi$ then $\Sigma \overset{\Delta}{\models} \varphi$.

Proof. $0\Delta^{\eta}_{\Sigma}\varphi$ means $(\Sigma_{\Gamma(\Delta)})^{\Gamma(\Delta),\eta} \vdash \varphi$, i.e. by Theorem 2.2(v), $((\Sigma_{\Gamma(\Delta)})^{\Gamma(\Delta),\eta} \cup \{\varphi\})_{\Gamma(\Delta)} = \Sigma_{\Gamma(\Delta)}$. The assumption $\Sigma \subseteq (\Sigma_{\Gamma(\Delta)})^{\Gamma(\Delta),\eta}$ entails $(\Sigma \cup \{\varphi\})_{\Gamma(\Delta)} = \Sigma_{\Gamma(\Delta)}$, which by theorem 3.7 gives $\Sigma \overset{\Delta}{\models} \varphi$. □

Lemma 3.11 $\boxed{C}$ $0\Delta^{\eta}_{\Sigma}\varphi$ iff $(\Sigma_{\Gamma(\Delta)})^{\Gamma(\Delta),\eta} \overset{\Delta}{\models} \varphi$.

Proof. Implication to the right follows from Lemma 3.5. For implication to the left assume $(\Sigma_{\Gamma(\Delta)})^{\Gamma(\Delta),\eta} \overset{\Delta}{\models} \varphi$. Using Theorem 3.7 we obtain $((\Sigma_{\Gamma(\Delta)})^{\Gamma(\Delta),\eta} \cup \{\varphi\})_{\Gamma(\Delta)} = \Sigma_{\Gamma(\Delta)}$, which by Theorem 2.2 (v) gives us $\varphi \in (\Sigma_{\Gamma(\Delta)})^{\Gamma(\Delta),\eta}$. □

Corollary 3.12 $\boxed{C}$ $(\Sigma_{\Gamma(\Delta)})^{\Gamma(\Delta),\eta} \vdash \varphi$ iff $(\Sigma_{\Gamma(\Delta)})^{\Gamma(\Delta),\eta} \overset{\Delta}{\models} \varphi$. □

The above Corollary, together with Lemma 3.10, gives a complete characterization of theories for which the logical entailment and the pragmatic entailment coincide. We put it in a form of

Theorem 3.13 $\Sigma \subseteq L$ satisfies for every $\varphi \in L$:

(i) $\Sigma \vdash \varphi$ iff $\Sigma \overset{\Delta}{\models} \varphi$

iff there exists η such that $\Sigma_L = (\Sigma_{\Gamma(\Delta)})^{\Gamma(\Delta),\eta}$.

Proof. Corollary 3.12 proves implication to the left. For the implication to the right assume (i), i.e. let $\{\varphi \in L \mid \Sigma\Delta\!\!\models \varphi\} = \Sigma_L$. Let, for some ordinal number $\alpha \in Dm(\eta)$, $\Sigma = Rg(\eta \restriction \alpha)$. Then

(ii) $\Sigma \subseteq (\Sigma_{\Gamma(\Delta)})^{\Gamma(\Delta),\eta}$,

and hence by Definition 3.2 and by Lemma 3.10,

$$(\Sigma_{\Gamma(\Delta)})^{\Gamma(\Delta),\eta} \subseteq \{\varphi \in L \mid \Sigma\Delta\!\!\models \varphi\}.$$

By (i) we obtain $(\Sigma_{\Gamma(\Delta)})^{\Gamma(\Delta),\eta} \subseteq \Sigma_L$, so by (ii) and by Theorem 2.2 (vi), $(\Sigma_{\Gamma(\Delta)})^{\Gamma(\Delta),\eta} = \Sigma_L$. □

Lemma 3.11 may be strengthened.

Lemma 3.14 $p\Delta\frac{\eta}{\Sigma}\varphi$ **iff** $(\Sigma_{\Gamma(\Delta)})^{\Gamma(\Delta),\eta} \cup p \,\Delta\!\!\models \varphi$.

Proof. $p\Delta\frac{\eta}{\Sigma}\varphi$ iff $0\Delta\frac{\eta}{\Sigma}\bigwedge p \supset \varphi$ iff (by Lemma 3.11) $(\Sigma_{\Gamma(\Delta)})^{\Gamma(\Delta),\eta}\Delta\!\!\models \bigwedge p \supset \varphi$ iff (by Lemma 3.6) $(\Sigma_{\Gamma(\Delta)})^{\Gamma(\Delta),\eta} \cup p \,\Delta\!\!\models \varphi$. □

This gives us a very important property of the relation $\Delta\frac{\eta}{\Sigma}$.

Lemma 3.15 [C] For each $\varphi \in L$ and $p \in Cond_\Delta(\Sigma)$,

$$p\Delta\frac{\eta}{\Sigma}\neg\varphi \text{ iff } \bigwedge_{q \in Cond_\Delta(\Sigma):p\subseteq q} \sim (q\Delta\frac{\eta}{\Sigma}\varphi).$$

Proof. $p\Delta\frac{\eta}{\Sigma}\neg\varphi$ iff (by Lemma 3.14) $(\Sigma_{\Gamma(\Delta)})^{\Gamma(\Delta),\eta} \cup p \,\Delta\!\!\models \neg\varphi$ iff (by Theorem 3.8) $(\bigwedge_{r \in Cond_\Delta((\Sigma_{\Gamma(\Delta)})^{\Gamma(\Delta),\eta}\cup p)}) \sim ((\Sigma_{\Gamma(\Delta)})^{\Gamma(\Delta),\eta} \cup p \cup r \vdash \varphi)$ iff (pose $q = p \cup r$; Δ is closed under $\cup$) $(\bigwedge_{q \in Cond_\Delta((\Sigma_{\Gamma(\Delta)})^{\Gamma(\Delta),\eta}):p\subseteq q}) \sim ((\Sigma_{\Gamma(\Delta)})^{\Gamma(\Delta),\eta} \cup q \vdash \neg\varphi)$ iff (by Lemma 3.1) $(\bigwedge_{q \in Cond_\Delta(\Sigma):p\subseteq q}) \sim (q\Delta\frac{\eta}{\Sigma}\varphi)$. □

Now we are ready to give a usual inductive syntactic characterization of generalized forcing.

Theorem 3.16 [C]

(i) If $\varphi \in \Gamma(\Delta)$ **then:** $p\Delta\frac{\eta}{\Sigma}\varphi$ iff $\Sigma \cup p \vdash \varphi$.

(ii) $p\Delta\frac{\eta}{\Sigma}\neg\varphi$ **iff** $(\bigwedge_{q \in Cond_\Delta(\Sigma):p\subseteq q}) \sim (q\Delta\frac{\eta}{\Sigma}\varphi)$.

(iii) $p\Delta\frac{\eta}{\Sigma}\varphi \wedge \psi$ **iff** $p\Delta\frac{\eta}{\Sigma}\varphi$ and $p\Delta\frac{\eta}{\Sigma}\psi$.

Proof. (i) follows from Theorem 2.2 (iv); (ii) is given by Lemma 3.15; (iii) follows from Definition 3.2. □

Negation clause (ii) in Theorem 3.16 makes it evident that *not* is interpreted here as *it cannot happen that*, provided $Cond_\Delta(\Sigma)$ contains all possible things that may happen, and only an increase of the increment is allowed. Theorem 3.16 has an interesting consequence: the set of quantifier-free forcible sentences does not depend on η but only on Σ.

Lemma 3.17 $\boxed{C}$ For every two enumerations η and ζ of L, every $\Sigma \subseteq L$, every $p \in Cond_\Delta(\Sigma)$, and every quantifier-free sentence φ of L:

$$p\Delta\frac{\eta}{\Sigma}\varphi \text{ iff } p\Delta\frac{\zeta}{\Sigma}\varphi.$$

Proof. Since every atomic sentence of L or its negation is in $\Gamma(\Delta)$, Theorem 3.16 determines unambiguously whether $p\Delta\frac{\eta}{\Sigma}\varphi$ or not for all non-atomic sentences φ of L. By Definition 3.2, $p\Delta\frac{\eta}{\Sigma}\varphi$ is equivalent to $p\Delta\frac{\eta}{\Sigma}\neg\neg\varphi$. Therefore for all atomic sentences φ of L, $p\Delta\frac{\eta}{\Sigma}\varphi$ is decided by Theorem 3.16 too. □

The following Lemma states that two different forcings cannot extend each other.

Lemma 3.18 If $\Delta\frac{\eta}{\Sigma} \subseteq \Delta\frac{\zeta}{\Sigma}$ then $\Delta\frac{\eta}{\Sigma} = \Delta\frac{\zeta}{\Sigma}$.

Proof. Follows from Theorem 2.2 (xii) and Definition 3.2. □

It should be noted that $\forall$-clause:

$$p\Delta\frac{\eta}{\Sigma}\forall x\varphi(x) \text{ iff } \bigwedge_{c\in C} p\Delta\frac{\eta}{\Sigma}\varphi(c), \tag{$*$}$$

which is valid in all instances of weak forcing defined in [Kei73], does not necessarily hold in our case, because otherwise $\Delta\frac{\eta}{\Sigma}$ would not depend on η, which is not true. On the other hand, since $\Gamma(\Delta) \subseteq \forall$, an appropriate choice of η enforces $(*)$. This fact will be implied by Theorem 5.8.

The following facts will be used in Section 5. in particular in the proof of Theorem 5.8, and in applications of Lemma 3.10.

Lemma 3.19 $\boxed{C}$ Let η and ζ be two enumerations consistent with Kleene-Mostowski hierarchy. If $\Delta\frac{\eta}{\Sigma}$ satisfies $\forall$-clause $(*)$ then $\Delta\frac{\eta}{\Sigma} = \Delta\frac{\zeta}{\Sigma}$.

Proof. We demonstrate by induction that for every $n \in \omega$, $\Delta\frac{\eta}{\Sigma} \restriction \forall_{n-1} = \Delta\frac{\zeta}{\Sigma} \restriction \forall_{n-1}$. Case $n = 0$ follows immediately from Lemma 3.17. Assume that for some $n \in \omega$, $\Delta\frac{\eta}{\Sigma} \restriction \forall_{n-1} = \Delta\frac{\zeta}{\Sigma} \restriction \forall_{n-1}$. By theorem 3.16 (ii), $\Delta\frac{\eta}{\Sigma} \restriction \exists_{n-1} = \Delta\frac{\zeta}{\Sigma} \restriction \exists_{n-1}$. Let $\varphi \in \forall_n$, i.e. $\varphi = \forall\vec{x}\psi(\vec{x})$, where $\psi(\vec{c}) \in \exists_{n-1}$. If $0\Delta\frac{\zeta}{\Sigma}\varphi$ then, by definition 3.2, $0\Delta\frac{\zeta}{\Sigma}\psi(\vec{c})$ holds for every $\vec{c} \in C^{|\vec{x}|}$, or, by induction hypothesis, $(\bigwedge_{\vec{c}\in C^{|\vec{x}|}})(p\Delta\frac{\eta}{\Sigma}\psi(\vec{c}))$. Hence, by $(*)$, $0\Delta\frac{\eta}{\Sigma}\varphi$. Let α and β satisfy $Rg(\eta \restriction \alpha) = Rg(\zeta \restriction \beta) = \forall_n$. We have $\Sigma^{\Gamma(\Delta),\zeta\restriction\beta} \subseteq \Sigma^{\Gamma(\Delta),\eta\restriction\alpha}$, so, by Theorem 2.2 (xii), $\Sigma^{\Gamma(\Delta),\zeta\restriction\beta} = \Sigma^{\Gamma(\Delta),\eta\restriction\alpha}$. Thus $\Delta\frac{\zeta}{\Sigma} \restriction \forall_n = \Delta\frac{\eta}{\Sigma} \restriction \forall_n$. □

Lemma 3.20 $\boxed{\forall}$ If $\Sigma \subseteq \forall$ then there exists η consistent with Kleene-Mostowski hierarchy such that $\Sigma^{\Gamma(\Delta),\eta} = (\Sigma_{\Gamma(\Delta)})^{\Gamma(\Delta),\eta}$.

Proof. Let η be consistent with Kleene-Mostowski hierarchy, and such that for some ordinal number $\beta \in Dm(\eta)$, $\Sigma = Rg(\eta \restriction \beta)$. Such η exists since $\Sigma \subseteq \forall$. If $\eta_\alpha \in \Sigma$ then $((\Sigma_{\Gamma(\Delta)})_\alpha^{\Gamma(\Delta),\eta} \cup \{\eta_\alpha\})_{\Gamma(\Delta)} \subseteq (\Sigma \cup \{\eta_\alpha\})_{\Gamma(\Delta)} = \Sigma_{\Gamma(\Delta)} = (\Sigma_{\Gamma(\Delta)})_{\Gamma(\Delta)}$, in other words,

$$((\Sigma_{\Gamma(\Delta)})_\alpha^{\Gamma(\Delta),\eta} \cup \{\eta_\alpha\})_{\Gamma(\Delta)} = (\Sigma_{\Gamma(\Delta)})_{\Gamma(\Delta)},$$

which implies $\eta_\alpha \in (\Sigma_{\Gamma(\Delta)})^{\Gamma(\Delta),\eta}$. We have shown that $\Sigma \subseteq (\Sigma_{\Gamma(\Delta)})^{\Gamma(\Delta),\eta}$. By Theorem 2.2 (vi), we get $\Sigma_\forall \subseteq (\Sigma_{\Gamma(\Delta)})^{\Gamma(\Delta),\eta}$. $\Gamma(\Delta) \subseteq \forall$ implies $\Sigma_{\Gamma(\Delta)} \subseteq \Sigma_\forall$. By Theorem 2.2 (x) we obtain $(\Sigma_{\Gamma(\Delta)})^{\Gamma(\Delta),\eta} = ((\Sigma_{\Gamma(\Delta)})^{\Gamma(\Delta),\eta})^{\Gamma(\Delta),\eta}$, which by theorem 2.2 (xi) gives $(\Sigma_\forall)^{\Gamma(\Delta),\eta} = (\Sigma_{\Gamma(\Delta)})^{\Gamma(\Delta),\eta}$. An application of $\Sigma \subseteq \forall$ yields the thesis. □

The negation clause (ii) of Theorem 3.16 has an important consequence: within the class of models for Σ there exists a semantics which is complete with respect to relation Δ^{η}_{Σ}.

Definition 3.21 *of generic set.*
Set G is called generic (relative to Δ^{η}_{Σ}) iff:
(i) each finite subset of G is in $Cond_\Delta(\Sigma)$
(ii) for each sentence $\varphi \in L$, $G\Delta^{\eta}_{\Sigma}\varphi$ or $G\Delta^{\eta}_{\Sigma}\neg\varphi$, where $G\Delta^{\eta}_{\Sigma}\vartheta$ means: $(\bigvee_{finite\ p\subseteq G})(p\Delta^{\eta}_{\Sigma}\vartheta)$. □

Definition 3.22 *of generic model.*
Let G be a generic set. A generic model $\mathcal{M}(G)$ (relative to Δ^{η}_{Σ}) corresponding to G is any structure such that for each sentence $\varphi \in L$, $\mathcal{M}(G) \models \varphi$ iff $G\Delta^{\eta}_{\Sigma}\varphi$. □

The following three results testify that our generalization of weak forcing preserves its basic properties.

Generic Set Theorem 3.23 $\boxed{C}$ For every $p \in Cond_\Delta(\Sigma)$ there is a generic set G with $p \subseteq G$.

Proof. As in [Bar78]: given an ω-enumeration φ_i of sentences of L, define an increasing sequence p_i of elements of $Cond_\Delta(\Sigma)$ as follows. $p_0 = p$. If $p_i\Delta^{\eta}_{\Sigma}\neg\varphi_i$ then $p_{i+1} = p_i$, otherwise $p_{i+1} = q$, where $p_i \subseteq q$ and $q\Delta^{\eta}_{\Sigma}\varphi_i$ (such q exists by Theorem 3.16 (ii)). $G = \bigcup\{p_i \mid i \in \omega\}$ is, by compactness theorem (cf e.g. [Bar78] Chap. A1, thm. 2.4), consistent with Σ, and for every φ_i, $G\Delta^{\eta}_{\Sigma}\varphi_i$ or $G\Delta^{\eta}_{\Sigma}\neg\varphi_i$. Therefore G is a generic set. □

Generic Model Theorem 3.24 $\boxed{C}$ For every generic set G there is a generic model $\mathcal{M}(G)$. Moreover, any two generic models $\mathcal{M}(G)$ and $\mathcal{M}'(G)$ are elementarily equivalent (i.e. they satisfy exactly the same sentences of L).

Proof. By Definition 3.21 (i), for each $p \subseteq G$, $p \in Cond_\Delta(\Sigma)$, therefore (by Lemma 3.1) $p \in Cond_\Delta((\Sigma_{\Gamma(\Delta)})^{\Gamma(\Delta),\eta})$. By the compactness theorem (cf e.g. [Bar78] Chap. A1, thm. 2.4) this means that $(\Sigma_{\Gamma(\Delta)})^{\Gamma(\Delta),\eta} \cup G$ is consistent. Hence, by Definition 3.21 (ii), $\Pi = \{\varphi \in L \mid G\Delta^{\eta}_{\Sigma}\varphi\}$ is complete.

Thus Π has a model, and every model $\mathcal{M}$ of Π satisfies $\Pi = \{\varphi \in L \mid \mathcal{M} \models \varphi\}$. □

Completeness Theorem 3.25 $\boxed{C}$ For every sentence $\varphi \in L$ and every $p \in Cond_\Delta(\Sigma)$:

$$p\Delta\frac{\eta}{\Sigma}\varphi \text{ iff } \bigwedge_{G \supseteq p : G \text{ is a generic set}} \mathcal{M}(G) \models \varphi.$$

Proof. Straightforward from Theorem 3.23 and 3.24. □

It may be easily observed that the consequence operation f^η_Δ defined by: $f^\eta_\Delta(\Sigma) = (\Sigma_{\Gamma(\Delta)})^{\Gamma(\Delta),\eta}$ is non-monotonic, i.e. $\Pi \subseteq \Sigma$ does not imply $f^\eta_\Delta(\Pi) \subseteq f^\eta_\Delta(\Sigma)$, even if $\Pi \cup \Sigma \subseteq \Gamma(\Delta)$. This property is an immediate consequence of the fact that for $\Sigma \subseteq \Gamma(\Delta)$, $Cn(\Sigma)$ may constitute a *proper* subset of $\Sigma^{\Gamma(\Delta),\eta}$.

Despite the fact that we have defined a unique class $\Gamma(\Delta)$ for each Δ, there may actually be many Λ's producing the same theory $(\Sigma_\Lambda)^{\Lambda,\eta}$. As we already have seen, one may close $\Gamma(\Delta)$ under $\wedge$ without changing $(\Sigma_{\Gamma(\Delta)})^{\Gamma(\Delta),\eta}$. One may also ask about the characterization of a class $\mathcal{K}_\Delta$ consisting of all such Λ's that satisfy for each η on L and $\Sigma \subseteq L$, $\Sigma_\Lambda \subseteq (\Sigma_{\Gamma(\Delta)})^{\Gamma(\Delta),\eta}$. We may easily demonstrate that $\mathcal{K}_\Delta$ is a filter, but it does not seem a lot.

$\bigcup \mathcal{K}_\Delta$ gives a generalization $\{\varphi \mid (\Sigma \cup \{\varphi\})_{\bigcup \mathcal{K}_\Delta} \subseteq \Sigma_{\bigcup \mathcal{K}_\Delta}\}$ of the Kaiser Hull of [Kai69].

4 Generalized $\forall$-Forcing

In this section we consider a special case $\overset{w}{\Vdash}$ of generalized forcing, namely the one which satisfies the $\forall$-clause $(*)$ of Section 3. We start from independent definition of this kind of forcing, without referencing to operation $^{\Gamma,\eta}$, investigating after that some of its properties. In Section 5 we will demonstrate (theorem 5.8) that there exists η, such that $\overset{w}{\Vdash}$ and $\Delta\frac{\eta}{\Sigma}$ coincide for every Σ and every Δ, that is to say, that $\overset{w}{\Vdash}$ is indeed a special case of the generalized forcing introduced before.

Because our weak generalized $\forall$-forcing $\overset{w}{\Vdash}$ is, as one can easily check, very close to weak forcing $\overset{w}{\Vdash}$ of [Rob71] and [Kei73] (actually, $\overset{w}{\Vdash}$ is a slight generalization of $\overset{w}{\Vdash}$), most of proofs of basic properties of $\overset{w}{\Vdash}$ strongly resemble the proofs of analogical properties of $\overset{w}{\Vdash}$.

Let *Atom* denote the set of atomic sentences of L.

Definition 4.1 $\boxed{C}$ The relation $\Vdash$ of *generalized $\forall$-forcing* is defined inductively for all $p \in Cond_\Delta(\Sigma)$ and sentences $\varphi \in L$ (Σ and Δ are implicit parameters of this relation).

(i) If $\varphi \in (Atom \cap \Gamma(\Delta)) \cup n(Atom \setminus \Gamma(\Delta))$ **then:**

$$p \Vdash \varphi \text{ iff } \Sigma \cup p \vdash \varphi.$$

(ii) If $\neg\varphi \notin n(Atom \setminus \Gamma(\Delta))$ **then:**

$$p \Vdash \neg\varphi \text{ iff } (\bigwedge_{q \in Cond_\Delta(\Sigma): p \subseteq q}) \sim (q \Vdash \varphi).$$

(iii) $p \Vdash \varphi \vee \psi$ **iff** $p \Vdash \varphi$ or $p \Vdash \psi$.

(iv) $p \Vdash \exists x \varphi(x)$ **iff** $(\bigvee_{c \in C})(p \Vdash \varphi(c))$.

(Other connectives we treat as appropriate abbreviations.)

The relation $\overset{w}{\Vdash}$ of *weak generalized* $\forall$*-forcing* is defined by:

(v) $p \overset{w}{\Vdash} \varphi$ **iff** $p \Vdash \neg\neg\varphi$. □

If $\Delta = Atom \cup nAtom$ then $Atom \subseteq \Gamma(\Delta)$. In this case $\overset{w}{\Vdash}$ coincides with weak forcing $\overset{w}{\Vdash}$ of [Rob71] (see [Bar78], Chap. A2 §8 for details).

Because for every atomic sentence φ, either φ of $\neg\varphi$ is in $\Gamma(\Delta)$, it is a routine induction to check that for every Σ and Δ there exists exactly one relation $\overset{w}{\Vdash}$ satisfying (i)...(v). Immediately from (ii) and (v) we get

Lemma 4.2 For every $p \in Cond_\Delta(\Sigma)$ and $\varphi \in L$,

$$p \overset{w}{\Vdash} \varphi \text{ iff } (\bigwedge_{q \in Cond_\Delta(\Sigma): p \subseteq q})(\bigvee_{r \in Cond_\Delta(\Sigma): q \subseteq r})(r \Vdash \varphi).$$ □

Both relations, $\Vdash$ and $\overset{w}{\Vdash}$, are monotonic with respect to their left arguments.

Lemma 4.3 For every $p, q \in Cond_\Delta(\Sigma)$ with $p \subseteq q$, and every $\varphi \in L$:

(i) $p \Vdash \varphi$ implies $q \Vdash \varphi$,

(ii) $p \overset{w}{\Vdash} \varphi$ implies $q \overset{w}{\Vdash} \varphi$.

Proof. (i) The inductional cases of $\varphi \in (Atom \cap \Gamma(\Delta)) \cup n(Atom \setminus \Gamma(\Delta))$, $\varphi \vee \psi$ and $\exists x \varphi(x)$ are easy. For the case of $\neg\varphi$ let us note that $p \Vdash \neg\varphi$ implies $(\bigwedge_{r \in Cond_\Delta(\Sigma): p \subseteq r}) \sim (r \Vdash \varphi)$ implies $(\bigwedge_{r \in Cond_\Delta(\Sigma): q \subseteq r}) \sim (r \Vdash \varphi)$ implies $q \Vdash \neg\varphi$.

(ii) Follows from (i) and Lemma 4.3. □

$\overset{w}{\Vdash}$ is stronger than $\Vdash$.

Lemma 4.4 For every $p \in Cond_\Delta(\Sigma)$ and $\varphi \in L$, $p \Vdash \varphi$ implies $p \overset{w}{\Vdash} \varphi$.

Proof. If $p \Vdash \varphi$ then by Lemma 4.3 (i) $(\bigwedge_{q \in Cond: p \subseteq q})(q \Vdash \varphi)$, and hence

$$(\bigwedge_{q \in Cond_\Delta(\Sigma): p \subseteq q})(\bigvee_{r \in Cond_\Delta(\Sigma): q \subseteq r})(r \Vdash \varphi),$$

that is to say (by Lemma 4.2), $p \overset{w}{\Vdash} \varphi$. □

Moreover, $\overset{w}{\Vdash}$ coincides with $\Vdash$ for negated sentences.

Lemma 4.5 For every $p \in Cond_\Delta(\Sigma)$ and $\varphi \in L$,

$$p \Vdash \neg\varphi \text{ iff } p \overset{w}{\Vdash} \neg\varphi.$$

Proof. Implication to the right follows from Lemma 4.4. For implication to the left assume $p \overset{w}{\Vdash} \neg\varphi$. If $\neg\varphi \in n(Atom \setminus \Gamma(\Delta))$ then by Definition 4.1 (i) and Theorem 4.6 (i), $p \Vdash \neg\varphi$. Otherwise, by Theorem 4.6 (ii), we obtain $(\bigwedge_{q \in Cond_\Delta(\Sigma): p \subseteq q}) \sim (q \overset{w}{\Vdash} \varphi)$, which by Lemma 4.4 implies $(\bigwedge_{q \in Cond_\Delta(\Sigma): p \subseteq q}) \sim (q \Vdash \varphi)$, i.e. $p \Vdash \neg\varphi$. □

The following Theorem provides a useful characterization of $\overset{w}{\Vdash}$.

Theorem 4.6 For every $p \in Cond_\Delta(\Sigma)$ and $\varphi \in L$, the following conditions are satisfied.

(i) If $\varphi \in (Atom \cap \Gamma(\Delta)) \cup n(Atom \setminus \Gamma(\Delta))$ **then** $p \overset{w}{\Vdash} \varphi$ iff $\Sigma \cup p \vdash \varphi$.

(ii) $p \overset{w}{\Vdash} \neg\varphi$ **iff** $(\bigwedge_{q \in Cond_\Delta(\Sigma): p \subseteq q}) \sim (q \overset{w}{\Vdash} \varphi)$.

(iii) $p \overset{w}{\Vdash} \varphi \wedge \psi$ **iff** $p \overset{w}{\Vdash} \varphi$ and $p \overset{w}{\Vdash} \psi$.

(iv) $p \overset{w}{\Vdash} \forall x \varphi(x)$ **iff** $(\bigwedge_{c \in C})(p \overset{w}{\Vdash} \varphi(c))$.

Proof. (i) Let $\varphi \in (Atom \cap \Gamma(\Delta)) \cup n(Atom \setminus \Gamma(\Delta))$. $p \overset{w}{\Vdash} \varphi$ iff (by Lemma 4.2) $(\bigwedge_{q \in Cond_\Delta(\Sigma): p \subseteq q})(\bigvee_{r \in Cond_\Delta(\Sigma): q \subseteq r})(r \Vdash \varphi)$ iff (by (i) of Definition 4.1)

(v) $(\bigwedge_{q \in Cond_\Delta(\Sigma): p \subseteq q})(\bigvee_{r \in Cond_\Delta(\Sigma): q \subseteq r})(\Sigma \cup r \vdash \varphi)$.

If $\Sigma \cup p \vdash \varphi$ then (v) is true. If $\Sigma \cup p \nvdash \varphi$ then (since $\{n\varphi\} \in \Delta$ and Δ is closed under $\cup$) $p \cup \{n\varphi\} \in Cond_\Delta(\Sigma)$ and $\Sigma \cup p \cup \{n\varphi\} \vdash \neg\varphi$, hence $(\bigvee_{q \in Cond_\Delta(\Sigma): p \subseteq q})(\bigwedge_{r \in Cond_\Delta(\Sigma): q \subseteq r})(\Sigma \cup r \vdash \neg\varphi)$, or in other words ($\Sigma \cup r$ is consistent), $\sim (\bigwedge_{q \in Cond_\Delta(\Sigma): p \subseteq q})(\bigvee_{r \in Cond_\Delta(\Sigma): q \subseteq r})(\Sigma \cup r \vdash \varphi)$. Thus in this case (v) is false.

(ii) $p \overset{w}{\Vdash} \neg\varphi$ iff (by Lemma 4.2)
$(\bigwedge_{q \in Cond_\Delta(\Sigma): p \subseteq q})(\bigvee_{r \in Cond_\Delta(\Sigma): q \subseteq r})(r \Vdash \neg\varphi)$, that is to say,
$(\bigwedge_{q \in Cond_\Delta(\Sigma): p \subseteq q}) \sim (\bigwedge_{r \in Cond_\Delta(\Sigma): q \subseteq r}) \sim (r \Vdash \neg\varphi)$, or in other words,
$(\bigwedge_{q \in Cond_\Delta(\Sigma): p \subseteq q}) \sim (q \Vdash \neg\neg\varphi)$, i.e. $(\bigwedge_{q \in Cond_\Delta(\Sigma): p \subseteq q}) \sim (q \overset{w}{\Vdash} \varphi)$.

(iii) $p \overset{w}{\Vdash} \varphi \wedge \psi$ iff ($\varphi \wedge \psi$ has been treated as an abbreviation for $\neg(\neg\varphi \vee \neg\varphi)$ in Definition 4.1) $p \overset{w}{\Vdash} \neg(\neg\varphi \vee \neg\psi)$, iff (by Lemma 4.5) $p \Vdash \neg(\neg\varphi \vee \neg\psi)$ iff (by Definition 4.1 (ii) and (iii) $(\bigwedge_{q \in Cond_\Delta(\Sigma): p \subseteq q})$ ($\sim (q \Vdash \neg\varphi)$ and $\sim (q \Vdash \neg\psi)$) iff $(\bigwedge_{q \in Cond_\Delta(\Sigma): p \subseteq q}) \sim (q \Vdash \neg\varphi)$ and $(\bigwedge_{q \in Cond_\Delta(\Sigma): p \subseteq q}) \sim (q \Vdash \neg\psi)$ iff (by Definition 4.1 (ii) and (v)) $p \overset{w}{\Vdash} \varphi$ and $p \overset{w}{\Vdash} \psi$.

(iv) $p \overset{w}{\Vdash} \forall x \varphi(x)$ iff ($\forall x \varphi(x)$ is an abbreviation for $\neg\exists x \neg\varphi(x)$ in Definition 4.1) $p \overset{w}{\Vdash} \neg\exists x \neg\varphi(x)$ iff (by Lemma 4.5) $p \Vdash \neg\exists x \neg\varphi(x)$ iff (by Definition 4.1 (ii) and (iv)) $(\bigwedge_{q \in Cond_\Delta(\Sigma): p \subseteq q})$ $(\bigwedge_{c \in C})(\bigvee_{r \in Cond_\Delta(\Sigma): q \subseteq r})(p \Vdash \varphi(c))$ iff (by Lemma 4.2) $(\bigwedge_{c \in C})(p \overset{w}{\Vdash} \varphi(c))$. □

One may observe that if $Atom \subseteq \Gamma(\Delta)$ then Theorem 4.6 may serve as an inductive definition of $\overset{w}{\Vdash}$. This case will take place in Section 7.

Similarly as for Δ^{η}_{Σ}, we introduce—respectively—notions of generic set and generic model. The notion of generic set (relative to $\overset{w}{\Vdash}$) remains the same as for generalized forcing Δ^{η}_{Σ} (see Definition 3.21). Theorem 4.6 (ii) ascertains the truthfulness of Generic Set Theorem 3.23 for $\overset{w}{\Vdash}$. The notion of generic model needs a strengthening: $\forall$-clause (iv) of Theorem 4.6 allows us, without loss generality, to restrict the class of generic models for Σ to canonic structures. (Recall that by a canonic structure for L we mean any first-order structure $\mathcal{N}$ such that the interpretation of C in $\mathcal{N}$ is *on* $\mathcal{N}$.)

Definition 4.7 Let G be a generic set relative to $\overset{w}{\Vdash}$. A generic $\forall$-model $\mathcal{M}_{\forall}(G)$ (relative to $\overset{w}{\Vdash}$) corresponding to G is any canonic structure for L such that for every sentence $\varphi \in L$,

$$\mathcal{M}_{\forall}(G) \models \varphi \text{ iff } G \overset{w}{\Vdash} \varphi. \qquad \square$$

Theorem 4.6, similarly to Theorem 3.16 allows us to derive a couple of classic properties of $\overset{w}{\Vdash}$.

Generic Model Theorem 4.8 $\boxed{C}$ For every generic set G there is a generic $\forall$-model $\mathcal{M}_{\forall}(G)$. If the equality symbol $=$ belongs to L then $\mathcal{M}_{\forall}(G)$ is unique (up to isomorphism).

Proof. Using the same argument as in proof of Theorem 3.24, we infer that $\Pi = \{\varphi \in L \mid G \overset{w}{\Vdash} \varphi\}$ is a complete consistent theory. On the other hand, $\Pi \vdash \forall x \varphi(x)$ iff $(\bigwedge_{c \in C})(\Pi \vdash \varphi(c))$. Therefore Π has a canonic model. Of course, if Π is a complete theory in a language with the equality symbol, and $\mathcal{M}$, $\mathcal{N}$ are its canonic models, then $\mathcal{M}$ and $\mathcal{N}$ are isomorphic. $\square$

Completeness Theorem 4.9 $\boxed{C}$ For every sentence $\varphi \in L$ and every $p \in Cond_{\Delta}(\Sigma)$:

$$p \overset{w}{\Vdash} \varphi \text{ iff } \bigwedge_{\substack{G \supseteq p:\ G \text{ is} \\ \text{a generic set}}} \mathcal{M}_{\forall}(G) \models \varphi.$$

Proof. Straightforward from theorems 3.23 and 4.8. $\square$

The Completeness Theorem 4.9 proves that $\overset{w}{\Vdash}$ defines a well behaved consequence operation.

Definition 4.10 The operation ^{S} is defined on a class of consistent subsets of $\forall$ ($\overset{w}{\Vdash}$ is the implicit parameter of this operation) by:

$$\Sigma^S = \{\varphi \mid 0 \overset{w}{\Vdash} \varphi\}. \qquad \square$$

The following lemma gives a semantic characterization of operation ^{S}.

Lemma 4.11 $\boxed{C}$

$$\Sigma^S = \{\varphi \in L \mid \bigwedge_{\mathcal{M} \models \Sigma : \mathcal{M} \text{ is a generic } \forall-\text{model}} \mathcal{M} \models \varphi\}.$$

Proof. Follows from Theorem 4.9. □

The operation ^{S} similarly to the operation $^{\Gamma,\eta}$ (cf. Definition 3.2) defines the relation $\overset{w}{\Vdash}$.

Lemma 4.12 For every $\Sigma \subseteq L$, every $p \in Cond_\Delta(\Sigma)$, and every $\varphi \in L$,

$$p \overset{w}{\Vdash} \varphi \text{ iff } \Sigma^S \cup p \vdash \varphi.$$

Proof. We have: $p \overset{w}{\Vdash} \varphi$ iff (by the completeness Theorem 4.9)

$$(\bigwedge_{G \supseteq p : G \text{ is a generic set}})(\mathcal{M}_\forall(G) \models \varphi)$$

iff, since $\mathcal{M}_\forall(G) \models G$,

$$(\bigwedge_{\text{generic set } G})(\mathcal{M}_\forall(G) \models \wedge p \supset \varphi)$$

iff (by the completeness theorem again) $0 \overset{w}{\Vdash} \wedge p \supset \varphi$ iff (by the Definition 4.10) $\Sigma^S \vdash \wedge p \supset \varphi$ iff $\Sigma^S \cup p \vdash \varphi$. □

The operation ^{S} has the following property.

Lemma 4.13 $\boxed{C}$ $\Sigma^S = (\Sigma_{\Gamma(\Delta)})^S$.

Proof. It follows from the Definition 3.21 of generic set and Theorem 4.9 that for Σ and Π satisfying $Cond_\Delta(\Sigma) = Cond_\Delta(\Pi)$, $\Sigma^S = \Pi^S$ holds. Therefore an application of Lemma 3.1 completes the proof. □

It is convenient for us, however, to postpone investigation of other properties of operation ^{S} until Section 5 (Thm. 5.12). The rest of this section is devoted to demonstrating that Theorem 3.16 (i) holds for $\overset{w}{\Vdash}$.

Lemma 4.14 $\boxed{C}$ For every quantifier-free sentence $\varphi \in L$, every $p \in Cond_\Delta(\Sigma)$, and every enumeration η of L:

$$p \Delta^{\eta}_{\Sigma} \varphi \text{ iff } p \overset{w}{\Vdash} \varphi.$$

Proof. Since for each $\varphi \in Atom$, either φ or $\neg\varphi$ is in $\Gamma(\Delta)$, the routine induction, with application of Theorems 3.16 and 4.6, shows that the thesis is true for all $\varphi \notin Atom \setminus \Gamma(\Delta)$. On the other hand, both $\{\psi \in L \mid p\Delta^{\eta}_{\Sigma}\psi\}$ and $\{\psi \in L \mid p \overset{w}{\Vdash} \psi\}$ are closed under $\vdash$ (by Definition 3.2 and Theorem 4.9), therefore for $\varphi \in Atom \setminus \Gamma(\Delta)$ we have: $p\Delta^{\eta}_{\Sigma}\varphi$ iff $p\Delta^{\eta}_{\Sigma}\neg\neg\varphi$ iff (already demonstrated) $p \overset{w}{\Vdash} \neg\neg\varphi$ iff $p \overset{w}{\Vdash} \varphi$. □

The following generalizes the analogous property of weak forcing.

Lemma 4.15 $\boxed{C}$ For every sentence $\varphi \in \Gamma(\Delta)$ and every $p \in Cond_{\Delta}(\Sigma)$,

$$p \overset{w}{\Vdash} \varphi \text{ iff } \Sigma \cup p \vdash \varphi.$$

Proof. The truthfulness of the thesis for quantifier-free $\varphi \in \Gamma(\Delta)$ follows from Theorem 3.16 (i) and Lemma 4.14. The rest of proof follows by induction on the number of quantified variables in φ. We have: $p \overset{w}{\Vdash} \forall x \varphi(x)$ iff $(\bigwedge_{c \in C})(p \overset{w}{\Vdash} \varphi(c))$ iff (by the definition of $\Gamma(\Delta)$ in Section 3, and induction hypothesis) $(\bigwedge_{c \in C})(\Sigma \cup p \vdash \varphi(c))$ iff (there are some constants in C not appearing in $\Sigma \cup p$) $\Sigma \cup p \vdash \forall x \varphi(x)$. □

This way we have proven the following

Lemma 4.16 $\boxed{C}$ For every $\varphi \in \Gamma(\Delta)$, every $p \in Cond_{\Delta}(\Sigma)$ and every enumeration η of L :

$$p \overset{w}{\Vdash} \varphi \text{ iff } p \Delta \frac{\eta}{\Sigma} \varphi.$$

Proof. Follows from Lemma 4.15 and Theorem 3.16. □

5 Incremental Models

Now we use the concept of generalized forcing to introduce a model for updating deductive data bases. In our approach, a consistent first-order theory Σ in language L constitutes the fixed part of the data base. Its changes are not, in general, captured by this model. The class Δ contains all virtual extensions of the fixed part Σ, called increments in this paper, so the actual content of a data base is represented by $\Sigma \cup p$, where $p \in Cond_{\Delta}(\Sigma)$.

To avoid unnecessary technical problems, in the first part of this section we assume that $\Sigma \subseteq \Gamma(\Delta)$, which gives us the following:

Lemma 5.1 $\boxed{\forall}$ Let $\Sigma \subseteq \Gamma(\Delta)$.

(i) $\Sigma \subseteq \Sigma_{\Gamma(\Delta)}$.

(ii) $\Sigma \subseteq (\Sigma_{\Gamma(\Delta)})^{\Gamma(\Delta),\eta}$.

Proof. (i) is obvious. (ii) follows from (i) and Theorem 2.2 (viii). □

The data base $\Sigma \cup p$ usually refers to a not necessarily completely known world, which is supposed to be a model of $\Sigma \cup p$, in our case a first-order one. So, the first-order consequences of $\Sigma \cup p$ are as true in this world as the data base itself. $(\Sigma \cup p)_L$ does not seem, however, to exhaust the sentences customarily accepted as its true consequences. The fact that not all sentences of L are allowed in p makes some other sentences practically nonverifiable. Therefore, there is some freedom with associating truth values with sentences of this kind. In particular, one may insist on using pragmatic $\Gamma(\Delta)$-truth instead of truth. This would never result in accepting the truthfulness of a sentence, which will be refuted after enlarging the dynamic increment p.

Having this motivation in mind, we require that a consistent syntactic system F of answers to queries satisfies the following postulates.

1. $\Sigma \subseteq F(\Sigma)$.

2. $\varphi \in F(\Sigma) \supset \Sigma \Delta\!\!\models \varphi$.

3. $\varphi \in F(\Sigma) \supset \neg\varphi \notin F(\Sigma)$.

4. $F(\Sigma)$ is closed under $\vdash$.

5. $F(\Sigma)$ is a maximal set of sentences of L which satisfies 1...4.

Obviously, postulates 3 and 4 make the operation F non-monotonic[4] in all cases it does not coincide with first-order consequence operation $_L$.

At this point we use forcing, putting $\Sigma^{\Gamma(\Delta),\eta}$ as $F(\Sigma)$. By selecting various η's one can obtain different F's. Of course, one can pick up different η's for different Σ's. What may be interesting here is:

Theorem 5.2 $\boxed{\forall}$ Operation $F : \mathcal{P}(\Gamma(\Delta)) \to \mathcal{P}(\Gamma(\Delta))$ satisfies postulates 1...5 iff for each $\Sigma \subseteq \Gamma(\Delta)$ there exists η such that $F(\Sigma) = \Sigma^{\Gamma(\Delta),\eta}$.

Proof. Take an enumeration η such that for some ordinal number $\beta \in Dm(\eta)$, $F(\Sigma) = Rg(\eta \restriction \beta)$. Because, by postulate 1, $\Sigma \subseteq F(\Sigma)$, we get $\Sigma_{\Gamma(\Delta)} \subseteq F(\Sigma)_{\Gamma(\Delta)}$ and, by Lemma 5.1 (i), $\Sigma \subseteq F(\Sigma)_{\Gamma(\Delta)}$. If $\varphi \in F(\Sigma)_{\Gamma(\Delta)}$ then, by postulate 4, $\varphi \in \Gamma(\Delta) \cap F(\Sigma)$, therefore $\varphi \in \Gamma(\Delta)$ and $\Sigma \Delta\!\!\models \varphi$. Hence, by Theorem 3.7, $\varphi \in \Gamma(\Delta)$ and $(\Sigma \cup \{\varphi\})_{\Gamma(\Delta)} = \Sigma_{\Gamma(\Delta)}$, that is to say, $\varphi \in \Sigma_{\Gamma(\Delta)}$. Thus $F(\Sigma)_{\Gamma(\Delta)} \subseteq \Sigma_{\Gamma(\Delta)}$. Taking into account $\Sigma_{\Gamma(\Delta)} \subseteq F(\Sigma)_{\Gamma(\Delta)}$, we obtain $\Sigma_{\Gamma(\Delta)} = F(\Sigma)_{\Gamma(\Delta)}$. Let $\eta_\alpha \in F(\Sigma)$. It is easily seen that $\Sigma_\alpha^{\Gamma(\Delta),\eta} \subseteq F(\Sigma)$, so $(\Sigma_\alpha^{\Gamma(\Delta),\eta} \cup \{\eta_\alpha\})_{\Gamma(\Delta)} \subseteq F(\Sigma)_{\Gamma(\Delta)} = \Sigma_{\Gamma(\Delta)}$, that is to say, $\eta_\alpha \in \Sigma^{\Gamma(\Delta),\eta}$. We obtained $F(\Sigma) \subseteq \Sigma^{\Gamma(\Delta),\eta}$. Now we check that $F(\Sigma) = \Sigma^{\Gamma(\Delta),\eta}$ satisfies postulates 1...4.
Postulate 1 follows from Theorem 2.2 (viii).
Postulate 2 follows from Lemma 5.1 (ii) and Lemma 3.10.
Postulate 3 follows from Theorem 2.2 (i).
Postulate 4 follows from Theorem 2.2 (vi).
Hence $F(\Sigma) \subseteq \Sigma^{\Gamma(\Delta),\eta}$ and the maximality of $F(\Sigma)$ gives us $F(\Sigma) = \Sigma^{\Gamma(\Delta),\eta}$. □

From the above theorem we easily conclude

Corollary 5.3 $\boxed{\forall}$ If $\Sigma \subseteq \Gamma(\Delta)$ then

$$\{\varphi \in L \mid \Sigma \Delta\!\!\models \varphi\} = \bigcup \{\Sigma^{\Gamma(\Delta),\eta} \mid Rg(\eta) = L\}.$$

□

[4] $\Sigma \subseteq \Pi$ does not imply $F(\Sigma) \subseteq F(\Pi)$.

It seems interesting to find a base (smallest collection of η's) for such a space. Perhaps the most important consequence of Theorem 5.2 is the existence of a narrow and well behaved semantics for which operation F is complete.

Corollary 5.4 $\boxed{\forall}$ If operation F satisfies postulates 1...5 then for every $\Sigma \subseteq \forall$ there exists η such that

$$F(\Sigma) = \{\varphi \in L \mid \bigwedge_{\substack{\mathcal{M} \models \Sigma : \mathcal{M} \text{ is a generic} \\ \text{model relative to } \Delta^{\eta}_{\Sigma}}} \mathcal{M} \models \varphi\}.$$

Proof. Immediately follows from theorems 3.25 and 5.2. □

Theorem 5.2 has a couple of other important consequences which we list below.

Theorem 5.5 $\boxed{\forall}$ $\boxed{C}$ Let $\Sigma \subseteq \Gamma(\Delta)$. Assume $\Pi \subseteq L$ satisfies for each $p \in Cond_{\Delta}(\Sigma)$ and $\varphi \in L$:

(i) if $\varphi \in \Gamma(\Delta)$ then: $\Pi \cup p \vdash \varphi$ iff $\Sigma \cup p \vdash \varphi$,

(ii) $\Pi \cup p \vdash \neg\varphi$ iff $(\bigwedge_{q \in Cond_{\Delta}(\Sigma) : p \subseteq q}) \sim (\Pi \cup q \vdash \varphi)$,

(iii) Π is closed under $\vdash$,

(iv) Π is a maximal subset of L satisfying (i)...(iii).

Then there exists η such that $\Pi = \Sigma^{\Gamma(\Delta),\eta}$.

Proof. We check that $F(\Sigma) = \Pi$ satisfies postulates 1...4.

Postulate 1 follows from (i) and $\Sigma \subseteq \Gamma(\Delta)$.

Postulate 2. By (ii), (iii) and Theorem 3.8 it suffices to prove that if

$$(\bigwedge_{p \in Cond_{\Delta}(\Sigma)}) \sim (\Pi \cup p \vdash \neg\varphi)$$

then

$$(\bigwedge_{p \in Cond_{\Delta}(\Sigma)}) \sim (\Sigma \cup p \vdash \neg\varphi).$$

The left side implies $(\bigwedge_{p \in Cond_{\Delta}(\Sigma)}) \sim (\Pi_{\Gamma(\Delta)} \cup p \vdash \neg\varphi)$, which by (i) gives $(\bigwedge_{p \in Cond_{\Delta}(\Sigma)}) \sim (\Sigma_{\Gamma(\Delta)} \cup p \vdash \neg\varphi)$, i.e. the right side (recall that $\Sigma \subseteq \Gamma(\Delta)$).

Postulate 3 follows from (ii), putting $p = q = 0$.

Postulate 4 explicitly assumed in (iii).

We have already demonstrated that postulates 1–4 are consequences of $F(\Sigma) = \Pi$ and (i)–(iii). Let

$$\mathbf{K} = \{\Lambda \subseteq L \mid \Pi \subseteq \Lambda \text{ and } F(\Sigma) = \Lambda \text{ satisfies postulates } 1 \ldots 4\}.$$

$\mathbf{K}$ is non-empty because $\Pi \in \mathbf{K}$. Let Ω be an increasing $\subseteq$-chain in $\mathbf{K}$. Let $\bigcup \Omega \vdash \varphi$. By the compactness theorem (cf e.g. [Bar78] Chap. A1, thm. 2.4) there exists a finite $\Phi \subseteq \bigcup \Omega$, such that $\Phi \vdash \varphi$. The finiteness of Φ implies that there exists $\Lambda \in \Omega$ such that $\Phi \subseteq \Lambda$. This proves that $\bigcup \Omega \vdash \varphi$ iff for some $\Lambda \in \Omega$, $\Lambda \vdash \varphi$. Therefore $F(\Sigma) = \bigcup \Omega$ satisfies postulates 1...4. From the Kuratowski-Zorn Lemma we conclude that $\mathbf{K}$ contains a maximal element, say Θ. $F(\Sigma) = \Theta$ satisfies postulates 1...5. Let $\Theta = \Sigma^{\Gamma(\Delta),\eta}$ (such an η exists by Theorem 5.2). By definition of $\mathbf{K}$ we have $\Pi \subseteq \Sigma^{\Gamma(\Delta),\eta}$. By Theorem 3.16 (i) and (ii) and Theorem 2.2 (vi), $\Pi = \Sigma^{\Gamma(\Delta),\eta}$ satisfies (i), (ii) and (iii) of the current theorem. Therefore (iv) and $\Pi \subseteq \Sigma^{\Gamma(\Delta),\eta}$ imply $\Pi = \Sigma^{\Gamma(\Delta),\eta}$. □

Theorem 5.6 $\boxed{\forall}$ $\boxed{C}$ Let $\Sigma \subseteq \Gamma(\Delta)$. Assume $\Pi \subseteq L$ satisfies for each $p \in Cond_{\Delta}(\Sigma)$ and $\varphi \in L$ conditions (i)...(iii) of Theorem 5.5 and moreover

(iv) $\Pi \cup p \vdash \forall x \varphi(x)$ **iff** $(\bigwedge_{c \in C})(\Pi \cup p \vdash \varphi(c))$.

Then there exists η such that $\Pi = \Sigma^{\Gamma(\Delta),\eta}$.

Proof. An easy induction shows that Π is a maximal subset of L satisfying (i)...(iii) of Theorem 5.5. □

The above Theorem provės that our generalized forcing covers all cases of [Kei73]. Since, in our case, condition (iv) of Theorem 5.6 need not be satisfied, it is a proper generalization. The following result demonstrates that the weak generalized $\forall$-forcing is covered as well.

Existence Lemma 5.7 $\boxed{\forall}$ $\boxed{C}$ For every $\Sigma \subseteq \Gamma(\Delta)$ there exists enumeration η of sentences of L such that:

$$\overset{w}{\Vdash} = \Delta_{\Sigma}^{\eta}.$$

Proof. Put $\Pi = \Sigma^{S}$. Using Definition 4.10 we check that assumptions of Theorem 5.6 are met: (i) is satisfied by lemma 4.15, (ii) by Theorem 4.6 (ii), (iii) follows from the completeness Theorem 4.9, and finally, (iv) follows from Theorem 4.6 (iv). □

It is possible to strengthen Lemma 5.7, extending Henrard's Theorem of [Hen73] over all cases of weak generalized $\forall$-forcing.

Theorem 5.8 $\boxed{\forall}$ $\boxed{C}$ If ζ is consistent with Kleene-Mostowski hierarchy then for every $\Sigma \subseteq \forall$,

$$\Delta_{\Sigma}^{\zeta} = \overset{w}{\Vdash}.$$

Proof. First, assume $\Sigma \subseteq \Gamma(\Delta)$. Let η be an enumeration of L satisfying Lemma 5.7, i.e. $\overset{w}{\Vdash} = \Delta_{\Sigma}^{\eta}$, and let ξ be an enumeration of L satisfying the following condition: for every $n \in \omega$ there exist ordinal numbers α_n and β_n with $Rg(\xi \restriction \alpha_n) = \forall_{n-1} \cap \Sigma^{\Gamma(\Delta),\eta}$ and $Rg(\xi \restriction \beta_n) = \forall_{n-1}$. It follows that ξ is consistent with Kleene-Mostowski hierarchy.

We show by induction that for every $n \in \omega$, $\Delta^{\eta}_{\Sigma} \restriction \forall_{n-1} = \Delta^{\xi}_{\Sigma} \restriction \forall_{n-1}$. The case $n = 0$ is given by Lemma 3.17. Assume for some $n \in \omega$ that $\Delta^{\eta}_{\Sigma} \restriction \forall_{n-1} = \Delta^{\xi}_{\Sigma} \restriction \forall_{n-1}$. Because $\Delta^{\eta}_{\Sigma} = \overset{w}{\Vdash}$, it satisfies the $\forall$-clause $(*)$ of Section 3. Therefore, using the same argument as in the proof of Lemma 3.19, one can demonstrate that $\Delta^{\xi}_{\Sigma} \restriction \forall_n \subseteq \Delta^{\eta}_{\Sigma} \restriction \forall_n$. On the other hand, all the sentences of $\forall_n \cap \Sigma^{\Gamma(\Delta),\eta}$ have smaller numbers in enumeration ξ than the sentences of $\forall_n \setminus \Sigma^{\Gamma(\Delta),\eta}$. Therefore $\Delta^{\eta}_{\Sigma} \restriction \forall_n \subseteq \Delta^{\xi}_{\Sigma} \restriction \forall_n$. This completes the induction.

We have $\overset{w}{\Vdash} = \Delta^{\eta}_{\Sigma} = \Delta^{\xi}_{\Sigma}$. Because ξ and ζ are consistent with Kleene-Mostowski hierarchy, Lemma 3.19 gives us $\Delta^{\xi}_{\Sigma} = \Delta^{\zeta}_{\Sigma}$. Thus $\overset{w}{\Vdash} = \Delta^{\zeta}_{\Sigma}$.

Second, assume $\Sigma \subseteq \forall$ instead of $\Sigma \subseteq \Gamma(\Delta)$. We have already demonstrated that (recall that $\Gamma(\Delta) \subseteq \forall$) for every η consistent with Kleene-Mostowski hierarchy, $(\Sigma_{\Gamma(\Delta)})^{\Gamma(\Delta),\eta} = (\Sigma_{\Gamma(\Delta)})^S$. So, let η be an enumeration satisfying Lemma 3.20. We have $\Sigma^{\Gamma(\Delta),\eta} = (\Sigma_{\Gamma(\Delta)})^{\Gamma(\Delta),\eta} =$ (already demonstrated) $(\Sigma_{\Gamma(\Delta)})^S =$ (by Lemma 4.13) Σ^S. Thus, by Lemma 4.12, $\Delta^{\eta}_{\Sigma} = \overset{w}{\Vdash}$. Hence Δ^{η}_{Σ} satisfies $(*)$, and therefore, by Lemma 3.19, $\Delta^{\eta}_{\Sigma} = \Delta^{\zeta}_{\Sigma}$, that is to say, $\overset{w}{\Vdash} = \Delta^{\zeta}_{\Sigma}$. □

Corollary 5.9 $\boxed{\forall}\,\boxed{C}$ For every $\Sigma \subseteq \forall$ and every ζ consistent with Kleene-Mostowski hierarchy,

$$\Sigma^S = \Sigma^{\Gamma(\Delta),\zeta}.$$

□

Corollary 5.10 $\boxed{\forall}$ For every $\Sigma \subseteq \forall$, if η is consistent with Kleene-Mostowski hierarchy then

$$\Sigma \subseteq (\Sigma_{\Gamma(\Delta)})^{\Gamma(\Delta),\eta}.$$

Proof. Follows from Theorem 2.2 (viii) and $(\Sigma_{\Gamma(\Delta)})^{\Gamma(\Delta),\eta} = \Sigma^{\Gamma(\Delta),\eta}$. □

Theorem 5.8 allows us to relate $\overset{w}{\Vdash}$ and $\Delta\!\!\models$ each other.

Theorem 5.11 $\boxed{\forall}\,\boxed{C}$ For each $\Sigma \subseteq \forall$ and each sentence $\varphi \in \forall$,

$$p \overset{w}{\Vdash} \varphi \text{ iff } \Sigma \cup p\Delta\!\!\models \varphi.$$

Proof. Let η be consistent with Kleene-Mostowski hierarchy. By Theorem 5.8 it suffices to prove that for each sentence $\varphi \in \forall$: $p\Delta^{\eta}_{\Sigma}\varphi$ iff $\Sigma \cup p\Delta\!\!\models \varphi$, that is to say (by Lemma 3.6) $0\Delta^{\eta}_{\Sigma} \wedge p \supset \varphi$ iff $\Sigma\Delta\!\!\models \wedge p \supset \varphi$, or ($\wedge p \supset \varphi$ is equivalent to a sentence $\psi \in \forall$) that $0\Delta^{\eta}_{\Sigma}\psi$ iff $\Sigma\Delta\!\!\models \psi$, for each $\psi \in \forall$.

Implication to the right follows from Corollary 5.10 and Lemma 3.10. We prove implication to the left by induction on the number of quantified variables in ψ. Assume ψ is quantifier-free and $\Sigma\Delta\!\!\models \psi$. By Theorem 3.7 we obtain $(\Sigma \cup \{\psi\})_{\Gamma(\Delta)} = \Sigma_{\Gamma(\Delta)}$. Let ζ be an enumeration of L, such that for some ordinal number $\beta \in Dm(\zeta)$, $\Sigma \cup \{\psi\} = Rg(\zeta \restriction \beta)$. It gives us $\psi \in \Sigma^{\Gamma(\Delta),\zeta}$, or in other words, $0\Delta^{\zeta}_{\Sigma}\psi$. Hence, by Lemma 3.17, $0\Delta^{\eta}_{\Sigma}\psi$. Assume

$\psi = \forall x \vartheta(x)$, where $\vartheta(c) \in \forall$. $\Sigma \mathrel{\Delta\!\!\!=} \forall x \vartheta(x)$ implies (by Definition 3.4) $(\bigwedge_{c \in C})(\Sigma \mathrel{\Delta\!\!\!=} \vartheta(c)$, i.e. (by induction hypothesis) $(\bigwedge_{c \in C})(0 \Delta^{\eta}_{\Sigma} \vartheta(c))$ or in other words (by theorems 5.8 and 4.6 (iv)), $0 \Delta^{\eta}_{\Sigma} \forall x \vartheta(x)$. □

The following theorem outlines the basic properties of forcing-based consequence operation ^{S}.

Theorem 5.12 $\boxed{\forall}$ $\boxed{C}$ For every $\Sigma \subseteq \forall$:

(i) $\Sigma \subseteq \Sigma^{S}$,

(ii) $\varphi \in \Sigma^{S} \supset \neg\varphi \notin \Sigma^{S}$,

(iii) Σ^{S} is closed under $\vdash$,

(iv) $(\Sigma^{S})_{\Gamma(\Delta)} = \Sigma_{\Gamma(\Delta)}$,

(v) ^{S} is a maximal operation which satisfies (i)...(iv).

Proof. (i) follows from Corollaries 5.10 and 5.9. Now, Lemma 4.13 allows us to assume $\Sigma \subseteq \Gamma(\Delta)$. By Theorem 3.7, (iv) implies

(vi) $\varphi \in \Sigma^{S} \supset \Sigma \mathrel{\Delta\!\!\!=} \varphi$.

On the other hand, (i) implies $\Sigma_{\Gamma(\Delta)} \subseteq (\Sigma^{S})_{\Gamma(\Delta)}$. Let $\varphi \in (\Sigma^{S})_{\Gamma(\Delta)}$. (iii) implies $\varphi \in \Sigma^{S} \cap \Gamma(\Delta)$, hence (vi) implies $\varphi \in \Gamma(\Delta)$ and $\Sigma \mathrel{\Delta\!\!\!=} \varphi$, which, by Theorem 3.7, means $\varphi \in \Gamma(\Delta)$ and $(\Sigma \cup \{\varphi\})_{\Gamma(\Delta)} = \Sigma_{\Gamma(\Delta)}$, or in other words, $\varphi \in \Sigma_{\Gamma(\Delta)}$. Thus (i), (iii), and (vi) imply $\Sigma_{\Gamma(\Delta)} = (\Sigma^{S})_{\Gamma(\Delta)}$. Therefore (i)...(v) are equivalent to (i)...(iii), (v), (vi). The truthfulness of these last ones is guaranteed by $\Sigma \subseteq \Gamma(\Delta)$, Theorem 5.2, and Corollary 5.9. □

Now we turn our attention to certain special cases of Δ.

6 Model 1: Universalism

The first case of generalized $\forall$-forcing, which we have found useful in data bases, corresponds to Δ containing all finite sets of atomic and negated atomic sentences of L, and $\Gamma(\Delta)$ (after closing it under $\wedge$) equal to $\forall$. Since in this case $Atom \setminus \Gamma(\Delta)$ is empty, the relation $\stackrel{w}{\Vdash}$ coincides with weak forcing $\stackrel{w}{\Vdash}$ of [Rob71]. In particular, Theorem 5.8 reduces to Henrard's Theorem of [Hen73].

If one interprets the elements of Δ as results of possible finite experiments then $0 \stackrel{w}{\Vdash} \varphi$ means that φ is experimentally irrefutable in any model for Σ or, in other words, that it is pragmatically $\forall$-true in all such models. It seems natural to adopt this kind of pragmatic truth when answering queries to a data base. This point of view we have advocated in [Suc84, Suc86,Suc89c]. It corresponds to an assumption that all possible updates of the data base are subject to direct experimental verification.

The degree of undecidability of $\Sigma^{f} = \{\varphi \in L \mid 0 \stackrel{w}{\Vdash} \varphi\}$ has been addressed in [Suc85]. In particular, it has been proven there that $\Sigma^{f} \cap \forall\exists$ is asymptotically decidable[5] relative to $Cond_{\Delta}(\Sigma)$.

7 Model 2: Positivism

From now on, we assume that the equality symbol = does not belong to L. Some of the facts we prove here do not use this assumption, however. For better clarity we precede every statement with symbol $\boxed{\neq}$ (no equality symbol) if this assumption is used (explicitly or implicitly) in its proof, or is needed for the correctness of the definition.

A formula is *positive* iff it is built up of atomic formulae using $\wedge, \vee, \forall$, and $\exists$. *Pos* denotes the set of all positive sentences of L.

The second case of generalized $\forall$-forcing we apply in this paper is the restriction of Δ to finite sets of *negated* atomic sentences of L (let us denote it by $nAtom^*$). In this case $\Gamma(\Delta)$ is equal to $Pos \cap \forall$. The assumption $\Sigma \subseteq \forall$ gives us $\Sigma_{Pos \cap \forall} \equiv \Sigma_{Pos}$.

Lemma 7.1 $\boxed{\forall}$ If $\Sigma \subseteq \forall$ then $(\Sigma_{Pos \cap \forall})_L = (\Sigma_{Pos})_L$.

Proof. Since $\Sigma_{Pos \cap \forall} \subseteq \Sigma_{Pos}$, it suffices to prove that $\Sigma_{Pos \cap \forall} \models \Sigma_{Pos}$. Let $\mathcal{A} \models \Sigma_{Pos \cap \forall} (= \Sigma_{Pos} \cap \forall)$. By the Łoś-Tarski Theorem (cf. [Bar78], Chap. A2, thm. 3.11), there exists $\mathcal{B} \models \Sigma_{Pos}$ with $\mathcal{A} \subseteq \mathcal{B}$. By the Lyndon Homomorphism Theorem ([Lyn59], thm. 5′), there exist $\mathcal{M} \models \Sigma$, $\mathcal{B}' \models \Sigma_{Pos}$, and $\mathcal{M} \models \Sigma$, such that $\mathcal{B} \subseteq \mathcal{B}'$, and $\mathcal{B}'$ is a homomorphic image of $\mathcal{M}$. Let A, B', and M be the universes of models $\mathcal{A}, \mathcal{B}'$, and $\mathcal{M}$, respectively. We have $A \subseteq B' = M$. Let $\mathcal{N} = \mathcal{M} \restriction A$. Because $\Sigma \subseteq \forall$, by the Łoś-Tarski Theorem, $\mathcal{N} \models \Sigma$. It is easily seen that $\mathcal{A}$ is a homomorphic image of $\mathcal{M}$. Using the Lyndon Homomorphism Theorem again, we obtain $\mathcal{A} \models \Sigma_{Pos}$. □

As we have noted previously (a remark after Theorem 4.6), the relation $\overset{w}{\Vdash}$ in case of $\Delta = nAtom^*$ is completely determined by 4.6 (i)...(iv), because $Atom \setminus \Gamma(\Delta)$ is empty. For the same reason 4.6 (i) may be restricted to $\varphi \in Atom$. This fact, similarly as in Section 6, enables an explicit inductive definition of this special case of generalized $\forall$-forcing.

Definition 7.2 $\boxed{C}$ The relation $\overset{w}{\dashv\Vdash}$ of *weak negative $\forall$-forcing* is defined inductively for all increments $p \in Cond_{nAtom^*}(\Sigma)$ and sentences $\varphi \in L$ (Σ is an implicit parameter of this relation).

(i) If $\varphi \in Atom$ **then** $p \overset{w}{\dashv\Vdash} \varphi$ **iff** $\Sigma \cup p \vdash \varphi$.

(ii) $p \overset{w}{\dashv\Vdash} \neg\varphi$ **iff** $(\bigwedge_{q \in Cond_{nAtom^*}(\Sigma): p \subseteq q}) \sim (q \overset{w}{\dashv\Vdash} \varphi)$.

(iii) $p \overset{w}{\dashv\Vdash} \varphi \wedge \psi$ **iff** $p \overset{w}{\dashv\Vdash} \varphi$ and $p \overset{w}{\dashv\Vdash} \psi$.

---

[5]i.e. Δ_2

(iv) $p \overset{w}{\neg\Vdash} \forall x\varphi(x)$ **iff** $(\bigwedge_{c\in C})(p \overset{w}{\neg\Vdash} \varphi(c))$.
(Other connectives are treated as appropriate abbreviations). □

Our assumption of $\Sigma \subseteq \forall$ is not necessary for the correctness of this definition. It also should be noted that $\overset{w}{\neg\Vdash}$ formally does not fall under the scheme of weak forcing of [Kei73].

We use $^{-S}$ instead of ^{S} (cf. Definition 4.10) in relation to weak negative $\forall$-forcing. Moreover, we use the term *negatively generic* $\forall$*-model* in sense of *generic* $\forall$*-model* in this case.

It follows from the previous discussion that $^{-S}$ is a maximal consequence operation preserving pragmatic $\forall \cap Pos$-truth, to which we refer as ***positive truth***. In this case, increments p of a data base may contain only negated atomic sentences. This means that the positive truth implies some kind of completeness of positive elementary information contained in the fixed part of a data base, which constitutes the essence of closed world assumptions. We will show that the restriction of $^{-S}$ to $\forall$ gives consistent version cwa_S (introduced in [Suc87]) of Reiter's closed world assumption cwa_R of [Rei78].

Definition 7.3 $\boxed{\neq}$

$$\varphi \in cwa_S(\Sigma) \text{ iff } \varphi \in L \cap \forall \text{ and } (\Sigma \cup \{\varphi\})_{Pos} = \Sigma_{Pos}. \qquad \square$$

cwa_S is stronger than cwa_R in all cases cwa_R is consistent. It is strictly stronger then generalized closed world assumption $GCWA$ of [Min82]. In Section 9 we will construct a quantifier-free sentence provable from cwa_S but not from $GCWA$, and a $\forall$-sentence provable from cwa_S but not from cwa_R. It has been shown in [Suc87], Theorem 6.1, that $cwa_S(\Sigma)$ is asymptotically decidable (actually: co-recursively enumerable) in the set of quantifier-free consequences of Σ.

The following theorem shows, among others, that the assumption of $\Sigma \subseteq \forall \cap Pos$ may be relaxed indeed to $\Sigma \subseteq \forall$ if used in the context of reasoning with cwa_S.

Theorem 7.4 $\boxed{\forall}$ $\boxed{\neq}$ Let $\Sigma \subseteq \forall$.
(i) $cwa_S(\Sigma)$ is consistent unless Σ is not.
(ii) If $\Pi \subseteq L \cap \forall$ and $\Pi_{Pos} = \Sigma_{Pos}$ then $\Pi \subseteq cwa_S(\Sigma)$.

Proof. In [Suc87], Theorem 4.1. □

The above result, a parallel of Kaiser's theorem of [Kai69], shows that cwa_S provides a powerful scheme of reasoning from positive fragments of universally axiomatizable data bases. This scheme, within domain of $\forall$-sentences, is a maximal one preserving positive truth, and, moreover, its conclusions never contradict the information contained in the data base. In the next section we will relate it to another widely adopted semantics of deductive data bases. Below we characterize cwa_S in terms of weak negative $\forall$-forcing.

Lemma 7.5 $\boxed{\forall}$ $\boxed{\neq}$ For every enumeration η, such that there exists an ordinal number $\alpha \in Dm(\eta)$ with $Rg(\eta \restriction \alpha) = \forall$, and for every $\Sigma \subseteq \forall$,

$$cwa_S(\Sigma) = \Sigma^{Pos \cap \forall, \eta} \cap \forall.$$

Proof. Let

$$\mathbf{K} = \{\Sigma^{Pos \cap \forall, \zeta} \mid \zeta \text{ is consistent with Kleene-Mostowski hierarchy}\}.$$

It follows from Definition 7.3 that $\bigcap \mathbf{K} \cap \forall \subseteq cwa_S(\Sigma) \subseteq \bigcup \mathbf{K} \cap \forall$. By Theorem 5.8 and Lemma 3.3, $\mathbf{K}$ consists of one element. Therefore the Lemma is true for every η consistent with Kleene-Mostowski hierarchy. On the other hand, if η is consistent with Kleene-Mostowski hierarchy then $\Sigma^{Pos \cap \forall, \eta} \cap \forall$ does not depend on how η behaves outside $\forall$. This observation completes the proof. □

An application of Corollary 5.9 to Lemma 7.5 gives us:

Theorem 7.6 $\boxed{\forall}$ $\boxed{C}$ $\boxed{\neq}$ For every $\Sigma \subseteq \forall$,

$$cwa_S(\Sigma) = \Sigma^S \cap \forall.$$ □

As an easy consequences of Theorem 7.6, Theorem 5.12 (iv), Theorem 4.6 (ii), and Lemma 7.1, we may observe that

Corollary 7.7 $\boxed{\forall}$ $\boxed{\neq}$ Let $\Sigma \subseteq \forall$.

(i) $cwa_S(\Sigma)_{Pos} = \Sigma_{Pos}$.

(ii) For each quantifier-free sentence φ of L,
$\neg\varphi \in cwa_S(\Sigma)$ iff $(\bigwedge_{p \in Cond_{nAtom^*}(\Sigma)}) \sim (\varphi \in cwa_S(\Sigma \cup p))$. □

Thus cwa_S is $Pos \cap \forall$-conservative, and monotone with respect to admissible increments.

8 Minimal Models

In this section we apply weak negative $\forall$-forcing to provide the minimal entailment with the $\forall$-complete syntactic characterization under the assumption that L is a purely relational[6] language. The reader is referred to [BS84], [EMR85], [BH86], [Hin88], and [She88], for other less successful trials in this aspect.

Definition 8.1 The binary relation $\preceq$ between first-order structures for language L is defined by

---

[6] i.e. no function symbols are allowed

> $\mathcal{M} \preceq \mathcal{N}$ **iff** $M = N$, and for every $c \in C$, the interpretation of c in $\mathcal{M}$ is the same as in $\mathcal{N}$, and for every atomic formula φ of L and every assignment s in $\mathcal{M}$, $\mathcal{M} \models \varphi[s]$ implies $\mathcal{N} \models \varphi[s]$;

where M and N denote the universes of $\mathcal{M}$ and $\mathcal{N}$, respectively. □

(The relation $\preceq$ was first introduced by Lyndon [Lyn59] under the name of *enlargement.*)

A structure $\mathcal{M}$ for the language L is called a *minimal* model of Σ iff it is canonic and $\preceq$-minimal in the class of models of Σ.

The main issue in this section is the following central problem of model theory for AI logic.

Problem 8.2 $\boxed{\forall}$ Given a $\forall$-theory Σ, find a complete syntactic characterization of the set $Cn_{min}(\Sigma)$ of all sentences of L true in each minimal model of Σ. □

We will provide here such a characterization for $Cn_{min}(\Sigma) \cap \forall$. For this purpose we need the following result.

Lemma 8.3 $\boxed{\neq}$ $\boxed{\forall}$ $\boxed{C}$ If L does not contain function symbols then for every sentence $\varphi \in \forall$,

$$\varphi \in Cn_{min}(\Sigma) \text{ iff } 0 \dashv\!\Vdash^{w} \varphi.$$

Proof. For infinite minimal models the proof is contained in [Suc89a], Lemma 4.10. Observe that in the absence of $=$ in L, every finite canonic model is elementarily equivalent to[7] an infinite minimal model, completes the proof. □

Lemma 8.3 allows us to relate minimalism, positivism, and closed world assumption to each other.

Theorem 8.4 $\boxed{\forall}$ $\boxed{C}$ $\boxed{\neq}$ If L does not contain function symbols then for each $\Sigma \subseteq \forall$ and each sentence $\varphi \in \forall$, the following conditions are equivalent:

(i) $\varphi \in cwas(\Sigma)$;
(ii) φ is true in all minimal models for Σ;
(iii) φ is positively true in all models for Σ;
(iv) φ is true in all negatively generic $\forall$-models for Σ;
(v) $0 \dashv\!\Vdash^{w} \varphi$.

Proof. The equivalency of (i) and (v) follows from Corollary 7.6. The equivalency of (ii) and (v) is given by Corollary 8.3. The equivalency of (iii) and (v) is given by Theorem 5.11. The equivalency of (iv) and (v) is given by the completeness theorem 3.25. □

---

[7] i.e. it satisfies exactly the same sentences as

Theorem 8.4 constitutes essential progress in characterizing minimal model semantics. In particular, it is strictly stronger then analogical characterization of [She88], Theorem 32, using Minker's *GCWA*. It is known that neither Reiter's cwa_R nor *GCWA* prove exactly those $\forall$-sentences which are true in all minimal models of $\forall$-theory Σ.

A combination of forcing and homomorphism techniques yields a version of Theorem 8.4 for functional languages with equality. The reader is referred to the forthcoming [Suc89b] and [ST] for details (also for a characterization of the case of not all relation symbols subjected to minimization).

The implication $\varphi \in Cn_{min}(\Sigma) \Rightarrow 0^{-}\overset{w}{\Vdash} \varphi$ holds for all $\varphi \in L$, not just for those from $\forall$, but outside $\forall$, ${}^{-}\overset{w}{\Vdash}$ is strictly stronger than the minimal entailment. Also the implication (cf. Corollary 3.10) $0 \dashv\!\overset{w}{\Vdash} \varphi \Rightarrow (\varphi$ is positively true in all models of $\Sigma)$ holds for arbitrary $\varphi \in L$, but not vice versa: $\dashv\!\overset{w}{\Vdash}$ is a maximal *consistent* operation which preserves positive truth. Finally, since $cwa_S(\Sigma)$ is co-recursively enumerable in the set of quantifier-free consequences of Σ, the same is true for $\forall$-fragment of minimal entailment.

9 Example

In this section we provide an example which shows that cwa_S is stronger than *cwa* and *GCWA*. We pick a language L with two unary relation symbols S and T. Naturally, L contains a countably infinite set C of constants, and the usual logical connectives: $\forall$, $\neg$, and $\wedge$ (others we treat as abbreviations). Moreover, we take $\Sigma = \{(\forall x)(S(x) \vee T(x))\}$.

$Cond_{-\Delta}(\Sigma)$ consists of unions of the form

$$\{\neg S(c) \mid c \in D_1\} \cup \{\neg T(c) \mid c \in D_2\},$$

where D_1 and D_2 are finite, disjoint subsets of C. For every $p \in Cond_{-\Delta}(\Sigma)$, $\Sigma \cup p$ has a minimal model of the form $\langle C, C_1, C_2\rangle$, where $C_1 \cup C_2 = C$, and $C_1 \cap C_2 = 0$, which is a minimal model of Σ too.

Let $\varphi = (\forall x)\neg(S(x) \wedge T(x))$. We show by direct calculations that $0 \dashv\!\overset{w}{\Vdash} \varphi$.

By Definition 7.2 (iv), $0^{-}\overset{w}{\Vdash} \varphi$ is equivalent to $(\bigwedge_{c\in C})(0^{-}\overset{w}{\Vdash} \neg(S(c) \wedge T(c)))$, that is to say, by (ii) of the same definition, to

$$(\bigwedge_{c\in C})(\bigwedge_{p\in Cond_{-\Delta}(\Sigma)}) \sim (p \dashv\!\overset{w}{\Vdash} S(c) \wedge T(c)),$$

or equivalently, by Theorem 5.12 (iv), to

$$(\bigwedge_{c\in C})(\bigwedge_{p\in Cond_{-\Delta}(\Sigma)}) \sim (\Sigma \cup p \vdash S(c) \wedge T(c)).$$

The last statement is true because $\neg(S(c) \wedge T(c))$ is true in certain models of $\Sigma \cup p$.

It follows from Definition 3.4 that φ is also positively true in every model of Σ. Also $\varphi \in cwas_S(\Sigma)$. As it follows from the definition of *GCWA* of [Min82], *GCWA* may add to $Cn(\Sigma)$ only negations of atomic sentences. Therefore, $\varphi \notin GCWA(\Sigma)$. Also $(\bigwedge_{c \notin C}) \sim (\neg(S(c) \wedge T(c)) \in GCWA(\Sigma))$. □

10 Open Problems

There are a number of open problems related both to the technical and applied sides of the theory of reasoning from updated data bases. As we have already seen, the operation $^{\Gamma,\eta}$ constitutes rather general scheme of non-monotonic logic. However, its relationships with other known non-monotonic logics, such as the Default Logic of [Rei80] with its subsequent modifications, require a systematic investigation. A relaxation of the implicit requirement $\Gamma(\Delta) \subseteq \forall$ (one of the more desirable possibilities would be $\Delta = Neg$ and $\Gamma(\Delta) = Pos$) in the definition of generalized $\forall$-forcing makes another problem. Also a weakening of the assumption $\Sigma \subseteq \forall$ (perhaps to $\Sigma \subseteq \forall\exists$) without violating $\Sigma \subseteq \Sigma^S$ seems interesting. The degree of undecidability of generalized $\forall$-forcing deserves a separate address.

Key words: closed world assumption, deductive data bases, machine reasoning, minimal entailment, model theory of non-monotonic logics, model-theoretic forcing, non-standard negation, pragmatic truth, theory of update.

AMS classification: 03C25, 03C40, 03C52, 68G99.

11 References

[Bar78] Jon Barwise, editor. *Handbook of Mathematical Logic*. North-Holland, Amsterdam, second edition, 1978.

[BH86] N. Bidoit and R. Hull. Positivism vs. minimalism in deductive databases. In A. Silberschatz, editor, *Proceedings of 5-th Symposium on Principles of Database Systems*, pages 123–132, A.C.M. SIGACT-SIGMOD, Association for Computing Machinery, New York, 1986.

[BS84] Genviéve Bossu and Pierre Siegel. Saturation, nonmonotonic reasoning, and the closed world assumption. *Artificial Intelligence*, 25(1):13–64, 1984.

[EMR85] David W. Etherington, Robert Mercer, and Raymond Reiter. On the adequacy of predicate circumscription for closed-world reasoning. *Computational Intelligence*, 1:11–15, 1985.

[Hen73] Paul Henrard. Le 'forcing-compagnon' sans 'forcing'. *C. R. Acad. Sc.*, 276(Ser. A):821–822, 1973.

[Hin88] Jaakko Hintikka. Model minimization - an alternative to circumscription. *Journal of Automated Reasoning*, 4:1–13, 1988.

[Kai69] Kurt Kaiser. Über eine verallgemeinerung der Robinsonschen Modellvervollstaändigung. *Zeitschrift für Mathematik Logik und Grundlagen der Mathematik*, 15:37–48, 1969.

[Kei73] H. Jerome Keisler. Forcing and the omitting types theorem. In M. Morley, editor, *Studies in Model Theory*, pages 96–133, Mathematical Association of America, 1973.

[Lyn59] Roger C. Lyndon. Properties preserved under homomorphism. *Pacific J. Math.*, 9:143–154, 1959.

[Min82] Jack Minker. On indefinite databases and closed world assumption. In *Proceedings of Sixth Conference on Automated Deduction*, pages 292–308, Springer Verlag, Berlin, New York, 1982.

[Rei78] Raymond Reiter. On closed world data bases. In Hervé Gallaire and Jack Minker, editors, *Logic and Data Bases*, pages 55–76, Plenum Press, 1978.

[Rei80] Raymond Reiter. A logic for default reasoning. *Artificial Intelligence*, 13(1–2):81–132, 1980.

[Rob71] Abraham Robinson. Forcing in model theory. In *Proc. Simp. Mat.*, pages 64–80, Institute Nationale di Alta Matematica, 1971.

[She88] John C. Shepherdson. Negation in logic programming. In Jack Minker, editor, *Foundations of Deductive Databases and Logic Programming*, Morgan Kaufmann Publ., Inc., Los Altos, 1988.

[ST] Marek A. Suchenek and James H. Thomas. Forcings and homomorphisms in deductive data bases. Submitted.

[Suc84] Marek A. Suchenek. Forcing treatment of incomplete information in data bases. In *International Symposium on Model Theory in Foundations of Computer Science*, Budapest, September 1984.

[Suc85] Marek A. Suchenek. *On asymptotic decidability of model-theoretic forcing.* Research Reports 63/85, Institute of Computer Science, Warsaw Technical University, Nowowiejska 15/19, 00-665 Warszawa, Poland, 1985.

[Suc86] Marek A. Suchenek. Non-monotonic derivations which preserve pragmatic truth. In Maria Zemankova and M. L. Emrich, editors, *Proceedings of the International Symposium on Methodologies for Intelligent Systems, Colloquia Program, ORNL-6362*, pages 69–74, Oak Ridge National Laboratory, Knoxville, October 1986.

[Suc87] Marek. A. Suchenek. Forcing versus closed world assumption. In Zbigniew W. Raś and Maria Zemankova, editors, *Methodologies for Intelligent Systems*, pages 453–460, North-Holland, 1987.

[Suc89a] Marek A. Suchenek. Minimal models for closed world data bases. In *Fourth International Symposium on Methodologies for Intelligent Systems*, North-Holland, Charlotte, N.C., October 12 - 14 1989. To appear in Proceedings of the Symposium.

[Suc89b] Marek A. Suchenek. A syntactic characterization of minimal entailment. In *North American Conference on Logic Programming*, MIT Press, Cleveland, OH, October 16–20 1989. To appear in Proceedings of the Conference.

[Suc89c] Marek A. Suchenek. Two applications of model-theoretic forcing to Lipski's data bases with incomplete information. *Fundamenta Informaticae*, 12:269–288, 1989.

Noncommutative Cylindric Algebras and Relativizations of Cylindric Algebras

Richard J. Thompson[1]

We obtain the class NA of noncommutative cylindric algebras from the class CA of cylindric algebras by weakening the axiom C_4 of commutativity of cylindrifications to C_4^* (see below), and we obtain NCA from CA by omitting C_4 completely.

1 Some motivation for studying noncommutative cylindric algebras

Noncommutative cylindric algebras (NA's) have the same "substitutional structure" as cylindric algebras (CA's from now on), where substitutional structure refers to the equational behaviour of the substitution operations, the s_j^i's. For certain technical reasons, NA's turn out to be useful in the study of the substitutional structure of CA's. See Theorem 3 below.

In a sense, NA's have a nicer representation theory than CA's. Namely, the class of NA's satisfying the merry-go-round identities (MGR) admits a geometric characterization in terms of certain "concrete" algebras (of sets of sequences), while the corresponding result for the class of CA's satisfying MGR is only a representation result but not a characterization result. (Some of the concrete algebras are not CA's. See the remarks between Theorems 6 and 7 below.) It is not clear to the present author how to obtain a similarly nice characterization ("if and only if") for a class of CA's satisfying "some finite schema".

A third motivation for NA's comes from the theory of relativations of CA's extensively investigated in **[HR75]**, **[HMT71]**, **[HMTAN81]**, **[HMT85]**, **[N80]**, and **[R75]**. See, for example, Theorem 8 and Corollary 9 below.

A fourth motivation comes from a strongly related development in relation algebra theory pursued in **[M82]**, **[TG87]**, **[N86]**, and **[N87]**. If we take the CA version of the symmetric relativization theory, then we get exactly the NA's from CA's.

We use the notation of **[HMT71]**, **[HMT85]**. Let α be any ordinal. An algebra $\mathfrak{A} = \langle A, +, \cdot, -, 0, 1, c_i, d_{ij}\rangle_{i,j\in\alpha}$, where $+$ and $\cdot$ are binary operations, $-$ and c_i are unary operations, and 0, 1, and d_{ij} are constants for every $i, j \in \alpha$, is an α-dimensional cylindric algebra (a CA_α) if it satisfies

[1](c/o I. Németi) Mathematical Institute of the Hungarian Academy of Sciences, Budapest, P.O. Box 127, H-1364 Hungary

C_0–C_7 for every $i,j,k \in \alpha$.

(C_0–C_3) $\langle A, +, \cdot, -, 0, 1, \mathsf{c}_i \rangle_{i\in\alpha}$ is a Boolean algebra with additive closure operators c_i such that the complements of c_i-closed elements are c_i-closed (*i.e.*, if $x = \mathsf{c}_i x$ then $\mathsf{c}_i(-x) = -x$),

(C_4) $\mathsf{c}_i \mathsf{c}_j x = \mathsf{c}_j \mathsf{c}_i x$,

(C_5) $\mathsf{d}_{ii} = 1$,

(C_6) $\mathsf{d}_{ij} = \mathsf{c}_k(\mathsf{d}_{ik} \cdot \mathsf{d}_{kj})$ if $k \notin \{i,j\}$,

(C_7) $\mathsf{d}_{ij} \cdot \mathsf{c}_i(\mathsf{d}_{ij} \cdot x) \leq x$ if $i \neq j$.

For all $i,j \in \alpha$, let $\mathsf{s}^i_j x = \mathsf{c}_i(\mathsf{d}_{ij} \cdot x)$ if $i \neq j$, else $\mathsf{s}^i_j x = x$. Let C_4^* be the following weaker version of C_4:

(C_4^*) $$\mathsf{c}_i \mathsf{c}_j x \geq \mathsf{c}_j \mathsf{c}_i x \cdot \mathsf{d}_{jk} \text{ if } k \notin \{i,j\}.$$

If Σ is a set of equations (in the language of CA_α's) then ModΣ denotes the class of all algebras that satisfy Σ (and which are similar to CA_α's).

DEFINITION 1. *By a non-commutative cylindric algebra of dimension α (an NA_α) we understand a model of C_0–C_3, C_4^*, and C_5–C_7, that is,* $\mathsf{NA}_\alpha = \text{Mod}\{C_0\text{–}C_3, C_4^*, C_5\text{–}C_7\}$.

DEFINITION 2. ([N86]) $\mathsf{NCA}_\alpha = \text{Mod}\{C_0\text{–}C_3, C_5\text{–}C_7\}$.

PROPOSITION 1. *In NCA_α, (validity of) C_4^* is equivalent with both (*) and (**) below.*

$$(*) \quad \mathsf{c}_k \mathsf{s}^i_j x \leq \mathsf{s}^i_j \mathsf{c}_k x \quad \textit{for} \quad k \notin \{i,j\}.$$

$$(**) \quad \mathsf{s}^k_j \mathsf{s}^i_j x = \mathsf{s}^i_j \mathsf{s}^k_j x \quad \textit{for distinct} \quad i,j,k \in \alpha.$$

Note that $\mathsf{CA}_\alpha \subset \mathsf{NA}_\alpha \subset \mathsf{NCA}_\alpha$ if $\alpha \geq 3$, else $\mathsf{NA}_\alpha = \mathsf{NCA}_\alpha$.

PROPOSITION 2. *All the results in [**HMT71**], §§1.2, 1.3, 1.5 are true for NCA_α as well, except Theorems 1.2.15–17, 1.3.18–20, 1.5.8(ii), 1.5.9(i), some parts of 1.5.10, and certain theorems among 1.5.11–23. In particular, 1.5.10(iii)(iv)(vi) hold in NA_α but not in NCA_α.*

The only embarrassment in the proof arises from Theorem 1.5.10: without C_4^* we can obtain (i), (v) and (ii); given (iii) and (iv) we can obtain (vi). However, C_4^* is required to obtain (iv): $\mathsf{s}^i_j \mathsf{s}^k_j x = \mathsf{s}^k_j \mathsf{s}^i_j x$, from which the more general (iii) can be obtained.

For the derivations mentioned, not using C_4^* (or C_4), the proofs in [**HMT71**] apply directly.

THEOREM 3.

(i) *Let τ, σ be terms (of CA_α's) built up exclusively from the s^i_j's as fundamental operations. Then*

$$\mathsf{CA}_\alpha \models \tau = \sigma \quad \textit{iff} \quad \mathsf{NA}_\alpha \models \tau = \sigma.$$

(ii) *If NA_α is replaced with NCA_α in (i), then (i) becomes false.*

Part (i) follows from Theorem 8(i) below (part $\subseteq$) together with (**) in the proof of Theorem 8(i).

Let $\mathsf{t}^i_j x = \mathsf{d}_{ij} \cdot \mathsf{c}_i x$. (Note that t^i_j is the conjugate of s^i_j in the sense of [**JT51**]). Let $\mathsf{C}_4^{*\partial}$ be

$$(\mathsf{C}_4^{*\partial}) \qquad \mathsf{s}^k_j \mathsf{s}^i_j x = \mathsf{t}^i_j \mathsf{s}^k_j x \quad \text{for } i, j, k \text{ distinct.}$$

PROPOSITION 4.

(i) $\mathsf{NCA}_\alpha \models (\mathsf{C}_4^* + \mathsf{C}_4^{*\partial}) \to \mathsf{C}_4$.
(ii) *Both* $(\mathsf{C}_4^* \to \mathsf{C}_4^{*\partial})$ *and* $(\mathsf{C}_4^* \leftarrow \mathsf{C}_4^{*\partial})$ *fail in* NCA_α *for* $\alpha \geq 2$.

The following is a useful fact about the connection between equations of t^i_j's and those of s^i_j's.

PROPOSITION 5. *Let* $\mathfrak{A} \in \mathsf{NCA}_\alpha$ *and* $n, r \in \omega$. *Then the following statements are equivalent.*

(i) $\mathfrak{A} \models \mathsf{s}^{i_1}_{j_1} \dots \mathsf{s}^{i_n}_{j_n}(x) = \mathsf{s}^{k_1}_{m_1} \dots \mathsf{s}^{k_r}_{m_r}(x)$.
(ii) $\mathfrak{A} \models \mathsf{t}^{i_n}_{j_n} \dots \mathsf{t}^{i_1}_{j_1}(x) = \mathsf{t}^{k_r}_{m_r} \dots \mathsf{t}^{k_1}_{m_1}(x)$.

The proof is a straightforward computation using the definitions of s^i_j and t^i_j (and relying heavily on C_7).

For a class K of algebras, EqK is the equational theory of K (the set of equations valid in K). MGR is the postulate scheme

$$(\mathrm{MGR}) \quad \mathsf{s}^k_i \mathsf{s}^i_j \mathsf{s}^j_m \mathsf{s}^m_k \mathsf{c}_k x = \mathsf{s}^k_m \mathsf{s}^m_i \mathsf{s}^i_j \mathsf{s}^j_k \mathsf{c}_k x \quad \text{for } k \notin \{i, j, m\} \text{ and } m \notin \{i, j\}$$

Henkin proved that $\mathsf{CA}_\alpha \not\models MGR$ if $\alpha > 2$ (see [**HMT85**]).

THEOREM 6. ([**N86**], [**N87**]).

(i) $Eq\mathsf{NCA}_\alpha$ *is decidable.*
(ii) $Eq(\mathsf{NA}_\alpha \cap \mathrm{Mod}(MGR))$ *is decidable.*

A geometric representation theorem, characterizing $\mathsf{NA}_\alpha \cap \mathrm{Mod}(MGR)$ as isomorphic copies of algebras of sets of α-sequences, is proved in [**AT88**] and [**M89**]. The theorem is due to Diane Resek and R. J. Thompson while its proof in [**AT88**] is due to H. Andreka. A stronger version is given in [**M89**]. ([**HMTAN81**] contains a detailed study of algebras of sets of α-sequences, but for introductory purposes the less detailed [**HMT85**] is more suitable.)

We do not know whether $Eq\mathsf{NA}_\alpha$ is decidable or not for $\alpha \geq 3$.

Recall from [**HMT71**] that if $b \in A$ and $\mathfrak{A} \in \mathsf{CA}_\alpha$ then $\mathfrak{Rl}_b\mathfrak{A}$ is the algebra obtained from $\mathfrak{A}$ by relativization to b. The universe of $\mathfrak{Rl}_b\mathfrak{A}$ is $\{x \in A : x \leq b\}$ and the operations are the natural ones. For a class K of algebras similar to CA_α's we define $\mathsf{Rl}K = \{\mathfrak{Rl}_b\mathfrak{A} : b \in \mathfrak{A} \in K\}$, and $\mathbf{S}K$ denotes the class of all subalgebras of elements of K.

THEOREM 7. (H. Andreka and R. J. Thompson)

(i) $\mathsf{RlNCA}_\alpha = \mathsf{SRlNCA}_\alpha$.
(ii) $\mathsf{RlNA}_\alpha \neq \mathsf{SRlNA}_\alpha$, *if* $\alpha \geq 3$.

For a class K similar to CA_α's define

$$\mathsf{RI}_{sym}K := \{\mathfrak{RI}_b\mathfrak{A} : \mathfrak{A} \in K, b \in A, \text{ and } b \leq \mathsf{s}^i_j b \text{ for all } i,j \in \alpha\}.$$

THEOREM 8.

(i) $\mathsf{NA}_\alpha = \mathsf{SRI}_{sym}\mathsf{CA}_\alpha$.

(ii) $\mathsf{NCA}_\alpha = \mathsf{SRI}_{sym}(\mathsf{NCA}_\alpha \cap \mathrm{Mod}(\mathrm{C}^{*\partial}_r))$.

So NCA_α is related to $\mathsf{NCA}_\alpha \cap \mathrm{Mod}(\mathrm{C}^{*\partial}_4)$ exactly as NA_α is related to $\mathsf{NA}_\alpha \cap \mathrm{Mod}(\mathrm{C}^{*\partial}_4) = \mathsf{CA}_\alpha$.

To prove the direction $\supseteq$, assume $b \in \mathfrak{A} \in \mathsf{CA}_\alpha$ with

$$(*) \qquad b \leq \mathsf{s}^i_j b \quad \text{for all } i,j \in \alpha \text{ (in } \mathfrak{A}).$$

Let $\mathfrak{R} = \mathfrak{RI}_b\mathfrak{A}$. By the proof of 2.2.3 of **[HMT71]**, $\mathfrak{R} \in \mathsf{NCA}_\alpha$. Let $rl_b : A \to R$ as in **[HMT85]**, p. 30. By Theorem 1(i) of [**N80**], p. 46, we have

$$(**) \qquad rl_b : \langle \mathfrak{BI}\,\mathfrak{A}, \mathsf{s}^i_j\rangle_{i,j\in\alpha} \to \langle \mathfrak{BI}\,\mathfrak{R}, \mathsf{s}^{i(\mathfrak{R})}_j\rangle_{i,j\in\alpha} \quad \text{is a homomorphism}$$

since $-b \geq -\mathsf{s}^i_j b = \mathsf{s}^i_j(-b)$, by $(*)$. Actually, [**N80**] implies that $(*)$ and $(**)$ are equivalent. Therefore all equations involving only the s^i_j's are preserved under rl_b whenever $(*)$ holds. By Proposition 1, $\mathfrak{R} \models (\mathrm{C}^*_4)$. Direction $\subseteq$ is harder to prove.

COROLLARY 9. $\mathsf{NA}_\alpha = \mathrm{Mod}\{\mathsf{c}_i\mathsf{d}_{ij} = 1 : i,j \in \alpha\} \cap \mathsf{SRICA}_\alpha$.

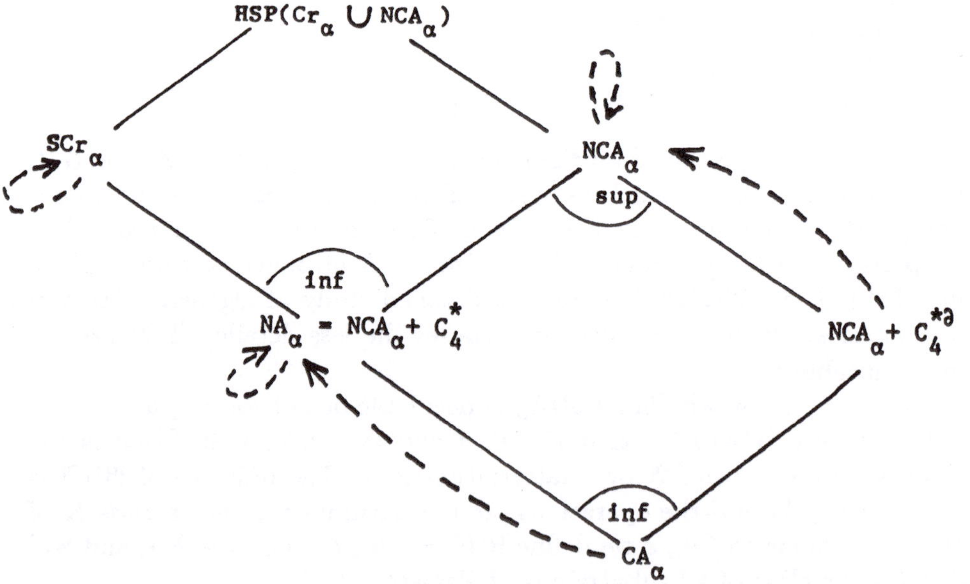

The above is a *sublattice* of the lattice of varieties where the broken arrow $- - \to$ represents the operator SRI_{sym}.

Recall that by Henkin's result, $\mathsf{NCA}_\alpha \not\models MGR$ for $\alpha \geq 3$.

PROPOSITION 10. $\mathsf{NCA}_\alpha \cap \mathrm{Mod}(MGR) \not\models (\mathsf{C}_4^*)$.

2 When do S and Rl commute?

If $\mathfrak{A} \in \mathsf{CA}_\alpha$ and $b \in A$, then we let $\mathsf{Rl}_b\mathsf{S}\mathfrak{A} := \{\mathfrak{Rl}_b\mathfrak{C} : b \in C,\ \mathfrak{C} \subseteq \mathfrak{A}\}$, and $\mathsf{Rl}_b\mathfrak{A} = \{\mathfrak{Rl}_b\mathfrak{A}\}$. For any $X \subseteq A$, $\mathfrak{Sg}^{\mathfrak{A}}X$ is the subalgebra of $\mathfrak{A}$ generated by X. $\mathfrak{Mn}(\mathfrak{A})$ denotes the subalgebra of $\mathfrak{A}$ generated by the constants of $\mathfrak{A}$, that is, $\mathfrak{Mn}(\mathfrak{A}) = \mathfrak{Sg}^{\mathfrak{A}}\emptyset$.

THEOREM 11. *Let $\alpha \geq \omega$, $\mathfrak{A} \in \mathsf{CA}_\alpha$ and $b \in A$. Assume $\mathsf{c}_i b \cdot \mathsf{c}_j b = b \leq \mathsf{s}_j^i b$ for all distinct $i, j \in \alpha$.*

(i) *Let $\mathfrak{B} \subseteq \mathfrak{Rl}_b\mathfrak{A}$. Then $\mathfrak{B} \in \mathsf{RlS}\mathfrak{A}$ iff*

$$b \cdot \mathsf{c}^{\partial}_{(\Delta)} \mathsf{c}_i(\bar{d}(\Delta \times \{i\}) \in B \text{ and } b \cdot \mathsf{c}^{\partial}_{(\Delta)} \mathsf{c}_i[\bar{d}(\Delta \times \{i\}) \cdot -\mathsf{c}_{(\Theta)} b] \in B$$

for all finite $\Delta \subseteq \Theta$ and $i \notin \Theta$.

(ii) *Then $\mathsf{SRl}_b\mathfrak{A} = \mathsf{Rl}_b\mathsf{S}\mathfrak{A}$ iff $\mathfrak{Mn}(\mathfrak{Rl}_b(\mathfrak{A}) \in \mathsf{RlS}\mathfrak{A}$.*

(iii) *If $Zd(\mathfrak{Rl}_b\mathfrak{Sg}^{\mathfrak{A}}\{b\}) \subseteq B$ then $\mathfrak{B} \in \mathsf{RlS}\mathfrak{A}$.*

The proof of part (i) goes by an elimination of cylindrifications argument, describing $\mathfrak{Sg}^{\mathfrak{A}}B$ whenever $(*)$ holds. Parts (ii) and (iii) are corollaries of part (i).

The following is a solution of Problem I.1 of **[HMTAN81]**, p. 127.

COROLLARY 14. *Let $\alpha \geq \omega$. Let $\mathfrak{B}$ be a normal Gws_α. Then there is $\mathfrak{A} \in \mathsf{Cs}_\alpha$ with $\mathrm{base}(\mathfrak{A}) = \mathrm{base}(\mathfrak{B})$ such that $\mathfrak{B} \in \mathsf{Rl}\mathfrak{A}$.*

Acknowledgments: Several of our proofs use results in **[R75]**. Thanks are due to I. Sain for extensive help in preparing this paper. This work was done while visiting the Mathematical Institute of Hungarian Academy of Science on an IREX fellowship.

References

[AT88] Andréka, H., Thompson, R. J., *A Stone type representation theorem for algebras of relations of higher rank*, Transactions of the American Mathematical Society **309 (2)** (October 1988), 671–682.

[HMT71] Henkin, L., Monk, J. D., Tarski, A., "Cylindric algebras, Part I," North-Holland, Amsterdam, 1971.

[HMT85] ——————, "Cylindric algebras, Part II," North-Holland, Amsterdam, 1985.

[HMTAN81] Henkin, L., Monk, J.D., Tarski, A., Andreka, H., Németi, I., "Cylindric Set Algebras," Lecture Notes in Mathematics **883**, Springer-Verlag, Berlin, 1981.

[HR75] Henkin, L., Resek, D., ***Relativization of cylindric algebras***, Fundamenta Mathematica **82** (1975), 363–383.

[JT51] Jónsson, B., Tarski, A., ***Boolean algebras with operators, Part I***, American Journal of Mathematics **73** (1951), 891–939.

[M82] Maddux, R.D., ***Some varieties containing relation algebras***, Transactions of the American Mathematical Society **272** (1982), 501–526.

[M89] ____________, ***Canonical relativized cylindric set algebras***, Proceedings of the American Mathematical Society **107** (1989), 465–478.

[N80] Németi, I., ***Some constructions of cylindric algebra theory applied to dynamic algebras of programs***, Computational Linguistics and Computer Languages **14** (1980), 43–65.

[N86] Németi, I., "Free algebras and decidability in algebraic logic," Doctoral Dissertation (B) for D.Sc. (or Dr. Rer. Nat.) with Hung. Acad. of Sci. Budapest, 1986..

[N87] Németi, I., ***Decidability of relation algebras with weakened associativity***, Proceedings of the American Mathematical Society **2** (1987), 340–344.

[R75] Resek, D., "Some results on relativized cylindric algebras," Doctoral Dissertation, University of California, Berkeley, 1975.

[TG87] Tarski, A., Givant, S., "A formalization of set theory without variables," Colloquium Publications in Mathematics **25**, American Mathematical Society, 1987.

On the λ-Definable Tree Operations

Marek Zaionc[1]

ABSTRACT A λ-language over simple type structures is considered. The type $\Upsilon = (0 \to (0 \to 0)) \to (0 \to 0)$ is called a binary tree type because of the isomorphism between binary trees and closed terms of this type. Therefore any closed term of type $\Upsilon \to (\Upsilon \to \cdots \to (\Upsilon \to) \cdots)$ represents an n-ary tree function. The problem is to characterize tree operations represented by the closed terms of the examined type. It is proved that the set of λ definable tree operations is the minimal set containing constant functions, projections and closed under composition and the limited version of recursion. This result should be contrasted with the results of Schwichtenberg and Statman (cf. [Sch75], [Sta79]) which characterize the λ definable functions over the natural number type $(0 \to 0) \to (0 \to 0)$ by composition only, as well as with the result of Zaionc (cf [Zai87]) for word λ definable functions over type $(0 \to 0) \to ((0 \to 0) \to (0 \to 0))$ which are also characterized by means of composition.

Introduction

The λ calculus introduced by Church is a calculus of expressions that naturally describes the notion of function. Functionals are considered dynamically like rules rather than set-theoretic graphs. The λ calculus mimics the procedure of computation of the program by the process called beta reduction. There is a natural way of expressing objects like numbers, words, trees and other syntactic entities in the λ calculus. Therefore dynamic operations on objects of this kind can be described by terms of λ calculus. All those objects are of considerable value for computer scientists. For example natural number "n" (Church's numeral) is represented by a functional that returns the n-fold composition of its argument. Therefore λ terms may be considered as algorithms (programs) on Church's numerals. This is a well known result (proved by Church and Kleene) relating all partial computable functions with λ terms. Of course, the notion of partial recursive function can be naturally extended to other structures like words, trees etc. It is natural that the Church-Kleene Theorem might be extended to these structure.

The typed version of λ calculus is obtained by imposing the simple types on the λ calculus. The problem of representing structures is basically the same in the typed λ calculus, however the rigid type structure in the syntax of λ calculus dramatically reduces expressiveness of functions on these structures. The first result concerning representability was proved by

[1]Instytut Informatyki Uniwersytet Jagielloński Kopernika 27, 31–501 Kraków, POLAND

Schwichtenberg in 1975 (see [Sch75]). The numerical functions represented in typed λ calculus are exactly the functions generated by the operation of composition from the constants 0 and 1 and functions ***addition, multiplication*** and ***conditional*** (extended polynomials). A similar result for word operations is the following: the word functions represented in typed λ calculus are exactly the functions generated by composition from the empty word and operations ***append, substitution*** and ***cut***. (Consult [Zai87] for details). Since any homogeneous free algebra may be represented in typed λ calculus in the same fashion as numbers or words, it seems interesting to investigate lambda definability on such algebras.

In this paper we investigate lambda definable operations on the algebra of binary trees. We prove the following characterization of tree functions represented in typed λ calculus: the λ definable tree functions are exactly the functions generated by composition and limited primitive recursion from the empty tree and operation join. It is not known to the author if the set of all λ definable tree operations can be generated by composition alone from some finite set of basic tree functions or if it is possible to reduce limited primitive recursion to composition only.

1 Binary Trees and Tree Operations

The set of binary trees T is defined recursively as follows: ε is a tree, and if t_1, t_2 are trees then $t_1 \wedge t_2$ is a tree. ε is called the empty tree and t_1, t_2 are left and right subtrees respectively of the tree $t_1 \wedge t_2$. We investigate tree operations, *i.e.*, functions $f : T^n \rightarrow T$. We begin by defining several tree operations.

DEFINITION 1.1. *The* ***tree constructor*** $\wedge$ *can be viewed as a function* $\wedge : T^2 \rightarrow T$. e^n *is the* n*-ary function which maps onto the empty tree* ε *constantly.* p_i^n *is the* n*-ary projection which extracts the* i*-th argument.*

$$\wedge(x_1, x_2) = x_1 \wedge x_2$$
$$e^n(x_1, \ldots, x_n) = \varepsilon$$
$$p_i^n(x_1, \ldots, x_n) = x_i.$$

We are going to investigate some classes of tree operations.

DEFINITION 1.2. *A class* X *of tree functions is* ***closed under primitive recursion*** *if the* $n+1$*-ary function* h *defined by*

$$h(\varepsilon, x_1, \ldots, x_n) = g(x_1, \ldots, x_n)$$
$$h(s \wedge s', x_1, \ldots, x_n) = f(h(s, x_1, \ldots, x_n), h(s', x_1, \ldots, x_n), x_1, \ldots, x_n)$$

belongs to X *whenever functions* g, f *are in* X.

Let us distinguish four classes of tree operations F, F_0, F_1, F_λ.

DEFINITION 1.3. *The class F of tree operations is defined as the minimal class containing the functions $\wedge$, e^n, p_i^n, and closed under composition and primitive recursion.*

The subclass $F_0 \subseteq F$ is the minimal class containing $\wedge$, e^n and p_i^n closed under composition.

F_1 is a class of all $n+2$-ary tree functions f such that for all trees $t_1, \ldots, t_n \in T$ the binary function p defined by $p(y,z) = f(y,z,t_1,\ldots,t_n)$ belongs to F_0.

The class F_λ is the minimal class containing $\wedge$, e^n and p_i^n, closed under composition and the following version of primitive recursion: if $g \in F_\lambda$ and $f \in F_\lambda \cap F_1$ then the function h defined below belongs to F_λ.

$$h(\varepsilon, x_1, \ldots, x_n) = g(x_1, \ldots, x_n)$$
$$h(s \wedge s', x_1, \ldots, x_n) = f\,(h(s, x_1, \ldots, x_n) h(s', x_1, \ldots, x_n), x_1, \ldots, x_n)$$

DEFINITION 1.4. *If f_1 and f_2 are two n-ary functions then by $f_1 \wedge f_2$ we mean the n-ary function defined by the formula $(f_1 \wedge f_2)(x_1, \ldots, x_n) = (f_1(x_1, \ldots, x_n)) \wedge (f_2(x_1, \ldots, x_n))$.*

2 Typed λ-Calculus

Our language is based on Church's [Chu41] simple theory of types. The set of types is introduced as follows: 0 is a type, and if τ and μ are types then $\tau \to \mu$ is a type. We will use the following notation: if $\tau_1, \ldots, \tau_n, \tau$ are types then by $\tau_1, \cdots, \tau_n \to \tau$ we understand the type $\tau_1 \to (\tau_2 \to \cdots (\tau_n \to \tau) \cdots)$. Therefore every type τ has a form $\tau_1, \cdots, \tau_n \to 0$. The type τ_i is called a component of τ and is denoted by $\tau[i]$. The type $\tau[i_1, \ldots, i_k]$ is defined inductively to be $\tau[i_k, \ldots, i_{k-1}])[i_k]$.

By $\tau^n \to \mu$ we mean the type $\tau, \cdots, \tau \to \mu$ with n occurrences of τ (with $\tau^0 \to \mu = \mu$). For any type τ we define numbers $\arg(\tau)$ and $\mathrm{rank}(\tau)$ as follows: $\arg(0) = \mathrm{rank}(0) = 0$ and if $\tau = \tau[1], \ldots \tau[n] \to 0$ then $\arg(\tau) = n$ and $\mathrm{rank}(\tau) = \max_{i=1 \ldots n}(\mathrm{rank}(\tau[i])) + 1$.

A denumerable set of variables $V(\tau)$ is given for any type τ. The set of terms is the minimal set containing the variables and closed under application and abstraction rules, *i.e.*, if T is a term of type $\tau \to \mu$ and S is a term of type τ then TS is a term of type μ, and if x is a variable of type τ and T is a term of type μ then $\lambda x \cdot T$ is a term of type $\tau \to \mu$. If T is a term of type τ we write $T \in \tau$. We shall use the notation $\lambda x_1 x_2 \ldots x_n \cdot T$ for the term $\lambda x_1 \cdot (\lambda x_2 \cdots (\lambda x_n \cdot T) \cdots)$ and $TS_1 S_2 \ldots S_n$ for $(\ldots((TS_1)S_2)\ldots S_n)$.

If T is a term and x is a variable of the same type as a term S then $T[x/S]$ denotes the term obtained by substituting the term S for each free occurrence of x in T. The axioms of equality between terms have the form of α, β, η conversions (see [Fri75] or [Bar81]), and the convertible terms will be written as $T =_{\beta\eta} S$. All terms are considered modulo α, β and η conversions. By $\mathrm{Cl}(\tau)$ we mean the set of all closed (without free variables) terms of type τ. If Y is a set of variables then $\mathrm{Cl}(\tau, Y)$ is a set of all terms of type τ with only free variables from Y. Obviously $\mathrm{Cl}(\tau, \emptyset) = \mathrm{Cl}(\tau)$ and $\mathrm{Cl}(\tau, \emptyset) \subseteq \mathrm{Cl}(\tau, Y)$.

The term T is in the *long normal form* iff $T = \lambda x_1 \ldots x_n \cdot yT_1 \ldots T_k$ where y is an x_i for $i \in [n]$ or y is a free variable, each T_j for $j \in [k]$ is in the long normal form and $yT_1 \ldots T_k$ is a term of type 0. It is easy to prove that long normal forms exist and are unique for $\beta\eta$ conversions (compare [Sta80] or Φ-normal form in [Cur68]).

Let us introduce a complexity measure π for closed terms. If T is a closed term written in the long normal form and $T = \lambda x_1 \ldots x_n \cdot x_i$ then $\pi(T) = 0$. If $T = \lambda x_1 \ldots x_n \cdot x_i T_1 \ldots T_k$ then $\pi(T) = \max_{j=1 \ldots k}(\pi(\lambda x_1 \ldots x_n \cdot T_j)) + 1$. For a closed term S, $\pi(S)$, is defined as $\pi(T)$ for T in long normal form such that $S =_{\beta\eta} T$.

3 Term Grammars

Let $NT = \{Y_1, Y_2, \ldots\}$ be a finite or denumerable set of variables (the elements of NT correspond to nonterminal elements in the classical grammars). A *production* is a pair (Y, T), denoted also by $Y \Longrightarrow T$, where Y is a variable in NT, Y and T have the common type τ and $T \in \mathrm{Cl}(\tau, NT)$.

DEFINITION 3.1. *A grammar is a finite or denumerable set of productions. The relation of indirect production* $\longrightarrow$ *in the grammar* G *is defined by induction as:*

$$\text{if } Y \Longrightarrow T \in G \text{ then } Y \longrightarrow T \text{ holds,}$$
$$\text{if } Y \longrightarrow T \text{ and } Z \longrightarrow S \text{ hold then } Y \longrightarrow R$$

where R *is any term obtained from* T *by substituting at most one free occurrence of* Z *by* S. *By* $L(G, Y)$ *we mean the set of all closed terms which are generated from* Y *by the grammar* G, *i.e., if* $Y \in \tau$ *then* $L(G, Y) = \{T \in \mathrm{Cl}(\tau) : Y \longrightarrow T\}$. *It is easy to see that if* $Y \longrightarrow T$ *and* $Z \longrightarrow S$ *hold then* $Y \longrightarrow T[Z/S]$ *holds.*

DEFINITION 3.2. *Let us assume that* $Y \Longrightarrow T$ *is a production and* $Y_1, \ldots, Y_n$ *are all free occurrences of nonterminal variables in the term* T. *Let* $Y \in \tau$, $Y_1 \in \tau_1$, $\ldots$, $Y_n \in \tau_n$. *We say that this production determines a function*

$$\alpha : \mathrm{Cl}(\tau_1) \times \mathrm{Cl}(\tau_2) \times \cdots \times \mathrm{Cl}(\tau_n) \to \mathrm{Cl}(\tau)$$

defined by: $\alpha(T_1, \ldots, T_n) = T[Y_1/T_1, \ldots, Y_n/T_n]$ *for all closed terms* $T_1 \in \mathrm{Cl}(\tau_1), \ldots, T_n \in \mathrm{Cl}(\tau_n)$. *If there are no nonterminal variables in the term* T *then the production* $Y \Longrightarrow T$ *determines a 0-ary function (constant)* T *which belongs to* $\mathrm{Cl}(\tau)$.

Let us define grammar $G(\tau)$ for a given type τ.

DEFINITION 3.3. *The construction of this grammar is analogous with the construction of the Huet matching tree for the unification problem (see [Hue75] chapter 3.4 page 37 with simplification for the β, η lambda-calculus in the chapter 4.5 page 51). Let Y be a nonterminal variable of type τ. For the type 0 the grammar $G(0)$ is $Y \Longrightarrow Y$. If $\tau = \tau[1], \ldots, \tau[n] \to 0$ then the grammar contains all productions which are of the form:*

(i) $\quad Y \Longrightarrow \lambda x_1 \ldots x_n \cdot x_i \qquad$ *if* $\arg(\tau[i]) = 0$

(ii) $\quad Y \Longrightarrow \lambda x_1 \ldots x_n \cdot x_i T_1 \ldots T_k \qquad$ *if* $\arg(\tau[i]) = k > 0$

where $T_j \in \tau[i,j]$ for $j \leq k$ are as follows

(ii1) $\quad T_j = Y x_1 \ldots x_n \qquad$ *iff* $\arg(\tau[i,j]) = 0$

(ii2) $\quad T_j = \lambda z_1 \ldots z_p \cdot Y' x_1 \ldots x_n z_1 \ldots z_p \qquad$ *iff* $\arg(\tau[i,j]) = p > 0$

where Y' is a new nonterminal variable of type $\tau[1], \ldots, \tau[n] \to \tau[i,j]$ and $z_s \in \tau[i,j,s]$ for $s \leq p$. This construction is repeated for all new nonterminal variables introduced at this step.

EXAMPLE 3.4. Let τ be the following type: $(0, 0 \to 0), 0, 0, 0 \to 0$. Let us consider the following grammar over $NT = \{Y\}$. Types of auxiliary variables are as follows: $p \in (0, 0 \to 0)$ and $x, v, z \in 0$.

$$Y \Longrightarrow \lambda pxvz \cdot x,$$
$$Y \Longrightarrow \lambda pxvz \cdot v,$$
$$Y \Longrightarrow \lambda pxvz \cdot z,$$
$$Y \Longrightarrow \lambda pxvz \cdot p(Ypxvz)(Ypxvz).$$

It is easy to observe that this grammar generates all closed terms of type τ (see Lemma 3.6).

THEOREM 3.5. *For every type τ the grammar $G(\tau)$ generates all closed terms of type τ.*

PROOF: By induction on the complexity measure. Let T be a closed term of type τ. If T is a projection written in normal form then T can be obtained by means of the production (i). Let $T = \lambda x_1 \ldots x_n \cdot x_i T_1 \ldots T_k$ where $\arg(\tau[i]) = k$ and $T_j \in \mathrm{Cl}(\tau[i,j], \{x_1, \ldots, x_n\})$ for $j \leq k$. Let S_j be the term $\tau x_1 \ldots x_n \cdot T_j$ for $j \leq k$. Every term S_j belongs to $\mathrm{Cl}(\tau[1], \ldots, \tau[n] \to \tau[i,j])$ for $j \leq k$. The complexity measure $\pi(S_j)$ is less then $\pi(T)$ for every $j \leq k$. So, from the inductive assumption every term S_j can be obtained by this grammar from the following nonterminal variables

if $\tau[i,j] = 0$ then $Y \longrightarrow S_j$ (case (ii1))

if $\tau[i,j] = \tau[i,j,1], \ldots, \tau[i,j,p] \to \mu$ then $Y' \longrightarrow S_j$ (case (ii2)).

Let α be the function determined by the production (ii) (see definition 3.2). Then the term T can be obtained by means of the production (ii) from the terms $S_1, \ldots, S_k$, and the following condition holds: $\alpha(S_1, \ldots, S_k) = T$.

Lemma 3.6. *For every type τ such that rank $(\tau) \leq 2$ there is a finite grammar which generates all closed terms of type τ.*

Proof: Let $\tau = \tau[1], \tau[2], \ldots, \tau[n] \to 0$. We will prove that the grammar $G(\tau)$ is finite and according to Theorem 3.5 produces all closed terms of type τ. The grammar contains all productions which are of the form:

$$Y \Longrightarrow \lambda x_1 \ldots x_n \cdot x_i \qquad \text{if } \arg(\tau[i]) = 0$$
$$Y \Longrightarrow \lambda x_1 \ldots x_n \cdot x_i T_1 \ldots T_k \qquad \text{if } \arg(\tau[i]) = k > 0$$

where $T_j = Y x_1 \ldots x_n$ for $j \leq k$.
Grammars described here can produce any free structure.

4 Representability

The type Υ is called a binary tree type because of the isomorphism between $\mathrm{Cl}(\Upsilon)$ and T. We define by induction that a closed term $T \in \mathrm{Cl}(\Upsilon)$ represents a tree t.

Definition 4.1. *The term $\lambda px \cdot x$ of type Υ represents the empty tree ε. If trees $t_1, t_2 \in T$ are represented by closed terms $T_1, T_2 \in \mathrm{Cl}(\Upsilon)$ then the tree $t_1 \wedge t_2$ is represented by the term $\lambda px \cdot p(T_1 px)(T_2 px)$. The unique (up to $\beta\eta$ conversions) term T which represents a tree t is denoted by $\underline{t}$. Thus, we have a 1–1 correspondence between $\mathrm{Cl}(\Upsilon)$ and T. For example the tree*

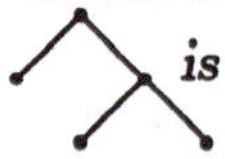

is represented by the term $\lambda px \cdot px(pxx)$.

Definition 4.2. *The function $h : T^n \to T$ is represented by a term $H \in \mathrm{Cl}(\Upsilon^n \to \Upsilon)$ iff, for all trees $t_1, \ldots, t_n \in T$, $H\underline{t_1}, \ldots, \underline{t_n} =_{\beta\eta} \underline{h(t_1, \ldots, t_n)}$. Any term which represents the function h is denoted by $\underline{h}$. The function $h : T^n \to T$ is called λ definable if there is a term $H \in \mathrm{Cl}(\Upsilon^n \to \Upsilon)$ which represents h.*

It is easy to see that any closed term H of type $\Upsilon^n \to \Upsilon$ uniquely defines the function $h : T^n \to T$ as follows: If $t_1, \ldots, t_n \in T$ are trees then the value of this function is the tree represented by the term $H\underline{t_1} \ldots \underline{t_n}$. On the contrary one function can be represented by many unconvertible terms.

Example 4.3. Let H be the closed term $\lambda TSpx \cdot Tp(Spx)$ of type $\Upsilon^2 \to \Upsilon$. It is easy to see that H represents the function append: $T^2 \to T$ defined by

$$\text{append}(\varepsilon, s) = s$$
$$\text{append}(t_1 \wedge t_2, s) = (\text{append}(t_1, s)) \wedge (\text{append}(t_2, s)).$$

The function append is obtained by the primitive recursion schema from the functions $g(s) = s$ and $f(y, z, s) = y \wedge z$. Since $g \in F_\lambda$ and $f \in F_\lambda \cap F_1$ the function append belongs to F_λ.

LEMMA 4.4. *For every closed term I of type $\Upsilon^n, (0,0 \to 0), 0,0,0 \to 0$ the closed term F of type $\Upsilon^{n+2} \to \Upsilon$ defined by*

$$\lambda Y Z T_1 \ldots T_n p x \cdot I T_1 \ldots T_n p x (Y p x)(Z p x)$$

represents a function from the class F_1.

PROOF: Let $f : T^{n+2} \to T$ be the function represented by F. Let $t_1, \ldots, t_n$ be fixed trees. Let us define the function $p(y,z) = f(y,z,t_1,\ldots,t_n)$ Let K be the term of type $(0,0 \to 0), 0,0,0 \to 0$ defined by $I\underline{t_1}\ldots\underline{t_n}$. The function p is represented by the term $\underline{p = \lambda Y Z p x \cdot K p x (Y p x)(Z p x)}$.

Now we prove the Lemma by induction on the construction of terms of the type $(0,0 \to 0), 0,0,0 \to 0$ (see Example 3.4). If $K =_{\beta\eta} \lambda p x v z \cdot x$ then $P =_{\beta\eta} \lambda Y Z p x \cdot x$, so $p(y,z) = \varepsilon$. Therefore $p \in F_0$ which means that $f \in F_1$. If K is $\beta\eta$ convertible to $\lambda p x v z \cdot v$ or $\lambda p x v z \cdot z$ then $P =_{\beta\eta} \lambda Y Z p x \cdot Y p x$ or $P =_{\beta\eta} \lambda Y Z p x \cdot Z p x$, respectively, so $p(y,z) = y$ or $p(y,z) = z$. In both cases $p \in F_0$ which means that $f \in F_1$. (Inductive step.) Let $K =_{\beta\eta} \lambda p x v z \cdot p(K_1 p x y z)(K_2 p x y z)$ where K_1, K_2 are closed terms of type $(0,0 \to 0), 0,0,0 \to 0$. Assuming that the Lemma is true for K_1 and for K_2 we assume that p_1, p_2 belong to F_0. But $p(y,z) = (p_1(y,z)) \wedge (p_2(y,z))$ so p belongs to F_0, therefore $f \in F_1$.

LEMMA 4.5. *Let I and G be closed terms of type $\Upsilon^n, (0,0 \to 0), 0,0,0 \to 0$ and $\Upsilon^n \to \Upsilon$, respectively. Let the closed term F of the type $\Upsilon^{n+2} \to \Upsilon$ be defined as $\lambda Y Z T_1 \ldots T_n p x \cdot I T_1 \ldots T_n p x (Y p x)(Z p x)$. Let $f : T^{n+2} \to T$ be a function represented by F and $g : T^n \to T$ be a function represented by G. Then the function $h : T^{n+1} \to T$ obtained by the primitive recursive schema from g and f is represented by the term H of type $\Upsilon^{n+1} \to \Upsilon$ defined by $\lambda S T_1 \ldots T_n p x \cdot S(\lambda y z \cdot I T_1 \ldots T_n p x y z)(G T_1 \ldots T_n p x)$.*

PROOF: Let $t_1, \ldots, t_n$ be fixed trees. Since I and G and $\underline{t_1}, \ldots, \underline{t_n}$ are closed then the terms $K = I\underline{t_1}, \ldots, \underline{t_n}$ and $L = G\underline{t_1}, \ldots, \underline{t_n}$ are also closed terms of types $(0,0 \to 0), 0,0,0 \to 0$ and Υ, respectively. We prove by induction on the construction of a tree s that $H\underline{s}\underline{t_1}, \ldots, \underline{t_n} =_{\beta\eta} \underline{h(s,t_1,\ldots,t_n)}$. If $s = \varepsilon$ then

$$H\underline{\varepsilon}\,\underline{t_1}, \ldots, \underline{t_n} =_{\beta\eta} G(\underline{t_1}, \ldots, \underline{t_n}) =_{\beta\eta} \underline{g(t_1,\ldots,t_n)} =_{\beta\eta} \underline{h(\varepsilon,t_1,\ldots,t_n)}.$$

Let us assume for induction that

$$H s \underline{t_1}, \ldots, \underline{t_n} =_{\beta\eta} \underline{h(s,t_1,\ldots,t_n)}$$
$$H s' \underline{t_1}, \ldots, \underline{t_n} =_{\beta\eta} \underline{h(s',t_1,\ldots,t_n)}.$$

This means that $\lambda p x \cdot \underline{s}(\lambda y z \cdot K p x y z)(L p x) =_{\beta\eta} \underline{h(s,t_1,\ldots,t_n)}$ and $\lambda p x \cdot \underline{s'}(y z \cdot K p x y z)(L p x) =_{\beta\eta} \underline{h(s',t_1,\ldots,t_n)}$. We are going to prove that

$H\underline{s \wedge s'}\underline{t_1}, \ldots, \underline{t_n} =_{\beta\eta} \underline{h(s \wedge s', t_1, \ldots, t_n)}$.

$$H\underline{s \wedge s'}\underline{t_1}, \ldots, \underline{t_n} =_{\beta\eta}$$
$$\lambda px \cdot [[\lambda qv \cdot q(\underline{s}qv)(\underline{s}'qv)](\lambda yz \cdot I\underline{t_1} \ldots \underline{t_n}pxyz)(G\underline{t_1} \ldots \underline{t_n}px)] =_{\beta\eta}$$
$$\lambda px \cdot [[\lambda qv \cdot q(\underline{s}qv)(\underline{s}'qv)](\lambda yz \cdot Kpxyz)(Lpx)] =_{\beta\eta}$$
$$\lambda px \cdot [[\lambda yz \cdot Kpxyz]\,(\underline{s}(\lambda yz \cdot Kpxyz)(Lpx))\,(\underline{s}'(\lambda yz \cdot Kpxyz)(Lpx))] =_{\beta\eta}$$
$$\lambda px \cdot Kpx\,(\underline{s}(\lambda yz \cdot Kpxyz)(Lpx))\,(\underline{s}'\,(\lambda yz \cdot Kpxyz)(Lpx)) =_{\beta\eta}$$
$$\lambda px \cdot Kpx \left(\underline{h(s, t_1, \ldots, t_n)}px\right) \left(\underline{h(s', t_1, \ldots, t_n)}px\right) =_{\beta\eta}$$
$$\lambda px \cdot F \left(\underline{h(s, t_1, \ldots, t_n)}\right) \left(\underline{h(s', t_1, \ldots, t_n)}\right) \underline{t_1} \ldots \underline{t_n}px =_{\beta\eta}$$
$$\underline{f\,(h(s, t_1, \ldots, t_n), h(s', t_1, \ldots, t_n), t_1, \ldots, t_n)} =_{\beta\eta}$$
$$\underline{h(s \wedge s', t_1, \ldots, t_n)}.$$

LEMMA 4.6. *Let H and $\widehat{H}$ be closed terms of types $\Upsilon^n \rightarrow \Upsilon$ and $\Upsilon^{n+1} \rightarrow \Upsilon$ which are of the forms*

$$H = \lambda T_1 \ldots T_n px \cdot T_i(\lambda yz \cdot IT_1 \ldots T_n pxyz)(GT_1 \ldots T_n px) \qquad \text{and}$$
$$\widehat{H} = \lambda ST_1 \ldots T_n px \cdot S(\lambda yz \cdot IT_1 \ldots T_n pxyz)(GT_1 \ldots T_n px)$$

for some closed terms I and G of types $\Upsilon^n, (0,0 \rightarrow 0), 0, 0, 0 \rightarrow 0$ and $\Upsilon^n \rightarrow \Upsilon$ respectively. Let $h\colon T^n \rightarrow T$, $\hat{h}\colon T^{n+1} \rightarrow T$ be functions represented by H, $\widehat{H}$ respectively. For all trees $t_1, \ldots, t_n$, $h(t_1, \ldots, t_n) = \hat{h}(t_i, t_1, \ldots, t_n)$.

PROOF: Let $t_1, \ldots, t_n$ be trees. The terms $H\underline{t_1} \ldots \underline{t_n}$ and $\widehat{H}\underline{t_i}\underline{t_1} \ldots \underline{t_n}$ are $\beta\eta$ convertible.

LEMMA 4.7. *Let M be a closed term of the type $\Upsilon^2 \rightarrow \Upsilon$ such that the function $m\colon T^2 \rightarrow T$ represented by M belongs to F_0. For all trees $t, t' \in T$ the terms*

$$A = \lambda px \cdot M(\lambda p'x' \cdot \underline{t}px)(\lambda p'x'\underline{t}'px)px \qquad \text{and}$$
$$B = \lambda px \cdot M\underline{t}\underline{t}'px$$

are $\beta\eta$ convertible.

PROOF: By induction on the construction of the function m. If $m(y, z) = \varepsilon$ for all y, z then $M = \lambda YZpx \cdot x$, so $A =_{\beta\eta} B =_{\beta\eta} \lambda px \cdot x$. If $m(y, z) = y$ or $m(y, z) = z$ then $M = \lambda YZpx \cdot Ypx$ or $M = \lambda YZpx \cdot Zpx$, so $A =_{\beta\eta} B =_{\beta\eta} \underline{t}$ or $\underline{t}'$. Suppose the terms are convertible for functions m_1 and m_2. The function $m_1 \wedge m_2$ (see definition 1.4) is represented by the term $M =_{\beta\eta} \lambda YZpx \cdot p(\underline{m_1}YZpx)(\underline{m_2}YZpx)$, therefore

$$A =_{\beta\eta} \lambda px \cdot M(\lambda p'x' \cdot \underline{t}px)(\lambda p'x' \cdot \underline{t}'px)px =_{\beta\eta}$$
$$\lambda px \cdot p\,(m_1(\lambda p'x' \cdot \underline{t}px)(\lambda p'x' \cdot \underline{t}'px))\,(m_2(\lambda p'x' \cdot \underline{t}px)(\lambda p'x' \cdot \underline{t}'px)) =_{\beta\eta}$$
$$\lambda px \cdot p(\underline{m_1}\,\underline{t}\,\underline{t}'px)(\underline{m_2}\,\underline{t}\,\underline{t}'px) =_{\beta\eta}$$
$$\lambda px \cdot \underline{(m_1 \wedge m_2)}\underline{t}\,\underline{t}'px =_{\beta\eta}$$
$$\lambda px \cdot \underline{m}\,\underline{t}\,\underline{t}'px =_{\beta\eta} B.$$

LEMMA 4.8. *If the functions* $g : T^n \to T$ *and* $f : T^{n+2} \to T$ *are both* λ *definable and* $f \in F_1$*, then the* $n+1$*-ary function* h *obtained from* g *and* f *by the primitive recursion rule is also* λ *definable.*

PROOF: Let G be a closed term of type $\Upsilon^n \to \Upsilon$ representing g. Let F be a closed term of type $\Upsilon^{n+2} \to \Upsilon$ representing f. The function $h : T^{n+1} \to T$ is defined by

$$h(\varepsilon, x_1, \ldots, x_n) = g(x_1, \ldots, x_n)$$
$$h(s \wedge s', x_1, \ldots, x_n) = f\left(h(s, x_1, \ldots, x_n), h(s', x_1, \ldots, x_n), x_1, \ldots, x_n\right).$$

We are going to prove that h is λ definable by the term

$$H = \lambda S T_1 \ldots T_n p x \cdot S(\lambda yz \cdot F(\lambda p'x' \cdot y)(\lambda p'x' \cdot z) T_1 \ldots T_n p x)(G T_1 \ldots T_n p x).$$

Let $t_1, \ldots, t_n$ be fixed trees. Let M and L be the terms of types $\Upsilon^2 \to \Upsilon$ and Υ defined by $M = \lambda YZpx \cdot FYZ\underline{t_1} \ldots \underline{t_n}px$ and $L = G\underline{t_1} \ldots \underline{t_n}$. Let $m \colon T^2 \to T$ be the function represented by M. Since the function $f \in F_1$, the function m belongs to F_0. The proof is by induction on the construction of a tree s. If $s = \varepsilon$ then $H\underline{\varepsilon}\underline{t_1} \ldots \underline{t_n} =_{\beta\eta} G\underline{t_1} \ldots \underline{t_n} =_{\beta\eta} \underline{g(t_1, \ldots, t_n)} =_{\beta\eta} \underline{h(\varepsilon, t_1, \ldots, t_n)}$. Let us assume that the Theorem is true for trees s and s'. Let t be the tree $h(s, t_1, \ldots, t_n)$ and t' be the tree $h(s', t_1, \ldots, t_n)$. Then we have

$$\underline{t} =_{\beta\eta} H\underline{s}\underline{t_1} \ldots \underline{t_n} =_{\beta\eta} \lambda px \cdot \underline{s}\,(\lambda yz \cdot M(\lambda p'x' \cdot y)(\lambda p'x' \cdot z)px)\,(Lpx)$$
$$\underline{t'} =_{\beta\eta} H\underline{s'}\underline{t_1} \ldots \underline{t_n} =_{\beta\eta} \lambda px \cdot \underline{s'}\,(\lambda yz \cdot M(\lambda p'x' \cdot y)(\lambda p'x' \cdot z)px)\,(Lpx).$$

We calculate the application of the term H to the arguments $\underline{s \wedge s'}$, $\underline{t_1}, \ldots, \underline{t_n}$.

$$\begin{aligned}
&H(\underline{s \wedge s'})\underline{t_1} \ldots \underline{t_n} =_{\beta\eta} \\
&\lambda px \cdot (\underline{s \wedge s'})\,(\lambda yz \cdot M(\lambda p'x' \cdot y)(\lambda p'x' \cdot z)px)\,(Lpx) =_{\beta\eta} \\
&\lambda px \cdot [\lambda qv \cdot q(\underline{s}qv)(\underline{s'}qv)]\,(\lambda yz \cdot M(\lambda p'x' \cdot y)(\lambda p'x' \cdot z)px)\,(Lpx) =_{\beta\eta} \\
&\lambda px \cdot [\lambda yz \cdot M(\lambda p'x' \cdot y)(\lambda p'x' \cdot z)px] \\
&\qquad (\underline{s}(\lambda yz \cdot M(\lambda p'x' \cdot y)(\lambda p'x' \cdot z)px)(Lpx)) \\
&\qquad (\underline{s'}(\lambda yz \cdot M(\lambda p'x' \cdot y)(\lambda p'x' \cdot z)px)(Lpx)) =_{\beta\eta} \\
&\lambda px \cdot M\,(\lambda p'x' \cdot \underline{s}(\lambda yz \cdot M(\lambda p'x' \cdot y)(\lambda p'x' \cdot z)px)(Lpx)) \\
&\qquad (\lambda p'x' \cdot \underline{s'}(\lambda yz \cdot M(\lambda p'x' \cdot y)(\lambda p'x' \cdot z)px)(Lpx)) =_{\beta\eta} \\
&\lambda px \cdot M(\lambda p'x' \cdot \underline{t}px)(\lambda p'x' \cdot \underline{t'}px).
\end{aligned}$$

Since M is representative of function m from F_0, by Lemma 4.7 we have

$$\begin{aligned}
&\lambda px \cdot M(\lambda p'x' \cdot \underline{t}px)(\lambda p'x' \cdot \underline{t'}px) =_{\beta\eta} \\
&\lambda px \cdot M\underline{t}\,\underline{t'}px =_{\beta\eta} \\
&\underline{F\underline{t}\,\underline{t'}\underline{t_1} \ldots \underline{t_n}} =_{\beta\eta} \\
&\underline{f(t, t', t_1, \ldots, t_n)} =_{\beta\eta} \\
&\underline{h(s \wedge s', t_1 \ldots, t_n)}.
\end{aligned}$$

Lemma 4.9. *If two functions f and g are λ definable then so is $f \wedge g$.*

Proof: $f \wedge g$ is λ definable by the term

$$\lambda T_1 \ldots T_n px \cdot (\underline{f} T_1 \ldots T_n px)(\underline{g} T_1 \ldots T_n px).$$

Lemma 4.10. *Any composition of λ definable functions is λ definable.*

Proof: If a k-ary function f and n-ary functions $g_1, \ldots g_k$ are λ definable then the composition

$$(x_1, \ldots, x_n) \to f\,(g_1(x_1, \ldots, x_n), \ldots, g_k(x_1, \ldots, x_n))$$

is λ definable by the term $\lambda T_1 \ldots T_n \cdot \underline{f}(\underline{g_1} T_1 \ldots T_n) \ldots (\underline{g_k} T_1 \ldots T_n)$.

Theorem 4.11. *(Soundness.) If $h \in F_\lambda$ then h is λ definable.*

Proof: By induction on the construction of the function f. The function e^n which maps onto the empty tree constantly is represented by $\lambda T_1 \ldots T_n px \cdot x$. The projection p_i^n is represented by $\lambda T_1 \ldots T_n px \cdot T_i px$. The function $\wedge$ is represented by $\lambda TSpx \cdot p(Tps)(Spx)$. Lemma 4.10 guarantees that the set of λ definable functions is closed under composition. If the functions $g \in F_\lambda$ and $f \in F_\lambda \cap F_1$ are λ definable then according to Lemma 4.8 the function obtained by primitive recursion is also λ definable.

Theorem 4.12. *(Completeness.) If h is λ definable then $h \in F_\lambda$.*

Proof: Let h be represented by the term H of the type $\Upsilon^n \to \Upsilon$. Let H be in long normal form. First we will construct the grammar $G(\Upsilon^n \to \Upsilon)$ which generates all closed terms of the type $\Upsilon^n \to \Upsilon$ (see Theorem 3.5) and then we will prove, by induction of the complexity of the term H, that H represents a function from F_λ. From the grammar construction it follows that every closed term H of the type $\Upsilon^n \to \Upsilon$ has one of three possible forms.

$$H = \lambda T_1 \ldots T_n px \cdot x,$$
$$H = \lambda T_1 \ldots T_n px \cdot p(H_1 T_1 \ldots T_n px)(H_2 T_1 \ldots T_n px)$$

where H_1 and H_2 are closed terms of the type $\Upsilon^n \to \Upsilon$,

$$H = \lambda T_1 \ldots T_n px \cdot T_i(\lambda yz \cdot I T_1 \ldots T_n pxyz)(G T_1 \ldots T_n px)$$

where I and G are closed terms of types $\Upsilon^n, (0, 0 \to 0), 0, 0, 0 \to 0$ and $\Upsilon^n \to \Upsilon$ respectively.

If $\pi(H) = 0$ then H must be in the first form, so H represents the function $h(t_1, \ldots, t_n) = \varepsilon$. Therefore h belongs to F_λ. Let $\pi(H) = n > 0$. Assume the Theorem is true for all terms H' such that $\pi(H') < n$. Suppose H is of the second form, *i.e.,* $H = \lambda T_1 \ldots T_n px \cdot p(H_1 T_1 \ldots T_n px)(H_2 T_1 \ldots T_n px)$ with $\pi(H_1) < n$ and $\pi(H_2) < n$. Then the functions h_1 and h_2 represented by H_1 and H_2 belong to F_λ. Since $h = h_1 \wedge h_2$ we have $h \in F_\lambda$ (see Lemma 4.10).

Suppose H is of the third form, *i.e.*,

$$H = \lambda T_1 \dots T_n px \cdot T_i(\lambda yz \cdot IT_1 \dots T_n pxyz)(GT_1 \dots T_n px)$$

where I and G are closed terms of appropriate types and $\pi(I) < n$ and $\pi(G) < n$. Let us define the term F of type $\Upsilon^{n+2} \to \Upsilon$ by $\lambda YZT_1 \dots T_n px \cdot IT_1 \dots T_n px(Ypx)(Zpx)$. Let f and g be the functions represented by F and G. Since $\pi(F) = \pi(I) < n$, f belongs to F_λ and since $\pi(G) < n$, $g \in F_\lambda$. From Lemma 4.4 it follows that $f \in F_1$. Therefore the function $\hat{h}$ obtained by primitive recursion from the functions g and f belongs to F_λ. According to Lemma 4.5 the function $\hat{h}$ is represented by the term $\widehat{H} = \lambda ST_1 \dots T_n px \cdot S(\lambda yz \cdot IT_1 \dots T_n pxyz)(GT_1 \dots T_n px)$. Because of Lemma 4.6 the function h can be expressed by means of $\hat{h}$, $h(t_1, \dots, t_n) = \hat{h}(t_i, t_1, \dots, t_n)$, so h belongs to F_λ.

EXAMPLE 4.13. This example illustrates the algorithm presented in Theorem 4.12 and Lemma 4.5. Let H be a term $\lambda Tpx \cdot T(\lambda yz \cdot p(Tpy)z)x$ of type $\Upsilon \to \Upsilon$. The problem is to find the function represented by H. The term H is of the form $\lambda Tpx \cdot T(\lambda yz \cdot ITpxyz)(Gpx)$ where I and G are terms $I = \lambda Tpxyz \cdot p(Tpy)z$ and $G = \lambda Tpx \cdot x$. Let us define the term F of type $\Upsilon^3 \to \Upsilon$ as $F = \lambda YZTpx \cdot ITpx(Ypx)(Zpx) =_{\beta\eta} \lambda YZTpx \cdot p(Tp(Ypx))(Zpx)$. Let f and g be represented by F and G. Then the function $\hat{h}$ defined by

$$\hat{h}(\varepsilon, y) = g(y),$$
$$\hat{h}(s \wedge s', y) = f\left(\hat{h}(s, y), \hat{h}(s', y), y\right)$$

is represented by the term $\widehat{H} = \lambda STpx \cdot S(\lambda yz \cdot p(Tpy)z)x$ (Lemma 4.5). By Lemma 4.6, $h(t) = \hat{h}(t, t)$. Therefore we have reduced the problem of finding the representation of H to the problem of finding the representation of the less complex terms G and F. G represents $g(t) = \varepsilon$. F is of the form $F = \lambda YZTpx \cdot p(F_1 YZTpx)(F_2 YZTpx)$ where $F_1 = \lambda YZTpx \cdot Tp(Ypx)$ and $F_2 = \lambda YZTpx \cdot Zpx$. F_1 represents $f_1(y, z, t) = \text{append}(t, y)$ (see Example 4.3), F_2 represents the projection $f_2(y, z, t) = z$ and we have $f = f_1 \wedge f_2$. Therefore the following system of recursive functions define the function h

$$h(t) = \hat{h}(t, t)$$
$$\hat{h}(\varepsilon, t) = \varepsilon$$
$$\hat{h}(s \wedge s', t) = f\left(\hat{h}(s, t), \hat{h}(s', t), t\right)$$
$$f(y, z, t) = \text{append}(t, y) \wedge z.$$

EXAMPLE 4.14. This example illustrates the algorithm used in the Soundness Theorem (4.11). Let the function $h : T^2 \to T$ be given by a system of recursive equations. The problem is to find a term representing h. Let h

be given as

$$h(\varepsilon, t) = g(t) = t$$
$$h(s \wedge s', t) = f\,(h(s,t), h(s',t), t)$$
$$f(y,z,t) = t \wedge y.$$

First it is easy to check that $f \in F_1$. The function g is represented by $G = \lambda Tpx \cdot Tpx$ and f is represented by $F = \lambda YZTpx \cdot p(Tpx)(Ypx)$. Therefore the function h is represented by (see Lemma 4.8)

$$H = \lambda STpx \cdot S(\lambda yz \cdot F(\lambda p'x' \cdot y)(\lambda p'x' \cdot z)Tpx)(GTpx) =_{\beta\eta}$$
$$\lambda STpx \cdot S\,(\lambda yz \cdot p(Tpx)y)\,(Tpx).$$

5 λ-Definability in other Free Structures

For some free structures similar results concerning representability are known. The first order type $\mathbf{N} = (0 \to 0) \to (0 \to 0)$ can be called the natural number type because of the existence of an isomorphism between positive integers and closed terms of this type (Church's integers). This correspondence is given in the following:

DEFINITION 5.1. *The term* $n = \lambda ux \cdot u\,(u(\ldots ux)\ldots) \in \mathbf{N}$ *with* n *occurrences of* u *represents the number* $n \in N$ *(with* $0 = \lambda ux \cdot x$*). Therefore any closed term of the type* $\mathbf{N}^k \to \mathbf{N}$ *represents a* k*-ary interger function.*

The λ representability of the first order integer functionals is characterized by the following Theorem (see [Sch75],[Sta79]).

THEOREM 5.2. *λ definable integer functions are just extended polynomials.*

DEFINITION 5.3. *The domain of all strings over a finite alphabet* $\Sigma_n = \{\alpha_1, \ldots, \alpha_n\}$ *can be identified with the type* $\mathbf{B}_n = (0 \to 0)^n \to (0 \to 0)$ *by the following. The empty word* Λ *is represented by the term* $\lambda u_1 \ldots u_n x \cdot x$. *If* $w \in \Sigma_n^*$ *is represented by the closed term* $W \in \mathbf{B}_n$ *then the word* $\alpha_i w$ *is represented by the closed term* $\lambda u_1 \ldots u_n x \cdot u_i(Wu_1 \ldots u_n x)$.

This constitutes a 1–1 correspondence between $\mathrm{Cl}(\mathbf{B}_n)$ and Σ_n^*. Therefore any closed term of the type $(\mathbf{B}_n)^k \to \mathbf{B}_n$ represents a k-ary word operation. The λ representability of the first-order word operations is characterized by the following Theorem (see [Zai87]).

THEOREM 5.4. *The λ definable word operations are those obtained by composition from* app, substitution, cut_{α_i}, …,cut_{α_n}, *where* app *is the* *append operation, and* substitution *and* cut_{α_i} *are defined by*

$$\text{substitution}(\Lambda, w_1, \ldots, w_n) = \Lambda,$$
$$\text{substitution}(\alpha_i w, w_1, \ldots, w_n) = \text{app}(w_i, \text{substitution}(w, w_1, \ldots, w_n)),$$
$$\text{cut}_{\alpha_i}(\Lambda) = \Lambda,$$
$$\text{cut}_{\alpha_i}(\alpha_j w) = \Lambda \quad \textit{iff} \quad i \leq j,$$
$$\text{cut}_{\alpha_i}(\alpha_i w) = \text{app}(\alpha_i, \text{cut}_{\alpha_i}(w)).$$

For finite domains λ definability is discussed in [Sta83] p. 24 and in [Zai87] p. 5. The finite set $\{1, \ldots, n\}$ can be identified with the type $\tau_n = 0^n \to 0$.

DEFINITION 5.5. *The natural number* i *is represented by the projection* $\lambda x_1 \ldots x_n \cdot x_i$. *Therefore any closed term of type* $(\tau_n)^k \to \tau^n$ *represents a function* $f : \{1, \ldots, n\}^k \to \{1, \ldots, n\}$.

THEOREM 5.6. *Every function* $f : \{1, \ldots, n\}^k \to \{1, \ldots, n\}$ *is representable.*

OPEN QUESTION. It is easy to observe that any free algebra can be identified with some first-order type (c.f [CoH85]). If A is a free algebra determined by the signature $I = (n_1, \ldots, n_k)$ where $n_i \in \mathbf{N}$ for $i \leq k$, then by the type τ^A we mean $\tau_1, \ldots, \tau_k \to 0$ where $\tau_i = 0^{n_i} \to 0$ for $i \leq k$. It can be seen that closed terms of type τ^A reflect constructions in the algebra A. Therefore any term of type $(\tau^A)^n \to \tau^A$ defines some n-ary mapping in this algebra. The problem is to find a characterization of all τ definable mappings in a free algebra given by signature I. Having two algebras A and B we may also ask about the characterization of all τ definable mappings from A to B, *i.e.*, the functions determined by closed terms of the type $\tau^A \to \tau^B$.

EXAMPLES. The free algebra given by the signature $(1, 0)$ (natural numbers) generates the type $(0 \to 0) \to (0 \to 0)$. The solution of the problem in this case is described in Theorem 5.2. The solution for the free algebra $(1, \ldots, 1, 0)$ (strings over n-ary alphabet) is sketched in Theorem 5.4. The free algebra $(2, 0)$ (binary trees) is discussed in this paper.

Acknowledgments: The research described in this paper was done while the author was at the Dept. of Computer & Information Sciences, University of Alabama at Birmingham, USA. I would like to thank the anonymous referee for many helpful suggestions and valuable comments.

References

[Bar81] H. P. Barendregt, *The Lambda calculus*, in "Studies in Logic and the foundations of Mathematics," North-Holland, 1981.

[CoH85] T. Coquand, G. Huet, *Constructions: A Higher Order Proof System for Mechanizing Mathematics*, Eurocol **85** (1985). Linz, Austria.

[Chu41] A. Church, "The Calculi of Lambda-Conversion," Princeton University Press, Princeton, NY, 1941.

[Cur68] H. B. Curry, R. Feys, "Combinatory Logic vol. 1," North-Holland, Amsterdam, 1968.

[Fri75] H. Friedman, *Equality between functionals*, in "Lecture Notes in Mathematics vol. 453," 1975, pp. 22–37.

[Hue75] G. Huet, *A unification algorithm for typed τ-calculus*, in "Theoretical Computer Science 1," 1975, pp. 27–58.

[Sch75] H. Schwichtenberg, *Definierbare Functionen im τ-Kalkuli mit Typen*, in "Arch Math. Logic Grundlagenforsch," 1975–76.

[Sta79] R. Statman, *The Typed λ-calculus is not elementary recursive*, in "Theoretical Computer Science 9," 1979.

[Sta80] R. Statman, *On the existence of closed terms in the type τ calculus I*, in "Combinatory logic, lambda calculus, and formal systems R. Hindley, J. Seldin (eds)," Academic Press, New York, 1980.

[Sta83] R. Statman, *λ-Definable functionals and $\beta\eta$ conversion*, Arch. Math. Logic **23** (1983).

[Zai85] M. Zaionc, *The set of unifiers in typed τ calculus as regular expression*, in "Lecture Notes in Computer Science 202," 1985, pp. 430–440.

[Zai87] M. Zaionc, *Word operations definable in the typed λ calculus*, Theoretical Computer Science **52** (1987), 1–14.

Vol. 379: A. Kreczmar, G. Mirkowska (Eds.), Mathematical Foundations of Computer Science 1989. Proceedings, 1989. VIII, 605 pages. 1989.

Vol. 380: J. Csirik, J. Demetrovics, F. Gécseg (Eds.), Fundamentals of Computation Theory. Proceedings, 1989. XI, 493 pages. 1989.

Vol. 381: J. Dassow, J. Kelemen (Eds.), Machines, Languages, and Complexity. Proceedings, 1988. VI, 244 pages. 1989.

Vol. 382: F. Dehne, J.-R. Sack, N. Santoro (Eds.), Algorithms and Data Structures. WADS '89. Proceedings, 1989. IX, 592 pages. 1989.

Vol. 383: K. Furukawa, H. Tanaka, T. Fujisaki (Eds.), Logic Programming '88. Proceedings, 1988. VII, 251 pages. 1989 (Subseries LNAI).

Vol. 384: G.A. van Zee, J.G.G. van de Vorst (Eds.), Parallel Computing 1988. Proceedings, 1988. V, 135 pages. 1989.

Vol. 385: E. Börger, H. Kleine Büning, M.M. Richter (Eds.), CSL '88. Proceedings, 1988. VI, 399 pages. 1989.

Vol. 386: J.E. Pin (Ed.), Formal Properties of Finite Automata and Applications. Proceedings, 1988. VIII, 260 pages. 1989.

Vol. 387: C. Ghezzi, J. A. McDermid (Eds.), ESEC '89. 2nd European Software Engineering Conference. Proceedings, 1989. VI, 496 pages. 1989.

Vol. 388: G. Cohen, J. Wolfmann (Eds.), Coding Theory and Applications. Proceedings, 1988. IX, 329 pages. 1989.

Vol. 389: D.H. Pitt, D.E. Rydeheard, P. Dybjer, A.M. Pitts, A. Poigné (Eds.), Category Theory and Computer Science. Proceedings, 1989. VI, 365 pages. 1989.

Vol. 390: J.P. Martins, E.M. Morgado (Eds.), EPIA 89. Proceedings, 1989. XII, 400 pages. 1989 (Subseries LNAI).

Vol. 391: J.-D. Boissonnat, J.-P. Laumond (Eds.), Geometry and Robotics. Proceedings, 1988. VI, 413 pages. 1989.

Vol. 392: J.-C. Bermond, M. Raynal (Eds.), Distributed Algorithms. Proceedings, 1989. VI, 315 pages. 1989.

Vol. 393: H. Ehrig, H. Herrlich, H.-J. Kreowski, G. Preuß (Eds.), Categorical Methods in Computer Science. VI, 350 pages. 1989.

Vol. 394: M. Wirsing, J.A. Bergstra (Eds.), Algebraic Methods: Theory, Tools and Applications. VI, 558 pages. 1989.

Vol. 395: M. Schmidt-Schauß, Computational Aspects of an Order-Sorted Logic with Term Declarations. VIII, 171 pages. 1989 (Subseries LNAI).

Vol. 396: T.A. Berson, T. Beth (Eds.), Local Area Network Security. Proceedings, 1989. IX, 152 pages. 1989.

Vol. 397: K.P. Jantke (Ed.), Analogical and Inductive Inference. Proceedings, 1989. IX, 338 pages. 1989 (Subseries LNAI).

Vol. 398: B. Banieqbal, H. Barringer, A. Pnueli (Eds.), Temporal Logic in Specification. Proceedings, 1987. VI, 448 pages. 1989.

Vol. 399: V. Cantoni, R. Creutzburg, S. Levialdi, G. Wolf (Eds.), Recent Issues in Pattern Analysis and Recognition. VII, 400 pages. 1989.

Vol. 400: R. Klein, Concrete and Abstract Voronoi Diagrams. IV, 167 pages. 1989.

Vol. 401: H. Djidjev (Ed.), Optimal Algorithms. Proceedings, 1989. VI, 308 pages. 1989.

Vol. 402: T.P. Bagchi, V.K. Chaudhri, Interactive Relational Database Design. XI, 186 pages. 1989.

Vol. 403: S. Goldwasser (Ed.), Advances in Cryptology – CRYPTO '88. Proceedings, 1988. XI, 591 pages. 1990.

Vol. 404: J. Beer, Concepts, Design, and Performance Analysis of a Parallel Prolog Machine. VI, 128 pages. 1989.

Vol. 405: C.E. Veni Madhavan (Ed.), Foundations of Software Technology and Theoretical Computer Science. Proceedings, 1989. VIII, 339 pages. 1989.

Vol. 406: C.J. Barter, M.J. Brooks (Eds.), AI '88. Proceedings, 1988. VIII, 463 pages. 1990 (Subseries LNAI).

Vol. 407: J. Sifakis (Ed.), Automatic Verification Methods for Finite State Systems. Proceedings, 1989. VII, 382 pages. 1990.

Vol. 408: M. Leeser, G. Brown (Eds.), Hardware Specification, Verification and Synthesis: Mathematical Aspects. Proceedings, 1989. VI, 402 pages. 1990.

Vol. 409: A. Buchmann, O. Günther, T.R. Smith, Y.-F. Wang (Eds.), Design and Implementation of Large Spatial Databases. Proceedings, 1989. IX, 364 pages. 1990.

Vol. 410: F. Pichler, R. Moreno-Diaz (Eds.), Computer Aided Systems Theory – EUROCAST '89. Proceedings, 1989. VII, 427 pages. 1990.

Vol. 411: M. Nagl (Ed.), Graph-Theoretic Concepts in Computer Science. Proceedings, 1989. VII, 374 pages. 1990.

Vol. 412: L.B. Almeida, C.J. Wellekens (Eds.), Neural Networks. Proceedings, 1990. IX, 276 pages. 1990,

Vol. 413: R. Lenz, Group Theoretical Methods in Image Processing. VIII, 139 pages. 1990.

Vol. 414: A. Kreczmar, A. Salwicki, M. Warpechowski, LOGLAN '88 – Report on the Programming Language. X, 133 pages. 1990.

Vol. 415: C. Choffrut, T. Lengauer (Eds.), STACS 90. Proceedings, 1990. VI, 312 pages. 1990.

Vol. 416: F. Bancilhon, C. Thanos, D. Tsichritzis (Eds.), Advances in Database Technology – EDBT '90. Proceedings, 1990. IX, 452 pages. 1990.

Vol. 417: P. Martin-Löf, G. Mints (Eds.), COLOG-88. International Conference on Computer Logic. Proceedings, 1988. VI, 338 pages. 1990.

Vol. 419: K. Weichselberger, S. Pöhlmann, A Methodology for Uncertainty in Knowledge-Based Systems. VIII, 136 pages. 1990 (Subseries LNAI).

Vol. 420: Z. Michalewicz (Ed.), Statistical and Scientific Database Management, V SSDBM. Proceedings, 1990. V, 256 pages. 1990.

Vol. 421: T. Onodera, S. Kawai, A Formal Model of Visualization in Computer Graphics Systems. X, 100 pages. 1990.

Vol. 422: B. Nebel, Reasoning and Revision in Hybrid Representation Systems. XII, 270 pages. 1990 (Subseries LNAI).

Vol. 423: L. E. Deimel (Ed.), Software Engineering Education. Proceedings, 1990. VI, 164 pages. 1990.

Vol. 424: G. Rozenberg (Ed.), Advances in Petri Nets 1989. VI, 524 pages. 1990.

Vol. 425: C.H. Bergman, R.D. Maddux, D.L. Pigozzi (Eds.), Algebraic Logic and Universal Algebra in Computer Science. Proceedings, 1988. XI, 292 pages. 1990.